ELASTOMERS

Edited by **Nevin Çankaya**

Elastomers
http://dx.doi.org/10.5772/66012
Edited by Nevin Çankaya

Contributors

Wirach Taweepreda, Na Ni, Ling Zhang, Abdul Kariem Arof, Li Na Sim, Magdalena Lipińska, Sandra Paszkiewicz, Iman Taraghi, Anna Szymczyk, Elzbieta Piesowicz, Zbigniew Roslaniec, Xufeng Dong, Chenguang Niu, Min Qi, Azanam Shah Hashim, Siew Kooi Ong, Dongli Fan, Xin Zhou, Yiming Zhang, Xiaohua Shi, Fabiula De Sousa, Weiyu Wang, Wei Lu, Nam-Goo Kang, Jimmy Mays, Kunlun Hong, Manuel Valero, Said Arévalo

Published by InTech

Janeza Trdine 9, 51000 Rijeka, Croatia

Publishing Process Manager Edi Lipovic
Technical Editor SPi Global
Cover InTech Design team

First published August, 2017
Printed in Croatia
Legal deposit, Croatia: National and University Library in Zagreb

Additional hard copies can be obtained from orders@intechopen.com

Elastomers, Edited by Nevin Çankaya
p. cm.
Print ISBN 978-953-51-3487-9
Online ISBN 978-953-51-3488-6

For EU product safety concerns:
IN TECH d.o.o., Prolaz Marije Krucifikse Kozulić 3, 51000 Rijeka, Croatia,
info@intechopen.com or visit our website at intechopen.com.

Meet the editor

Ass. Prof. Nevin Çankaya is associated with Usak University since 2011. She received her PhD, master's, and bachelor's degrees from Fırat University, Turkey. She is an expert in polymer composite science, working with bio-nanocomposites, polymeric materials, and polymer-clay composites. She has published over 20 scientific articles, over 30 conference papers, a book chapter entitled "Cellulose Grafting by Atom Transfer Radical Polymerization Method," *Cellulose - Fundamental Aspects and Current Trends*, another book edited by InTech Publication. Also, she has been a manager and researcher in many research projects. The editor is married and is a mother of two.

Contents

Preface IX

Section 1 Mechanical Properties of Elastomers 1

Chapter 1 **Electrorheological Elastomers 3**
Xufeng Dong, Chenguang Niu and Min Qi

Chapter 2 **Elastomers and Their Potential as Matrices in Polymer Electrolytes 21**
Li Na Sim and Abdul Kariem Arof

Chapter 3 **Nitrile Elastomer/LDH Composites with Varying Mg/Al Ratio, Curing, Nanoparticles Dispersion and Mechanical Properties 39**
Magdalena Lipińska

Chapter 4 **Nanocomposites Based on Thermoplastic Polyester Elastomers 75**
Sandra Paszkiewicz, Iman Taraghi, Anna Szymczyk, Elżbieta Piesowicz and Zbigniew Rosłaniec

Chapter 5 **Thermoplastic Elastomers Based on Block, Graft, and Star Copolymers 97**
Weiyu Wang, Wei Lu, Nam-Goo Kang, Jimmy Mays and Kunlun Hong

Chapter 6 **Fabrication and Properties of Rubber Nanofiber from Electrospinning 121**
Wirach Taweepreda

Section 2 Elastomers for Natural and Medical Applications 135

Chapter 7 **Castor Oil Polyurethanes as Biomaterials 137**
Said Arévalo-Alquichire and Manuel Valero

Chapter 8 **Natural Rubber and its Derivatives 159**
Azanam Shah Hashim and Siew Kooi Ong

Chapter 9 **Preparation, Characterization, and Preliminary Biocompatibility Evaluation of Carbon Ion-Implanted Silicone Rubber 189**
Xin Zhou, Yiming Zhang, Xiaohua Shi and Dongli Fan

Chapter 10 **Devulcanization of Elastomers and Applications 209**
Fabiula Danielli Bastos de Sousa

Chapter 11 **Dielectric Elastomer Sensors 231**
Na Ni and Ling Zhang

Preface

The purpose of this book is to concentrate on recent developments on elastomers. The articles collected in this book are contributions by invited researchers with a long-standing experience in different research areas. I hope that the material presented here will be understandable to a broader audience, not only scientists but also people with many different disciplines. The book contains eleven chapters in two sections: (1) "Mechanical Properties of Elastomers" and (2) "Elastomers for Natural and Medical Applications." The book provides detailed and current reviews in these different areas written by experts in their respective fields. This book will be useful for polymer workers and other scientists alike and will contribute to the training of current and future researchers, academics, PhD students, and other scientists.

Mechanical Properties of Elastomers:

- Electrorheological Elastomers
- Elastomers and Their Potential as Matrix in Polymer Electrolytes
- Nitrile Elastomer/LDH Composites with Varying Mg/Al Ratio, Curing, Nanoparticle Dispersion, and Mechanical Properties
- Nanocomposites Based on Thermoplastic Polyester Elastomers
- Thermoplastic Elastomers Based on Block, Graft and Star Copolymers
- Fabrication and Properties of Rubber Nanofiber from Electrospinning

Elastomers for Natural and Medical Applications:

- Castor Oil Polyurethanes as Biomaterials
- Natural Rubber and Its Derivatives Medical Değil
- Preparation, Characterization, and Preliminary Biocompatibility Evaluation of Carbon Ion-Implanted Silicone Rubber
- Devulcanization of Elastomers and Applications
- Dielectric Elastomer Sensors

Nevin Çankaya, PhD
Assistant Professor
Department of Chemistry, Faculty of Science
Usak University, Turkey

Section 1

Mechanical Properties of Elastomers

Chapter 1

Electrorheological Elastomers

Xufeng Dong, Chenguang Niu and Min Qi

Additional information is available at the end of the chapter

http://dx.doi.org/10.5772/intechopen.68396

Abstract

Electrorheological elastomers (EREs) are smart elastomers with tunable dynamic properties by applying electric field. They are composed by polarizable particles dispersing within elastomers matrix. In comparison with electrorheological fluids, the particles in EREs do not undergo aggregation and sedimentation. Without the requirement of large-sized electromagnetic coils, the smart devices based on EREs have simpler structures than those based on magnetorheological (MR) elastomers. The unique properties and merits of EREs give rise to their bright future in various fields. During the past decades, great progress has been made by scholars around the world. In this chapter, the conception, preparation, physical mechanisms, classification, development, influence factors, and applications of electrorheological elastomers is reviewed.

Keywords: electrorheological (ER), elastomers, dynamic properties, smart materials, storage modulus

1. Introduction

Smart materials are reactive materials. Their properties can be changed by exposure to stimuli, such as magnetic and electric fields, temperature, pH value, and moisture [1, 2]. Shape memory alloys and polymers [3], piezoelectric materials [4], magnetostrictive materials [5], and magnetorheological fluids and elastomers [6] are typical smart materials. Electrorheological (ER) materials have been considered as another promising smart material. The pioneering example of ER materials is electrorheological fluids, whose rheological properties, such as viscosity and yield stress, can be reversibly changed by applying external electric field. Electrorheological fluids are usually composed with polarizable particles dispersing within an insulating carrier liquid [7–9]. Under an applied electric field, ER fluids will transfer from liquid-like state to semisolid-like state. Compared with their magnetorheological counterparts, MR fluids, ER fluids present faster response (within a millisecond) and simpler design, but their field-induced yield stress cannot compete with MR fluids, and much higher electric field (several kV/mm) is required for notable changes. Besides, just like MR fluids, the aggregation of particles and the leakage of carrier liquid are two crucial shortcomings that prevent ER fluids from widely applications [10, 11].

To overcome those shortcomings, electrorheological elastomers (EREs) have developed in recent years. Typical ER elastomers consist of polarizable particles and elastomer matrix [12, 13]. According to the distribution of polarizable particles in the elastomer matrix, two kinds of ER elastomers can be distinguished as isotropic EREs and anisotropic EREs. The particles in the isotropic EREs are randomly distributed without order, while they are form ordered structures as chains or columns in the anisotropic ones. To get the anisotropic EREs, an electric field has to be applied during the crosslinking or curing or vulcanization process [14]. The ordered structures of particles are locked in the matrix. Compared with ER fluids, which are used in their postyield region, EREs are used in the preyield region. The viscoelastic properties, such as storage modulus and loss modulus, are electric field responsive, as shown in **Figure 1**. The most outstanding advantage of EREs with respect to ERFs is that the particles in ER elastomers do not undergo sedimentation. EREs also present obvious merits in comparison with magnetorheological elastomers (MREs). The EREs are driven by electric field rather than magnetic field, which is much easier to apply without the need of large-sized electromagnetic coils [15–17]. Therefore, the EREs can be used to design smart devices with simple structure and low weight. Due to those advantages, ER elastomers have been extensively studied in recent years and have presented promising future in applications.

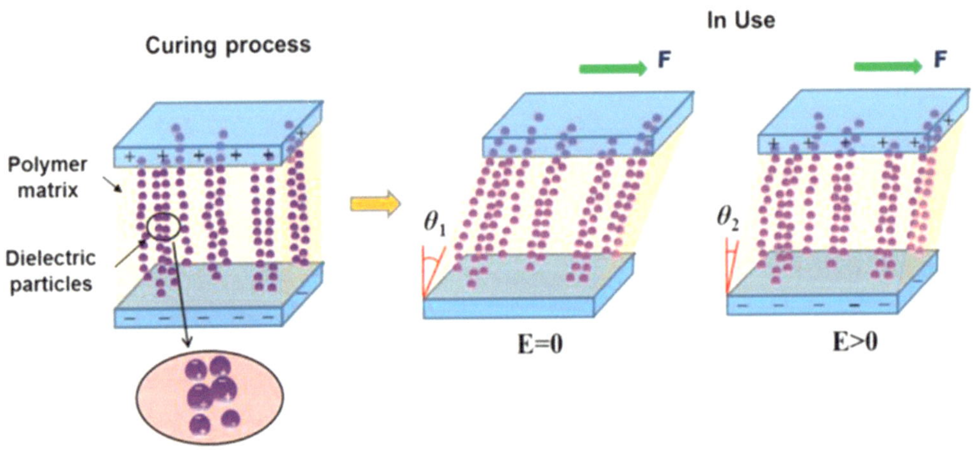

Figure 1. Schematic of EREs in curing process and in use.

2. Formulations and preparation of EREs

EREs are generally consisted of dielectric or semiconducting filler particles and polymer matrix. To prepare the EREs, filler particles are dispersed in the unvulcanized polymer and then locked in the matrix with the curing process. As mentioned above, the EREs will have isotropic structures if the particles randomly distribute in the matrix. In this case, the particles are dispersed at random and cured in the polymer without applying electric field. By curing the mixture of

http://dx.doi.org/10.5772/intechopen.68396

polymer and particles in the presence of an electric field, field-induced interaction between adjacent particles promote them to form anisotropic microstructures like chains or columns aligned paralleling to the field direction. These structures will be retained in the matrix and that have been proved to be helpful for improving the field-dependence of EREs' mechanical performance when they are exposed to extra electric field. SEM photographs have been used to show the presence of chain structures of particles in the EREs, besides, wide-angle X-ray scattering technique is also used to observe the anisotropic structures in clay/gelatin EREs [18, 19].

For anisotropic EREs adopting gels or rubbers as matrix, the mixtures are cured in the mold made of glass or other insulating materials with two electrodes, which is shown as **Figure 2**. The electric field should be maintained during the entire curing process of the matrix, and the strength of the field is at least 1–2 kV/mm to ensure effective polarization of the particles. The appearance of EREs will be strongly influenced by the characteristics of the matrix raw materials. Some bubbles may be left in the mixture during the stirring process to achieve a ERE with voids, which will degrade the performance of the material. Thus, the mixture should be performed degassing to remove the bubbles within it.

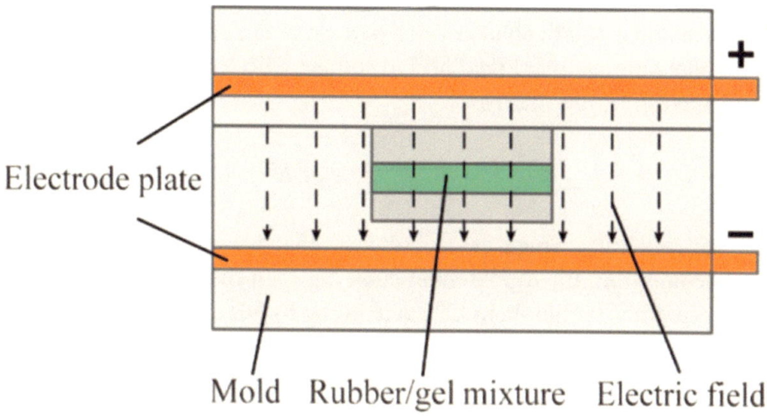

Figure 2. Schematic of ERE vulcanizing mold with DC electric field [18].

3. Physical mechanisms of EREs

The electroviscoelastic characteristics of EREs are mainly caused by the electro attraction between dispersed particles. Thus, the nature of the forces within adjacent particles is one of outstanding issues. Although some theories have been proposed, it is widely accepted that particle polarization forces play a dominant role in the ER effect [20, 21]. Klingenberg et al. first proposed the theoretical model to describe some aspects of ER suspensions with the one point dipole approximation model [22]. In this model, real ER solids could be considered as monodisperse suspensions of dielectric hard spheres in a continuous polymer matrix. The polarization in the one point dipole approximation model that occurs inside the particles instead of on their surface, and the dipoles form within adjacent particles will be stronger as the dipoles come close to each other. Based on the classical theory, the polarization force of the particles under extra electric field can be given by [12]

$$F(R,\theta) = 12\pi r^2 \varepsilon_0 \varepsilon_m \kappa^2 E^2 (r/R)^4 [(3\cos^2\theta - 1)e_r + \sin(2\theta)e_\theta] \quad (1)$$

$$\kappa = (\varepsilon_p - \varepsilon_m)/(\varepsilon_p + 2\varepsilon_m) \quad (2)$$

where R is the center-to-center distance between the adjacent particles, θ is the angle between the applied electric field and the chain axis (**Figure 3**), r is the radius of the spheroidal particle, E is the strength of electric field, ε_0 is the permittivity in vacuum (= 8.854×10^{-12} F/m), ε_p and ε_m are the dielectric constants of the filler particles and matrix, respectively, e_r and e_θ are unit vectors in r and chain axis directions, respectively.

Eq. (1) is an expression obtained from solving Laplace's equation about a pair of spheroidal particles in the continuous unbounded medium. By assuming the spheres interaction was replaced by dipoles with the same intensity as the dielectric spheres in an electric field, they further obtained the point dipole moment, which can be written as

$$\mu = 4\pi r^3 \varepsilon_0 \varepsilon_m \kappa E \quad (3)$$

Eq. (2) shows the volume polarizability of one particle under applied field. If the adjacent two particles are parallel to the applied field and in contact with each other, the dipole interaction force between them can be rewritten as

$$F = (3/2)\pi r^2 \varepsilon_0 \varepsilon_m \kappa^2 E^2 \quad (4)$$

Based on the above results, Shiga et al. assumed that interactions between two adjacent particles mainly come from the dipole interaction between these two particles instead of the interaction between paths within them [23]. According to this assumption, macroscopic visco-elastic properties such as storage and loss modulus can be estimated by calculating the overall

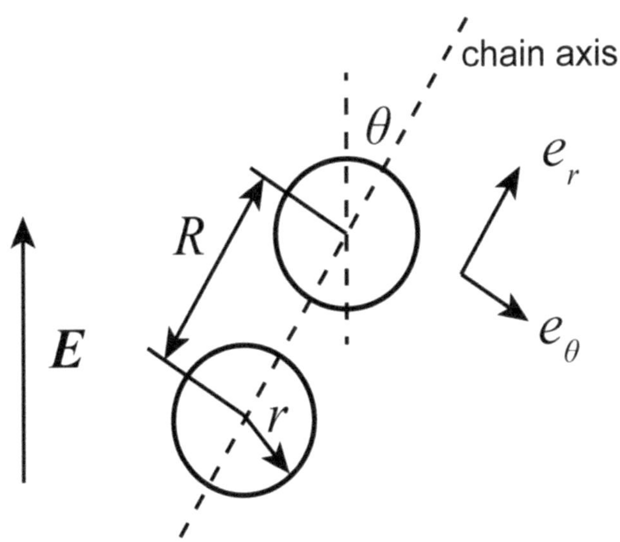

Figure 3. Two adjacent particles in electric field.

http://dx.doi.org/10.5772/intechopen.68396

electrostatic interaction force within the adjacent particles in the matrix. Thus, the following equation can be used to calculate the increase in storage modulus with varied extra electric field:

$$\Delta G = (9/4)\phi\varepsilon_m\kappa^2E^2 \tag{5}$$

where ΔG is the change of storage modulus, φ is the volume fraction of dielectric particles. It can be inferred from Eq. (5) that ΔG is proportional to the particle content and electric field intensity.

Liu et al. investigated the change of electrorheological and mechanical properties of EREs with anisotropically filled dielectric particles by numerical simulations [24]. Their simulations mainly focus on the distribution of electrostatic energy between the tips of two adjacent particles in a chain, as shown in **Figure 4**. According to their research, when the dielectric constant of filler particles is much greater than the matrix, the intensity of electric field will be concentrated between these two particles. This implied that electrostatic stress is located on the particle surface and the local fields between the tips of particles play an important role in determining the strength of restoring force.

The importance of local field strength between the particles on ER materials was further proved in the study of Lu et al., and a polar molecule (PM) dominated model had been proposed by them to explain the enhanced ER performance of the novel ER suspension [25, 26]. For this kind of ER suspensions, the polar molecules may orientate along the applied field in the tips of adjacent particles caused by the local field, which is three orders of magnitude larger than external electric field under certain condition. Once the polar molecules on particle surface are orientated under electric field, the PM-ER effect mainly comes from the attractive force f_{m-e} of polar molecules and polarization charges on the nearby particle (**Figure 5**). The overall attractive force F_{m-e} of molecule-charge in a unit area that perpendicular to applied electric field is given by

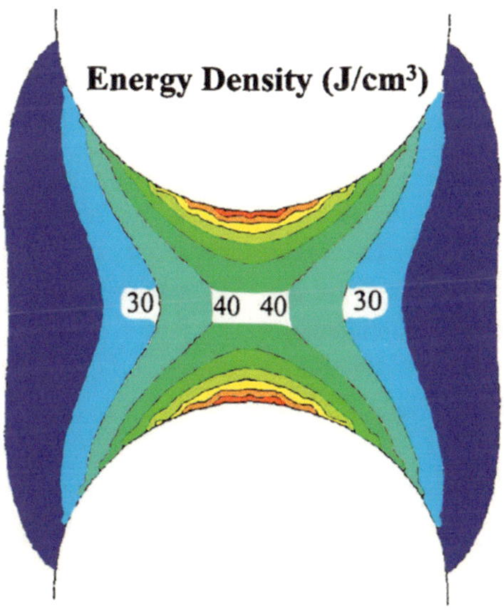

Figure 4. Contour of electrostatic energy distribution for rod-like filler particles in a 2-D axisymmetric simulation [24].

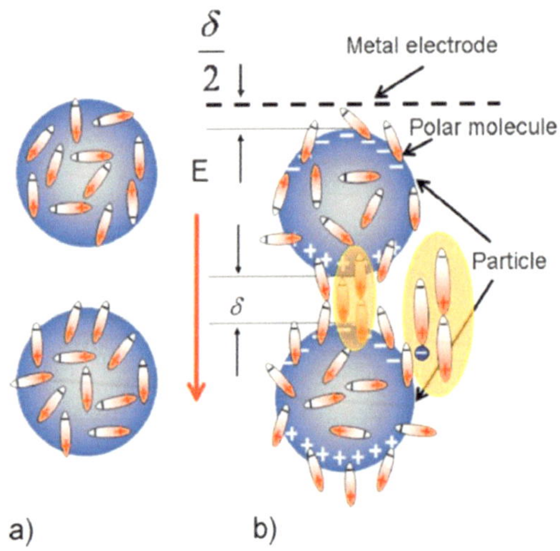

Figure 5. Schematic of the polar molecules aligned in the gap of adjacent particles. (a) The polar molecules adsorbed on particle surface are arbitrarily oriented when $E = 0$. (b) The polar molecules in the gap of the particles are oriented parallel to the field when E is high enough [26].

$$F_{m-e} = \frac{3\phi}{2\pi r^2} N f_{m-e} = A \frac{3\phi \rho_m e \mu^2 E}{\pi r \varepsilon_0 \varepsilon_m d^2} \tag{6}$$

in which

$$f_{m-e} = \frac{e\mu}{2\pi \varepsilon_0 \varepsilon_m d_{m-e}^3} \tag{7}$$

where A is a factor that related to the properties of the particle and oil, φ is the volume fraction of dielectric particles, ρ_m is the area density of the polar particles on the particle surface, e is the fundamental charge, r is the radius of filler particle, μ and d are the dipole moment and size of the polar molecules, respectively. Polar molecule-dominated model was initially used to explain the mechanical properties of PM-ER fluids, while it also opened the door to the innovation of novel ER elastomers filled with polar molecule modified particles.

One recent study by Zhao et al. proposed a theory on the mesostructured design of EREs for gaining giant tunable stiffness [27]. The EREs in this theory consist of two kinds of layers with defined thickness and dielectric constant, and the effective dielectric constant of the layered structure $\overline{\varepsilon}$ can be regarded as a function of variable θ with coefficient ε, n, and m, as shown in **Figure 6**. By connecting the thermodynamic system with the geometrical parameters of EREs, they can calculate the electrorheological coefficient as

$$K_{ER} = 4\alpha \cos(2\theta)\left(\frac{\partial\theta}{\partial\Delta\theta}\right)^2 + 2\alpha \sin(2\theta)\frac{\partial^2\theta}{\partial\Delta\theta^2} \tag{8}$$

$$f(\theta) = \alpha \cos(2\theta) + \beta \tag{9}$$

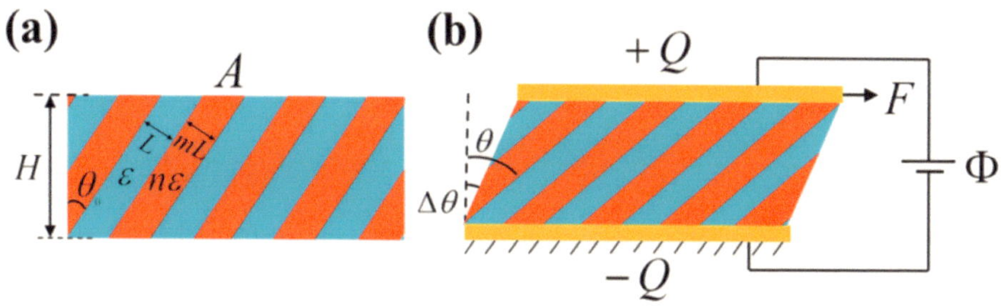

Figure 6. Schematic of one ERE with layered mesostructures at undeformed state (a) and undergoing simple shear deformation by an applied force and a voltage (b) [27].

where $\alpha = [f(0) - f(\pi/2)]/2$ and $\beta = [f(0) + f(\pi/2)]/2$. The EREs will have giant tunable stiffness when the coefficient K_{ER} reaches the maximum value. Finally, they found when $m = \sqrt{n}$, K_{ER} reaches its maximum with a given m. The K_{ER} can be expressed as

$$K_{\mathrm{ER}}^{\max}(0) = 2(\sqrt{n} - 1)^2 \tag{10}$$

In their conclusion, the EREs with a layered mesostructured arranged in parallel configuration will have the most widely tunable stiffness among all the structures with different orientations.

4. Development and diversification of EREs

4.1. Inorganic particles filled EREs

Inorganic particles are most frequently employed as the dispersion phase as their outstanding properties, including high permittivity, easy synthesis, and good environmental stability. A range of inorganic micro/nanosize particles, such as $PbTiO_3$ [28], silica [29], montmorillonite clay [19], titanium dioxide [30], and barium titanate [31] have been applied as the candidates of dispersed phase in EREs. Gao and Zhao investigated the mechanical properties of hydrogel elastomers containing barium titanate particles [32]. They found the aligned particle chain effect could induce large compression modulus that can be controlled by applied field strength. While in the research of Mitsumata on swollen silicone gel containing barium titanate, a negative electrorheological effect appeared for microscopic phase separation in the elastomers [33]. To enhance the electric responses, Gao et al. also prepared $BaTiO_3$ particles coated with many kinds of polymers, such as polyimide, chitosan, polystyrene, poly acrylic acid, and poly acrylamide [34]. They found the hydrous elastomer filled with $BaTiO_3$/polyimide particles show strongest electric response, which is closely related to hydrophilic surface and permittivity of the particles.

In order to improve the permittivity and conductivity of the particles, Kossi et al. prepared the TiO_2 doped with acetylacetone dipolar molecules (Acac), and the EREs prepared by dispersing

TiO_2-Acac in PDMS achieved a relative storage modulus change higher than 500 kPa at 2 kV/mm [30]. The relative storage modulus change is defined by the following equation

$$\frac{\Delta G'}{G'_0} = \frac{G' - G'_0}{G'_0} 100\% \tag{11}$$

where G' and G_0' are the storage modulus with and without an electric field, $\Delta G'$ is the change of storage modulus. Inspired by the study of Wen et al. on polar molecules dominated ER materials, Dong et al. proposed a kind of elastomer containing polar molecule-modified particles, which is expected to have much enhanced ER activities [18]. The elastomers synthesized by dispersing urea-coated titanium dioxide (TiO_2/urea) particles in silicon rubber presented high ER performance, the relative ER effects of which achieves 266% at 3 kV/mm (**Figure 7**).

4.2. Organic particles filled EREs

Organic hydrous particles, such as starch and cellulose, have been widely used in ER materials and shown strong ER performance. When Shiga et al. prepared the earliest elastomers with ER activity, the filler particles they adopted were polymethacrylic acid cobalt (II) salt with a small amount of absorbed water [21]. This kind of organic particles will present stable dispersion properties in polymer matrix and relatively good ER behaviors. However, the EREs containing organic hydrous particles will suffer from some defects. First, the leakage current density of EREs would be large if too much water is absorbed on the

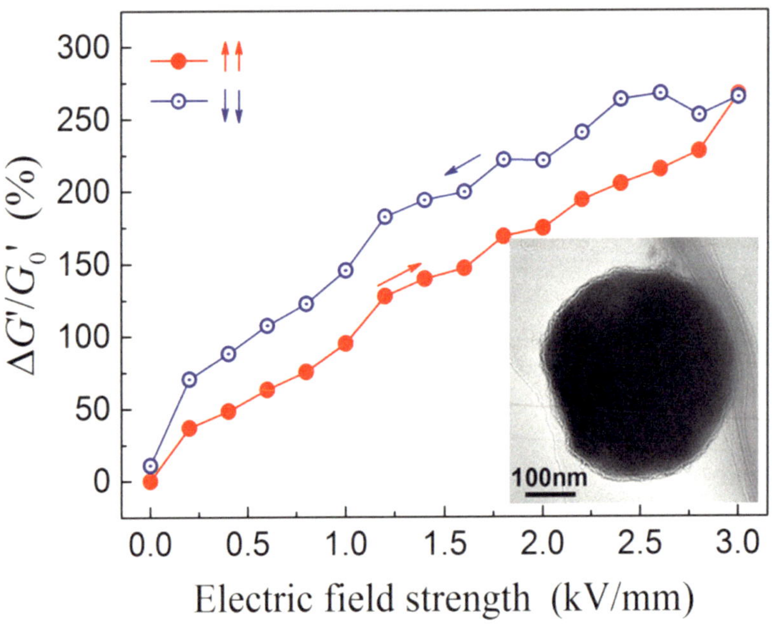

Figure 7. Relative ER effect of TiO_2/urea particles filled elastomers as a function of electric field strength, inset figure is the TEM images of TiO_2/urea particles.

particles, which will be unable to apply high voltage on the samples. In addition, the EREs containing hydrous particles are sensitive to working conditions, especially the operating temperature. Besides, the moisture in the EREs is also possible to cause corrosion problems on the electrode [35].

To overcome the shortcomings of the EREs filled with organic hydrous particles, some systems with anhydrous polymer particles have been investigated in recent years. Conducting polymers that contains conjugated π bonds have been used as ER materials due to their polarization properties under applied electric field. Besides, polymers with polarizable functional groups on the molecules backbone belong to another kind of anhydrous organics that adopted as ER materials. PANI is one kind of conducting polymers that has been regarded as an appropriate candidate for EREs. It has many excellent properties such as high conductivity and good environment stability, which is important for application in ER materials. Hiamtup and Sirivat investigated the electrorheological response of EREs filled with PANI particles, and they found the storage modulus of this kind of ERE will increase about 10–50% with the ascending field strength [36]. Other conducting polymers, such as polythiophene [37], poly(p-phenylene) [38], and polypyrrole [39] have also been studied as the dispersed phase of EREs. Compared with the inorganic particles, the polymeric dispersion particles are softer and have better compliance with the matrix.

5. Influence factors of ER performance

Since EREs are usually made of two constituent materials with different properties, the influence factors mainly come from three parts in them: the filler particles, matrix, and interface of these two phases.

5.1. Effects of particle properties

As the dispersion phase of EREs, particles play the most important role in determining the ER performance. According to the theory in previous study, the dielectric constant mismatch is closely related to polarization of EREs [40]. The polarization force is positively related to the coefficient κ in Eq. (2), which infers the difference between the dielectric constant of particles and matrix. When the permittivity of filler particle is much larger than that of polymer matrix, the ER materials will have stronger ER activity.

The content of particles filled in the EREs also has a great influence on their ER performance, and it has been investigated by many researchers theoretically and experimentally. As described in Eq. (5), the change of storage modulus of EREs is $\Delta G'$ is proportional to the content of filler particles [23]. In the study of Lu et al., the total interaction force between polarized molecule and charge in a unit area is also proportional to the particle content [25]. In 2006, Hiamtup and Sirivat investigated the viscoelastic properties of EREs with the particle volume fraction ranging from 0 to 20%, and they got the conclusion that both storage and loss moduli increased with the particle concentration, as shown in **Figure 8** [36].

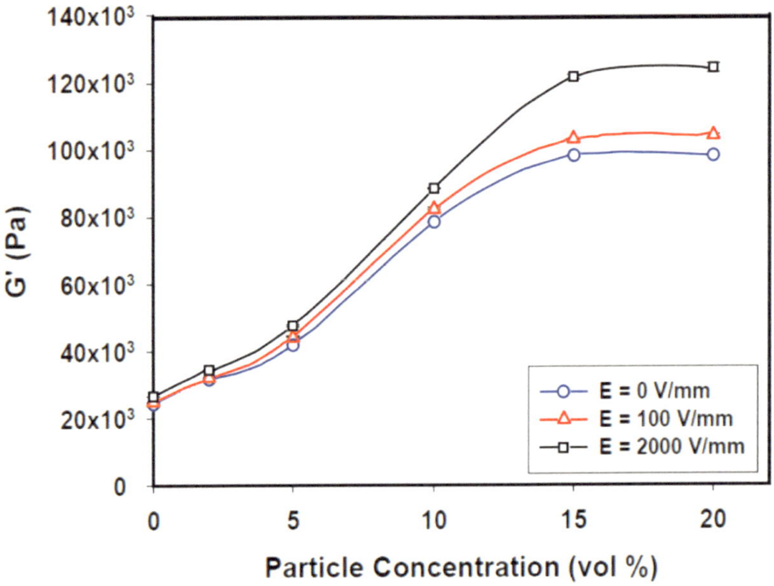

Figure 8. Particle concentration dependence of storage modulus of PANI/PDMS EREs under various electric field [36].

The morphology of particles is another important factor influences the viscoelastic properties of EREs under electric field. Kanu and Shaw systematically studied the role of particle geometry on ER response based on the dipole theory [41]. They found the polarization of the particles with larger aspect ratios will be stronger and these particles will participate in enhanced dipole-dipole interactions. Liu et al. computed the electrostatic forces between the particles with three kinds of geometries (sphere, capsule, and pointed capsule) in the EREs using numerical simulations [14]. Their calculation inferred the EREs containing the particles with a higher aspect ratio will present higher ER response for the formation of more particle chains. Jang et al. investigated the ER activity of the suspension containing pure silica or PANI coated mesoporous silica particles with different aspect ratio [42, 43]. According to their research, the ER materials with higher aspect ratios will have improved dielectric properties of large polarizability and short relaxation time. Therefore, the synergistic contribution of geometric effects and dielectric properties resulted in stronger ER activity.

5.2. Effects of polymer matrix

The tunable stiffness range is one of the most important properties for EREs, which can be evaluated by the relative ER effect as Eq. (11). Many research studies have focused on the materials used for the matrix of EREs due to their roles in determining the initial elastic modulus. Sakurai et al. adopted silicone elastomers with different moduli as the matrix of EREs and investigated how the ER properties depend on the elasticity of the elastomers [44]. They found the largest relative increase in modulus under electric field occurs when silicone

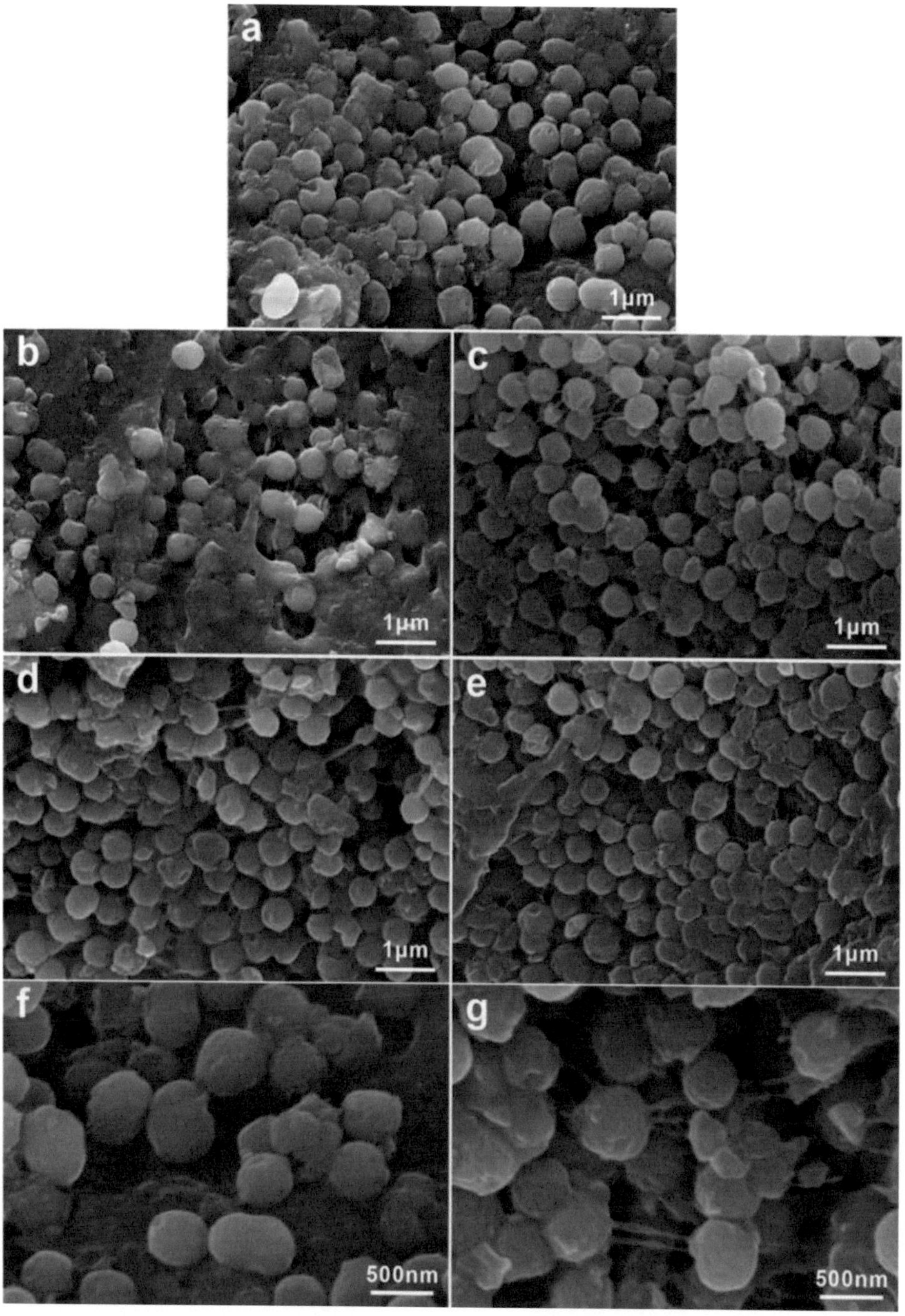

Figure 9. Section SEM microphotographs of (a) TiO_2, (b) TiO_2/A1, (c) TiO_2/A3, (d) TiO_2/V1, and (e) TiO_2/V3 particles filled elastomers have been soaked in toluene and (f and g) the magnification images of (a and d). A1 means the weight fraction of A174 is 1% to the TiO_2 particles, the rest symbols represent similar meaning [48].

elastomer with low elasticity is used. Except for silicone rubbers, many elastomers with low initial moduli have been used as the matrix of EREs, such as acrylic and polyisoprene elastomer [37, 45]. To further decrease the initial moduli of the EREs, Gao et al. also tried to use the gels like gelatin and chitosan as the matrix [46, 47]. However, most of the EREs used gels as the matrix can only work in compression mode due to their mechanic properties. Besides, it is mentioned in the study of Liu et al. that the conductivity of polymer matrix can play an important role in the electric field distribution of EREs under high field strength [14]. It is implied that the polarization of filler particles in high field may be affected by the conductivity of the matrix.

5.3. Effect of particles/matrix interface

As a composite, the interface between filler particles and matrix plays an important role in the ER performance of EREs, which has been overlooked in previous studies. To improve the interfacial bond strength in the EREs and find out how does the interface influence the ER performance, Dong et al. adopted two coupling agents, 3-(trimethoxysilyl)propyl methacrylate (A174), and triethoxyvinylsilane (VTEO), to modify the surface of the TiO_2 particles [48]. Compared with the bare TiO_2 particle filled sample, the modified TiO_2 particle filled elastomers present bridging structures between the particles and the matrix, enhanced mechanical properties, and a higher dielectric constant, indicating enhanced interfacial bond strength (**Figure 9**).

The viscoelastic properties of the bare and the modified TiO_2 particle filled elastomers were tested and compared in this study. The results indicated that the ER performance, including the field-induced storage modulus and the storage modulus sensitivity, can be promoted by improving the bond strength between the TiO_2 particles and the silicon rubber, as shown in **Figure 10**.

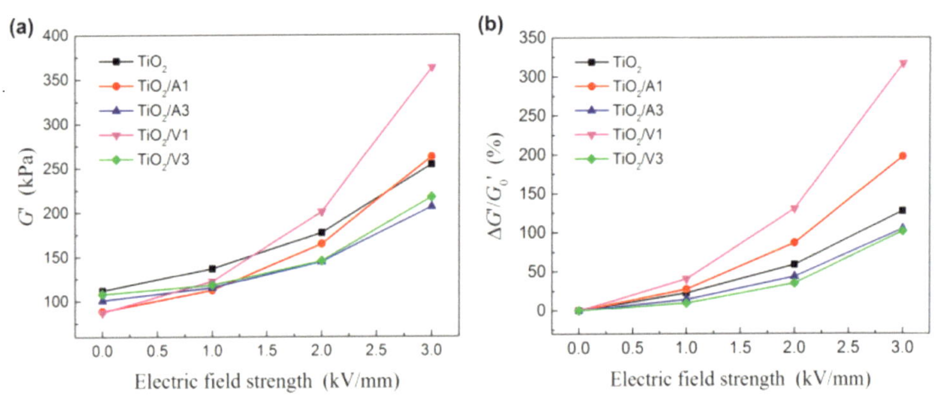

Figure 10. Storage modulus G' (a) and relative ER effect (b) of five types of particle filled elastomers as a function of electric field strength [48].

6. Applications of EREs

On the basis of the ER effect and special mechanic properties, EREs can be applied in several fields, including dampers, automotive industries, building base-isolation, and smart skins in bionic technology [12, 49]. Although commercialization of ERE-based devices has not been implemented, the tentative application of EREs has been done in some laboratory research studies. Biggerstaff and Kosmatka used the silicone gel based EREs as active dampers [50]. It is shown in the harmonic tests that the EREs reveal a factor six increase in stiffening and a factor of three decreases in damping under electric field. Koyanagi et al. created a novel prototype linear actuator with ERE drum and found the response time of the EREs was fast enough for mechatronics or robotics applications as physical supporting in welfare situations, while some problems like the relation between the pressure on the surface of the EREs and generative force were left for further investigation [51]. Besides, Wei et al. investigated the vibration characteristics and control capabilities of a cantilever sandwich beam at different electric fields [52]. The results indicated that the natural frequencies of the ERE sandwich beam increase and the vibration amplitudes at natural frequencies of the beam decrease with the ascending electric field intensity. Since the vibration characteristics of ERE beams can be controlled by changing the strength of applied field, the EREs are useful for applications in engineering structures where variable performance is desired. In 2013, Zhu et al. designed a shear mode ERE shock absorber and evaluated its vibration response performance under various excitation frequencies with and without external electric field [17]. Their experimental results showed that the stiffness and damping properties of the shock absorber could be controlled by changing the strength of applied electric field, whose macrofeatures was that the damping coefficient increased with the ascending electric field strength, and the damping effect in high frequency was better comparing with that in low frequency.

7. Summary and outlook

With respect to the scientific papers on EREs published in the past few years, methods of preparing new EREs and improving the dynamic mechanical properties, and the modeling approach for simulating the electrorheological and viscoelastic mechanical behavior of these EREs are of increasing scientific and technological importance. Based on the investigations about ER fluids, the point dipole approximation model and its modified model are most widely used to describe the viscoelastic properties controlled by electric field, and the meso-scale structure of EREs is also analyzed with micromechanics of composites. In addition, researchers have made considerable progress in developing new materials in this field, including inorganic particles, organic hydrous, and organic anhydrous particles filled EREs. For different types of EREs, the influencing factors on ER performance related to the filler particles, polymer matrix, and the interface between particles and matrix are studied experimentally or theoretically. In recent years, some preliminary explorations in the application of EREs have been carried out and some valuable conclusions are obtained.

Even though many reports focus on the mechanical and ER performance of EREs, some defects still prevent them from practical applications. The most important problem of EREs is the rather high voltage must be applied for obtaining large relative storage modulus, which will increase the security risks. The relatively low elastic modulus of the EREs makes them unable to sustain something of great weight, which is another defect that will narrow the application fields for EREs. Apart from that, considerable work is needed to broaden the practical application of EREs and make them more reliable under long service period. The EREs containing polar molecules modified particles have presented enhanced ER activities under relatively low field strength in previous studies, which indicates that to increase thc polarization abilities of particles under the guidance of polar molecules dominated ER effect model is an effective method to improve ER performance of the composites. Therefore, research studies of EREs are still widely open for extensive works on theoretical analysis of basic mechanisms on electroresponsive behavior of the particles based on their dielectric characteristics, and dynamic viscoelastic mechanical properties. Besides, new generation of EREs should have both large relative storage modulus and mechanical properties. On the other hand, it is important to design the devices with full insulation protection and safety power supply aiming at commercial applications. When these problems are solved, EREs should be one kind of promising smart materials that serve for modern society.

Acknowledgements

This work was supported by the National Natural Science Foundation of China under the grant number of 51478088.

Author details

Xufeng Dong*, Chenguang Niu and Min Qi

*Address all correspondence to: dongxf@dlut.edu.cn

Dalian University of Technology, China

References

[1] Kang I, Heung YY, Kim JH. Introduction to carbon nanotube and nanofiber smart materials. Composites Part B-Engineering. 2006;**37**(6):382–394. DOI: 10.1016/j.compositesb.2006.02.011

[2] Liu YD, Choi HJ. Electrorheological fluids: Smart soft matter and characteristics. Soft Matter. 2012;**8**(48):11961–11978. DOI: 10.1039/c2sm26179k

[3] Leng JS, Lu HB, Liu YJ. Shape-memory polymers-A class of novel smart materials. Mrs Bulletin. 2009;**34**(11):848–855. DOI: 10.1557/mrs2009.235

[4] Wang ZL, Song JH. Piezoelectric nanogenerators based on zinc oxide nanowire arrays. Science. 2006;**312**(5771):242–246. DOI: 10.1126/science.1124005

[5] Dong XF, Qi M, Guan XC. Microstructure analysis of magnetostrictive composites. Polymer Testing. 2010;**29**(3):369–374. DOI: 10.1016/j.polymertesting.2009.12.012

[6] Xu YG, Gong XL, Xuan SH. A high-performance magnetorheological material: Preparation, characterization and magnetic-mechanic coupling properties. Soft Matter. 2011;**7**(11):5246–5254. DOI: 10.1039/c1sm05301a

[7] Havelka KO, Filisko FE, editors. Progress in electrorheology: science and technology of electrorheological materials. New York: Springer US; 1995. DOI:10.1007/978-1-4899-1036-3

[8] Liu XH, Guo JJ, Cheng YC. Synthesis and electrorheological properties of polar molecule-dominated TiO_2 particles with high yield stress. Rheologica Acta. 2010;**49**(8):837–843. DOI: 10.1007/s00397-010-0452-y

[9] Dong XF, Huo S, Qi M. Comparison of electrorheological performance between urea-coated and graphene oxide-wrapped core-shell structured amorphous TiO_2 nanoparticles. Smart Materials and Structures. 2016;**25**(1):015033. DOI: 10.1088/0964-1726/25/1/015033

[10] Zhao XP, Yin JB. Preparation and electrorheological characteristics of rare-earth-doped TiO_2 suspensions. Chemistry of Materials. 2002;**14**(5):2258–2263. DOI: 10.1021/cm011522w

[11] Dong YZ, Yin JB, Zhao XP. Microwave-synthesized poly(ionic liquid) particles: A new material with high electrorheological activity. Journal of Materials Chemistry A. 2014;**2**(25):9812–9819. DOI: 10.1039/C4TA00828F

[12] Shiga T, Okada A, Kurauchi T. Electroviscoelastic effect of polymer blends consisting of silicone elastomer and semiconducting polymer particles. Macromolecules. 1993;**26**(25):6958–6963. DOI: 10.1021/ma00077a038

[13] Hao LM, Shi ZH, Zhao XP. Mechanical behavior of starch/silicone oil/silicone rubber hybrid electric elastomer. Reactive & Functional Polymers. 2009;**69**(3):165–169. DOI:10.1016/j.reactfunctpolym.2008.12.014

[14] Liu B, Boggs SA, Shaw MT. Electrorheological properties of anisotropically filled elastomers. IEEE Transactions on Dielectrics and Electrical Insulation. 2001;**8**(2):173–181. DOI: 10.1109/94.919919

[15] Gordaninejad F, Wang XJ, Mysore P. Behavior of thick magnetorheological elastomers. Journal of Intelligent Material Systems and Structures. 2012;**23**(9):1033–1039. DOI: 10.1177/1045389X12448286

[16] Li YC, Li JC, Li WH. A state-of-the-art review on magnetorheological elastomer devices. Smart Materials and Structures. 2014;**23**(12):123001. DOI: 10.1088/0964-1726/23/12/123001

[17] Zhu SS, Qian XP, He H. Experimental research about the application of ER elastomer in the shock absorber. In: Sung WP, Zhang CZ, Chen R, editors. Biotechnology, Chemical and Materials Engineering Ii, Pts 1 and 2. Stafa-Zurich: Trans Tech Publications Ltd; 2013. pp. 371–376. DOI: 10.4028/www.scientific.net/AMR.641-642.371

[18] Niu CG, Dong XF, Qi M. Enhanced electrorheological properties of elastomers containing TiO_2/Urea core-shell particles. ACS Applied Materials & Interfaces. 2015;**7**(44):24855–24863. DOI: 10.1021/acsami.5b08127

[19] Wang B. Rozynek Z, Zhou M. Wide angle scattering study of nanolayered clay/gelatin electrorheological elastomer. Journal of Physics: Conference Series. 2009;**149**(1):012032 (6 pp.)–012032 (6 pp.). DOI: 10.1088/1742-6596/149/1/012032

[20] Parthasarathy M, Klingenberg DJ. Electrorheology: Mechanisms and models. Materials Science & Engineering R-Reports. 1996;**17**(2):57–103. DOI: 10.1016/0927-796X(96)00191-X

[21] Shiga T, Ohta T, Hirose Y. Electroviscoelastic effect of polymeric composites consisting of polyelectrolyte particles and polymer gel. Journal of Materials Science. 1993;**28**(5):1293–1299. DOI: 10.1007/BF01191967

[22] Klingenberg DJ, Vanswol F, Zukoski CF. Dynamic simulation of electrorheological suspensions. Journal of Chemical Physics. 1989;**91**(12):7888–7895. DOI: 10.1063/1.457256

[23] Shiga T. Deformation and viscoelastic behavior of polymer gels in electric fields. Proceedings of the Japan Academy Series B-Physical and Biological Sciences. 1998;**74**(1):6–11. DOI:10.2183/pjab.74.6

[24] Liu B, Boggs SA, Shaw MT. Numerical simulation of electrorheological properties of anisotropically filled elastomers. In: 1999 Annual Report Conference on Electrical Insulation and Dielectric Phenomena (Cat. No.99CH36319); 1999; IEEE; Piscataway, NJ, USA. pp. 82–85 vol.1. DOI: 10.1109/ceidp.1999.804598

[25] Lu KQ, Shen R, Wang XZ. Polar molecule dominated electrorheological effect. Chinese Physics. 2006;**15**(11):2476–2480. DOI: 10.1088/1009-1963/15/11/002

[26] Shen R, Wang XZ, Lu Y. Polar-molecule-dominated electrorheological fluids featuring high yield stresses. Advanced Materials. 2009;**21**(45):4631–4635. DOI: 10.1002/adma.200901062

[27] Cao CY, Zhao XH. Tunable stiffness of electrorheological elastomers by designing mesostructures. Applied Physics Letters. 2013;**103**(4):041901. DOI: 10.1063/1.4816287

[28] Wen WJ, Tam WY, Sheng P. Electrorheological fluids using bidispersed particles. Journal of Materials Research. 1998;**13**(10):2783–2786. DOI: 10.1557/JMR.1998.0381

[29] Liu B, Shaw MT. Electrorheology of filled silicone elastomers. Journal of Rheology. 2001;**45**(3):641–657. DOI: 10.1122/1.1366716

[30] Kossi A, Bossis G, Persello J. Electro-active elastomer composites based on doped titanium dioxide. Journal of Materials Chemistry C. 2015;**3**(7):1546–1556. DOI: 10.1039/c4tc02535k

[31] Li RJ, Wei WX, Hai JL. Preparation and electric-field response of novel tetragonal barium titanate. Journal of Alloys and Compounds. 2013;**574**:212–216. DOI: 10.1016/j.jallcom.2013.04.203

[32] Gao LX, Zhao XP. Electrorheological behaviors of barium titanate/gelatin composite hydrogel elastomers. Journal of Applied Polymer Science. 2004;**94**(6):2517–2521. DOI: 10.1002/app.21212

[33] Mitsumata T, Sugitani K, Koyama K. Electrorheological response of swollen silicone gels containing barium titanate. Polymer. 2004;**45**(11):3811–3817. DOI: 10.1016/j.polymer.2004.03.056

[34] Gao LX, Li L, Qi XR. Enhancement on electric responses of $BaTiO_3$ particles with polymer-coating. Polymer Composites. 2013;**34**(6):897–903. DOI: 10.1002/pc.22507

[35] Fang FF, Choi HJ. Electrorheological Fluids: Materials and Rheology. In: Kontopoulou, Marianna, editors. Applied Polymer Rheology: Polymeric Fluids with Industrial Applications. 1st ed. John Wiley & Sons, Inc. Hoboken, NJ, USA; 2011. pp. 285–302. DOI: 10.1002/9781118140611.ch10

[36] Hiamtup P, Sirivat A. Soft and flexible actuator based on electromechanical response of polyaniline particles embedded in cross-linked poly(dimethyl siloxane) networks—art. no. 61680X. In: BarCohen, Y., editors. Smart Structures and Materials 2006: Electroactive Polymer Actuators and Devices. Bellingham: Spie-Int Soc Optical Engineering; 2006; pp. X1680–X1680. DOI: 61680x10.1117/12.658549

[37] Puvanatvattana T, Chotpattananont D, Hiamtup P. Electric field induced stress moduli in polythiophene/polyisoprene elastomer blends. Reactive & Functional Polymers. 2006;**66** (12):1575–1588. DOI: 10.1016/j.reactfunctpolym.2006.06.001

[38] Kunanuruksapong R, Sirivat A. Poly(p-phenylene) and acrylic elastomer blends for electroactive application. Materials Science and Engineering a-Structural Materials Properties Microstructure and Processing. 2007;**454**:453–460. DOI: 10.1016/j.msea.2006.12.033

[39] Ludeelerd P, Niamlang S, Kunaruksapong R. Effect of elastomer matrix type on electromechanical response of conductive polypyrrole/elastomer blends. Journal of Physics and Chemistry of Solids. 2010;**71**(9):1243–1250. DOI: 10.1016/j.jpcs.2010.05.003

[40] Chen Y, Sprecher AF, Conrad H. Electrostatic particle-particle interactions in electrorheological fluids. Journal of Applied Physics. 1991;**70**(11):6796–6803. DOI: 10.1063/1.349855

[41] Kanu RC, Shaw MT. Enhanced electrorheological fluids using anisotropic particles. Journal of Rheology. 1998;**42**(3):657–670. DOI: 10.1122/1.550944

[42] Noh J, Yoon CM, Jang J. Enhanced electrorheological activity of polyaniline coated mesoporous silica with high aspect ratio. Journal of Colloid and Interface Science. 2016;**470**:237–244. DOI: 10.1016/j.jcis.2016.02.061

[43] Yoon CM, Lee K, Noh J. Electrorheological performance of multigram-scale mesoporous silica particles with different aspect ratios. Journal of Materials Chemistry C. 2016;**4** (8):1713–1719. DOI: 10.1039/c5tc04124d

[44] Sakurai R, See H, Saito T. Effect of matrix viscoelasticity on the electrorheological properties of particle suspensions. Journal of Non-Newtonian Fluid Mechanics. 1999;**81**(3):235–250. DOI: 10.1016/S0377-0257(98)00099-8

[45] Kunanuruksapong R, Sirivat A. Electrical properties and electromechanical responses of acrylic elastomers and styrene copolymers: effect of temperature. Applied Physics a-Materials Science & Processing. 2008;**92**(2):313–320. DOI:10.1007/s00339-008-4513-3

[46] Gao LX, Zhao XP. The response of starceugelatin/glycerin aqueous electrorheological elastomer to applied electric field. International Journal of Modern Physics B. 2005;**19**(7–9):1449–1455. DOI: 10.1142/s0217979205030438

[47] Gao LX, Chen JL, Han XW. Electric-field response behaviors of chitosan/barium titanate composite hydrogel elastomers. Journal of Applied Polymer Science. 2015;**132**(25):42094 (1–6). DOI: 10.1002/app.42094

[48] Dong XF, Niu CG, Qi M. Enhancement of electrorheological performance of electrorheological elastomers by improving TiO_2 particles/silicon rubber interface. Journal of Materials Chemistry C. 2016;**4**(28):6806–6815. DOI: 10.1039/c6tc01447j

[49] Kakinuma Y, Aoyama T, Anzai H. Application of the electro-rheological gel to fixture devices for micro milling processes. Journal of Advanced Mechanical Design Systems and Manufacturing. 2007;**1**(3):387–398. DOI: 10.1299/jamdsm.1.387

[50] Biggerstaff JM, Kosmatka JB. Electroviscoelastic materials as active dampers. In: BarCohen Y, editors. Smart Structures and Materials 2002: Electroactive Polymer Actuators and Devices. Bellingham: Spie-Int Soc Optical Engineering; 2002. pp. 345–350. DOI:10.1117/12.475182

[51] Koyanagi K, Yamaguchi T, Kakinuma Y. Basic research of electro-rheological gel drum for novel linear actuator. Journal of Physics: Conference Series. 2009;**149**(1):012020 (4 pp.)–012020 (4 pp.). DOI: 10.1088/1742-6596/149/1/012020

[52] Wei KX, Bai Q, Meng G. Vibration characteristics of electrorheological elastomer sandwich beams. Smart Materials and Structures. 2011;**20**(5):055012. DOI: 10.1088/0964-1726/20/5/055012

Chapter 2

Elastomers and Their Potential as Matrices in Polymer Electrolytes

Li Na Sim and Abdul Kariem Arof

Additional information is available at the end of the chapter

http://dx.doi.org/10.5772/intechopen.69068

Abstract

Elastomers refer to natural or synthetic amorphous polymers which exhibit elastic properties of rubber. Elastomers are able to deform under stress and return to its original state upon removal of stress. As elastomers exhibit glass transition temperatures (T_g) below room temperature (r.t.), elastomers are soft and rubbery at r.t, and segmental motion exists. Elastomers, which contain polar atoms, such as oxygen (O), fluorine (F) and nitrogen (N), can be suitably employed as matrices in polymer electrolytes (PEs). The electronegative atoms serve as electron donors and are able to coordinate with cations from inorganic salts to form a complex. Among elastomers that have been employed in PEs include modified and copolymerised polyisoprene or natural rubber, polyurethanes, and polysiloxanes. This chapter focuses on the progress of natural rubber and its derivatives in the field of polymer electrolytes, and discussesc their interactions with other components of the PEs and ion conduction. Area percentage of ionic species, as well as ion transport parameters, such as number density, mobility and diffusion coefficient of lithium ions as obtained from deconvolution of infrared (IR) spectra, are also discussed in this book chapter.

Keywords: elastomers, polymer electrolytes, natural rubber, ion transport

1. Introduction

Polymer electrolytes (PEs) are ionic conductors, which consist of inorganic salts dissolved in polymer(s). Since the discovery of PEs by Fenton et al. [1], PEs are widely investigated by researchers with the aim to design flexible, light-weight, and safer solid-state ionic conductors as alternatives to the commercial leakage-prone liquid electrolytes. Excellent performing polymer electrolytes are aimed to be employed in electrochemical devices, such as fuel cells, rechargeable batteries, solar cells, sensors, supercapacitors, and electrochromic devices. The main components of a basic PE system include a polymer matrix, which serves as medium to hold the conducting ions; an inorganic salt, which acts as source of ions or charge carriers; and a solvent, which helps to dissolve the polymer and dissociate the ions from the salt. It is widely known that ion conduction occurs predominantly in the amorphous regions of the polymer electrolyte [2]. Hence, elastomers are suitable media for ion conduction to occur as their low glass transition temperature (T_g) below room temperature will allow high degree of polymer segmental motion to aid ion transport in polymer electrolytes.

Apart from possessing polar electron donating atoms, an elastomer has to meet certain criteria for it to be suitable for use as a polymer matrix in PEs. The elastomer also has to exhibit desirable mechanical and thermal properties, which enables it to be applicable in electrochemical devices. Complexation of cations to the polymer matrices is labile to allow ions to be transported via ion hopping and polymer segmental motion. Elastomers that have been used as polymer matrix in PEs are modified and copolymerised polyisoprene or natural rubber, polyurethanes, and polysiloxanes, which will be described briefly below. The chemical structures of natural rubber, polyurethane, and polysiloxane are depicted in **Figure 1**.

Natural rubber is widely used in many applications due to its characteristics such as flexibility, high tensile strength, being waterproof, and its availability from the rubber tree. Among applications that utilise natural rubber include tyres for vehicles, gloves and other industrial, consumer, and medical products. One of the reasons that sets natural rubber apart from the other elastomers as a preferred polymer matrix in polymer electrolytes mentioned in this chapter is that it is a natural or biopolymer, hence its sustainability and lower cost.

One setback of natural rubber is that it does not contain polar atoms that can complex with cations from salt. Hence, researchers have carried out modifications such as grafting and epoxidation to incorporate polymers possessing polar groups to allow ion transport. The use of natural rubber ($T_g \approx -70°$) in polymer electrolytes involves various kinds of chemical modification, that is, epoxidation and grafting, in order to introduce polar atoms into the polymer. Among all natural rubber-based polymer electrolytes, epoxidized natural rubber (ENR) (e.g. ENR-25, ENR-50; the number indicates the mole % of epoxidized group) and poly(methyl methacrylate) (PMMA)-grafted natural rubber (MG, e.g. MG30, MG45; the number indicates the percentage of PMMA-grafted) have been extensively studied [3–5]. The O atom of the epoxide group in ENR, and the carbonyl (C=O) and methoxy (C–O–CH_3) groups in PMMA-grafted natural rubber are able to complex with the cations, for example, Li^+ or H^+ ions, from inorganic salts to produce ion conduction. The soft and good elastic characteristics of natural rubber also help to produce good electrode-electrolyte contact.

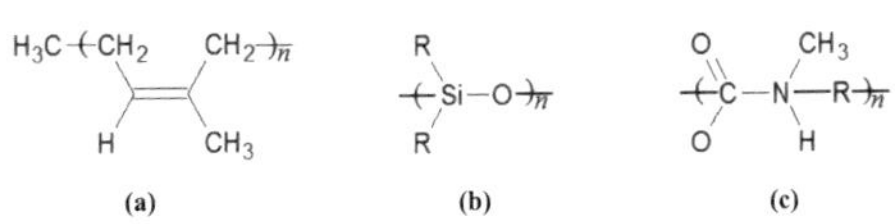

Figure 1. Chemical structure of (a) natural rubber, (b) polyurethane and (c) polysiloxane, where R represents an alkyl or aryl group.

Polyurethane is known for its high tensile strength and elasticity, which can be found in gaskets, cushioning, wheels and tyres for roller coasters and escalators, hoses, surface coatings and sealants, and biomedical devices. The uniqueness of polyurethane is found in its structure which is made up of alternating soft segments and hard segments connected by urethane linkages (NH—C=O—O). The soft segment of polyurethane is made up of long, flexible chains contributed by polyols, while the hard segment is made up of hydrogen bonded C=O and N—H groups contributed by isocyanates to form rigid physically cross-linked domains distributed in the soft segment matrix. In short, the soft segment provides medium for ion conduction, while the hard segment acts as reinforcing filler to provide dimensional stability. The O atoms in the soft segment of the polymer can coordinate with cations from salt to produce ion conduction. Its good adhesive property also allows good electrolyte-electrode contact during electrochemical device fabrication. Polyurethane-based polymer electrolytes have been reported to exhibit conductivity (σ) around 10^{-5} S cm^{-1}. Various modified and polymer blends of polyurethane have been reported, such as cross-linked poly(urethane acrylate) [6], as well as polyurethane blended with poly(vinylidene fluoride) (PVdF) [7], poly(vinylidenefluoride-co-hexafluoropropylene) (PVdF-HFP) [8], poly(ethylene oxide) (PEO) [9, 10], and polyacrylonitrile (PAN) [11].

Another elastomer that has been employed as polymer, matrix is polysiloxane, also known as silicones. Polysiloxanes are commonly found in silicone oil, grease, rubber, resin, plastics, and so on. Polysiloxanes are inorganic comb polymers, which are composed of substituents connected to the silicon with alternating silicon and oxygen atoms. This group of polymer is known to exhibit very low T_g (−127°C for poly(dimethylsiloxane)) [12], highly flexible backbone, chemical inertness, good thermal stability, and low toxicity. Due to their very low T_g, polysiloxanes exhibit low mechanical strength, and often exist as liquids or gums at room temperature. In order to impart mechanical strength into polysiloxanes, they are often cross-linked or combined with other polymers to design hybrid organic-inorganic polysiloxane-based electrolytes with good ion transport and mechanical stability. Polysiloxanes have been reported to be cross-linked with oligo(ethylene oxide)) [13], oligo(ethylene glycol) [14], oligo(ethylene oxide)-co-acrylate [15], and propyl(triethylene oxide) [16], and these ether-modified polysiloxanes are seen as alternatives to poly(ethylene oxide) (PEO). Other modifications include blending of ether-modified polysiloxanes with PVdF-HFP [16, 17], and grafting of polysiloxanes with tyrenesulfonyl(phenylsulfonyl)imide [18].

In this chapter, we focus on natural rubber–based polymer matrices, that is, modified natural rubber, namely PMMA-grafted natural rubber (MG) in polymer electrolytes. The conductivity, interactions between components, crystallinity as well as glass transition temperature of MG-based polymer electrolytes containing inorganic salt, plasticizer, and filler are discussed.

2. PMMA-grafted natural rubber (MG)

PMMA is a promising candidate as a polymer matrix which has been widely studied in polymer electrolytes since 1985 [19]. Advantages of PMMA include its high stability towards lithium electrode, which translates to higher life cycle in lithium batteries. Being an amorphous polymer, PMMA contains oxygen atoms in the carbonyl (C=O) and (C–O–C) groups, which can act as electron donor to complex with cations from salt. The presence of polar groups in PMMA in addition to its amorphous nature, high transparency, weatherability and good dimensional stability makes it suitable to be grafted with natural rubber. However, PMMA suffer from limitations in terms of mechanical strength and tends to be brittle [20]. Hence, combining the conducting ability of PMMA and the elasticity of natural rubber would counteract the limitations of the two polymers. The ability of PMMA to transport ions along its polymer chains combined with the resilience and flexibility of natural rubber allows PMMA-grafted natural rubber (MG), such as MG30 and MG49, to be employed as polymer matrices for ionic conductors. **Figure 2** illustrates the chemical structure of MG polymer matrix.

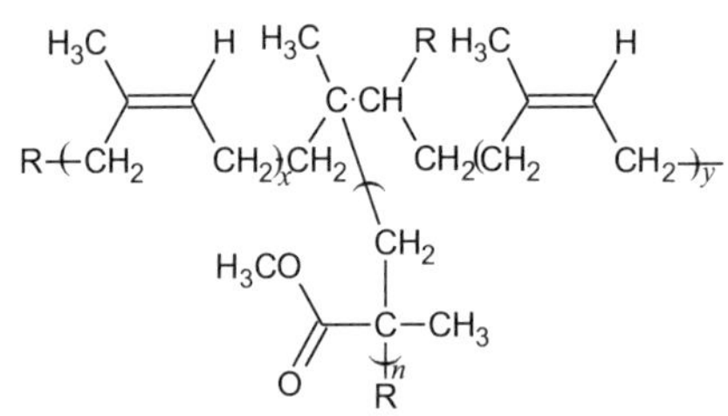

Figure 2. Chemical structure of MG polymer matrix. MG30 contains 30% of PMMA grafted onto natural rubber, whereas MG49 contains 49% of PMMA.

The chemical structure of MG30 and MG49 are similar and differs in the composition of PMMA which is grafted on natural rubber, where 30 and 49 represent the percentage of the methacrylate polymer present. Research on MG49 as a polymer matrix in polymer electrolytes has received more attention as compared to MG30, probably due to the higher amount of PMMA, and hence, more electron donors in the former are available for complexation with cations from salt.

2.1. MG-based polymer electrolytes incorporated with inorganic salts

MG-based PEs are easily prepared into films by dissolving MG polymer matrices in either toluene [21, 22] or tetrahydrofuran [5, 23] together with inorganic salt(s), and then solution casted to dry. In order to ease dissolution, the MG polymers are usually sliced into smaller pieces prior to use.

The ambient ionic conductivity of MG30 without addition of salt was reported to be in the order between 10^{-11} and 10^{-10} S cm^{-1} [5, 23], which is comparable to that of other synthetic polymers. Examples of salts added into MG30 and/or MG49 are lithium triflate ($LiCF_3SO_3$) [4, 5], lithium perchlorate ($LiClO_4$) [3], lithium tetrafluoroborate ($LiBF_4$) [24], and ammonium triflate ($NH_4CF_3SO_3$) [25], which provide ions for conduction to occur in the polymer matrix system.

Upon the incorporation of salt in MG polymer matrices, maximum conductivity is usually achieved with salt contents in the range of 15–35 wt.%. The increase in conductivity before the maximum point has been attributed to the increase in free ions [5] resulting from the salt dissociation. This is achieved by the complexation of salt cation with the electron donor atoms in the polymer matrix or solvent, while the separated anions will move freely around in the polymer. With increasing salt content, the amount of available complexation sites on the polymer matrices will be saturated, and hence, the cations will tend to associate with anions of salt to form less mobile ion pairs or ion aggregates [5, 23]. The reduction in free ions in combination with other factors, such as lower ion mobility and increased viscosity, can be used to explain the drop in the ionic conductivity beyond the maximum point [23].

Yap et al. [5] reported the best room temperature conductivity of 1.7 μS cm^{-1} obtained in MG30 sample containing 30 wt.% $LiCF_3SO_3$. The ion mechanism was found to obey the Arrhenius law, which suggested that ion transport occurs predominantly through ion hopping. The conductivity trend is in agreement with the trends of activation energy (E_a) and amounts of free ions, whereby the best conducting sample required the lowest E_a of 0.24 eV to hop from one complexation site to another and exhibits the highest amount of free ions.

The conductivity values of MG30- and MG49-based polymer electrolytes are listed in **Table 1**. Most MG-salt complexes reported an ambient ionic conductivity around 10^{-8} to 10^{-7} S cm^{-1}, which is considered insufficient for application in electrochemical devices.

Composition (wt.%)	Ambient conductivity, σ (S cm^{-1})	Reference
MG30: $LiCF_3SO_3$ (70:30)	1.7×10^{-6}	[5]
MG30:$LiCF_3SO_3$ (65:35)	$^a8.4 \times 10^{-4}$	[4]
MG30:$LiCF_3SO_3$:EC (15:9:76)	$^a9.0 \times 10^{-3}$	[4]
MG30:$LiCF_3SO_3$:PC (26:14:60)	$^a3.1 \times 10^{-3}$	[23]
MG30:$LiCF_3SO_3$:PEG200 (63:27:10)	3.7×10^{-4}	[29]
MG30:$NH_4CF_3SO_3$ (65:35)	2.0×10^{-4}	[25]
MG30:$NH_4CF_3SO_3$:EC (32.5:17.5:50)	8.9×10^{-4}	[25]
MG30:$LiCF_3SO_3$:SiO_2 (60:33:7)	$\approx 2.3 \times 10^{-4}$	[28]
MG30:$LiCF_3SO_3$:SiO_2:DMC (18:9.9:2.1:70)	$\approx 3.5 \times 10^{-4}$	[28]
MG30:$LiCF_3SO_3$:Al_2O_3 (59.8:32.2:8)	1.1×10^{-6}	[31]
MG49:$LiBF_4$ (80:20)	2.3×10^{-7}	[35, 36]
MG49:$LiClO_4$ (85:15)	4.0×10^{-8}	[35]
MG49:$LiClO_4$:TiO_2 (72:18:10)	3.5×10^{-6}	[27]
MG49:$LiClO_4$:TiO_2:EC (49:14:7:30)	1.1×10^{-3}	[27]
MG49:$NH_4CF_3SO_3$-PC (3:97)	1.2×10^{-2}	[30]
MG49:$NH_4CF_3SO_3$-PC:SiO_2 (2.8:89.2:8.0)	7.6×10^{-3}	[30]
MG49:$LiClO_4$: [HNO_3-THF-TiO_2-SiO_2] (69:23:8)	5.86×10^{-7}	[3]

^{a}At 30°C.

Table 1. Ionic conductivity values of MG-based polymer electrolytes incorporated with salt, plasticizer and filler.

2.2. MG-based polymer electrolytes incorporated with plasticizers

Plasticization is a common method used by researchers to enhance the ionic conductivity of polymer electrolytes. This is because plasticizers can increase the amorphousness of the polymer, and enhance polymer segmental motion and ion mobility [25]. The high dielectric constant of plasticizers can also help to dissociate the salt into free ions and prevent ion association [23]. Overall, plasticizer addition helps to improve the ionic conductivity of polymer electrolytes, which is an important factor that influences the device performance.

As mentioned earlier, most MG-based solid polymer electrolytes suffer from low conductivity ($\sigma = 10^{-8}$ to 10^{-7} S cm^{-1}). Hence, researchers have incorporated plasticizers into MG30 and MG49 to enhance ionic conductivity. Plasticizers such as ethylene carbonate (EC) [4, 20, 21, 25–27], propylene carbonate (PC) [23], dimethyl carbonate (DMC) [28], and poly(ethylene glycol) with molecular weight of 200 (PEG200) [29] have been introduced into MG polymer matrices. **Figure 3** illustrates some examples of plasticizers that have been added into MG polymer matrices.

One of the pioneering works on MG-based polymer electrolytes is by Kumutha et al. [21] who investigated on MG30-$LiCF_3SO_3$-EC-Al_2SiO_5 system. Results from infrared (IR) and differential scanning calorimetry (DSC) studies showed no distinct change in their wavenumbers and T_g value (which lie in the range of −39 to −42°C) upon addition of EC as plasticizer or Al_2SiO_5 as filler. The authors reported that there is no interaction between EC, and Al_2SiO_5, and the polymer.

In separate works, Ali and co-workers [4, 23] investigated on MG30-$LiCF_3SO_3$-EC and MG30-$LiCF_3SO_3$-PC systems. The ionic conductivity increased from 8.4×10^{-4} S cm^{-1} to achieve maximum conductivity at 9.0×10^{-3} S cm^{-1} for the EC system, and 3.1×10^{-3} S cm^{-1} for the system with PC on incorporation of 76 wt.% EC and 61 wt.% PC. For the MG30-$LiCF_3SO_3$-PC system, Ali et al. [23] reported that the sample containing 61 wt.% PC lost its dimensional stability despite exhibiting the best conductivity in the polymer electrolyte system. Increase in conductivity is attributed to several factors, namely increased polymer-cation interaction, high dielectric constant (ε) of EC at 89.6 (at 40°C) and PC at 64.4, which help to lower transient cross-linking between Li^+ ions and MG30, increased ion mobility and reduced intermolecular forces in MG30, that increased the amorphousness of the polymer in the presence of the plasticizer. The latter is manifested in the lowering of the T_g value of MG30-$LiCF_3SO_3$ (65:35)

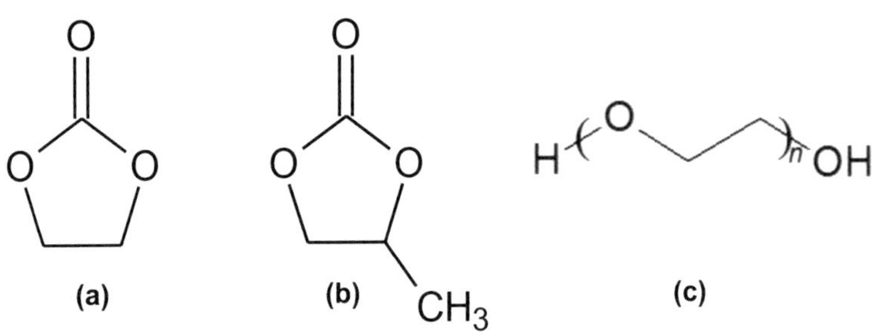

Figure 3. Chemical structure of (a) EC, (b) PC and (c) PEG200.

initially found between −66.6 and −59.5°C, and −77.1 and −74.6°C when 76 wt.% EC and 61 wt. of PC were added. In another work [26], gel polymer electrolytes consisting of MG30-$LiCF_3SO_3$ plasticized with 76 wt.% EC and PC were reported to exhibit lithium transference number (t_{Li}^{+}) below 0.3 and voltage breakdown at 4.2 V.

Zaki et al. [25] achieved considerably high ionic conductivity of 2.0×10^{-4} S cm^{-1} for MG30-$NH_4LiCF_3SO_3$ (65:35) sample. Ionic conductivity increased to 8.9×10^{-4} S cm^{-1} upon introduction of 50 wt.% EC. The authors attributed the conductivity enhancement to the high dielectric constant of EC, which helped to dissociate the salt and prevent ion association. Ion conduction was found to follow Vogel-Tammann-Fulcher (VTF) behaviour, whereby ion transport occurs in the amorphous phase via polymer segmental motion. Results from XRD showed reduction in crystallinity from the gradual broadening of amorphous hump located from $2\theta = 12°$ to 22° with addition of EC.

In the work of Yap et al. [29], PEG was added into MG30-$LiCF_3SO_3$ system. Conductivity improved from 1.7×10^{-6} S cm^{-1} to a maximum value at 3.7×10^{-4} S cm^{-1}, and this was exhibited in the sample containing 10 wt.% PEG. X-ray diffraction (XRD) analysis confirmed the sample to exhibit the lowest degree of crystallinity.

2.3. MG-based polymer electrolytes incorporated with fillers and other additives

Although polymer electrolytes added with plasticizer exhibit conductivity enhancement, they suffer from low mechanical stability [30]. Hence, researchers have attempted using inorganic fillers to improve the physical and mechanical properties of polymer electrolytes. For MG-based polymer electrolytes, fillers that have been reported are aluminium silicate (Al_2SiO_5) [20], silica (SiO_2) [28, 30], alumina (Al_2O_3) [31] and titania-silica (TiO_2-SiO_2) [3], which are all passive fillers, and therefore do not participate directly in the Li^+ ion transport process. Filler-added polymer electrolytes are known as composite polymer electrolytes (CPEs).

In the work of Hashim et al. [28], SiO_2 filler was incorporated into MG30-$LiCF_3SO_3$, followed by the addition of dimethyl carbonate (DMC) plasticizer. The ionic conductivity initially increased from ≈6.0×10^{-5} S cm^{-1} to 2.3×10^{-4} S cm^{-1} upon addition of 7 wt.% SiO_2. The rise in conductivity is attributed to the formation of conducting pathways for ion transport. Decreased conductivity above 7 wt.% SiO_2 was caused by the blocking effect of the nanofillers. After that, the authors managed to further enhance the conductivity value to ≈3.5×10^{-4} S cm^{-1} with addition of 7 wt.% DMC. The DMC plasticizer enhanced polymer segmental motion.

MG49-$NH_4CF_3SO_3$-PC-SiO_2 CPE system was reported by Kamisan et al. [30]. The authors first determined the best conducting $NH_4CF_3SO_3$ molarity in PC at 0.7 M. Gel polymer electrolyte containing 3 wt.% MG49 in the $NH_4CF_3SO_3$-PC liquid electrolyte exhibited high conductivity in the order of 10^{-2} S cm^{-1}. However, they lacked mechanical stability due to the high amount of liquid electrolyte. The addition of SiO_2 initially decreased the conductivity to 10^{-3} S cm^{-1}, but increased to achieve two maxima at 1.2×10^{-2} S cm^{-1} (3 wt.% SiO_2) and 7.6 10^{-3} S cm^{-1} (8 wt.% SiO_2). The activation energy values calculated from the Arrhenius equation displayed two minima at 0.11 and 0.12 eV for samples with 3 and 8 wt.% SiO_2 content, respectively. The CPE viscosity increased from 10.0 to 68.2 cP upon addition of 8 wt.% SiO_2, which authors took as indication of increased mechanical stability.

Kang and co-workers [3] incorporated fixed amount of TiO_2-SiO_2 filler into MG49-$LiClO_4$ system containing different solvents, that is, HNO_3-THF and $ClHNO_2$-THF. The conductivities of both MG49-$LiClO_4$-HNO_3-THF-TiO_2-SiO_2 and MG49-$LiClO_4$-$ClHNO_2$-THF-TiO_2-SiO_2 systems increased from 1.2×10^{-4} S cm^{-1} to achieve a maximum at 5.9×10^{-7} S cm^{-1} and 4.8×10^{-7} S cm^{-1}, respectively, upon incorporation of 25 wt.% $LiClO_4$, before decreasing at higher salt contents. Above 25 wt.% $LiClO_4$, the ion transport process is hindered by ionic aggregation and the formation of transient cross-links. The two different polymer electrolyte systems exhibited different ion conduction behaviors: the system containing HNO_3-THF followed the Arrhenius rule, whereby ion transport occurs by ion hopping, whereas the $ClHNO_2$-THF system obeyed VTF rule, which indicates polymer segmental relaxation. Both polymer electrolyte systems are thermally stable up to around 270–290°C.

Thermal oxidation of MG-based electrolytes has also been a subject of study. Aging of polymers, which in this case is natural rubber, is a topic of concern as the unsaturated hydrocarbons will degrade on exposure to oxygen and elevated temperatures. Consequently, the physical, electrical, mechanical, and other characteristics of PEs will deteriorate upon aging. Aziz and Ali [32] studied the effect of varied contents of an antioxidant, N-(1,3-dimethylbutyl)-N′-phenyl-p-phenylenediamine (6PPD) on the thermal oxidation resistance of MG30. Using IR analysis, the absorbance/intensity ratio of the C=O band at 1726 cm^{-1} to the =C–H band located at ≈835 cm^{-1} of different MG30-6PPD samples were calculated to determine the degree of oxidation [33, 34]. Only the sample with 0.5 wt.% 6PPD displayed anti-oxidation effects as the intensity ratio of the bands 1726 cm^{-1}:835 cm^{-1} decreased. At higher contents of 6PPD (1, 2, 4, 6 and 8 wt.%), no anti-oxidation effect was observed. However, the MG30 sample containing 0.5 wt.% 6PPD with the best anti-oxidation effect exhibited a lower ambient conductivity of 5.50×10^{-6} S cm^{-1} than the untreated sample which had conductivity of 3.10×10^{-5} S cm^{-1}. The authors suggested incorporating plasticizer to increase the conductivity of MG30-6PPD sample in the future.

2.4. Blends of PMMA-grafted natural rubber

In order to improve the properties of MG30- and MG49-based polymer electrolytes, researchers have attempted blending of the MG polymer matrices with other polymers. MG polymer matrices were simply mixed together with another polymer to produce polymer blend(s) which possess properties that are superior than individual polymers. MG49 has been blended with PMMA [24], PEO [36], PVdF-HFP [37], and PVdF [38]. The chemical structures of PMMA, PEO, PVdF-HFP, and PVdF are shown in **Figure 4**.

Ahmad et al. [36] found improvement in the conductivity of MG49 with and without $LiBF_4$ salt upon blending of the polymer with 50 wt.% PMMA. Pure MG49 exhibited ambient conductivity of 1.0×10^{-12} S cm^{-1}, whereas pure MG49-PMMA blend possessed conductivity of 1.7×10^{-12} S cm^{-1}. Upon incorporation of $LiBF_4$, MG49-$LiBF_4$ and MG49-PMMA-$LiBF_4$ samples exhibited maximum conductivity of 2.3×10^{-7} S cm^{-1} (at 20 wt.%) and 8.3×10^{-6} S cm^{-1} (at 25 wt.%), respectively. The more superior conductivity in the polymer blend electrolyte is attributed to the higher ratio of number of oxygen atoms to lithium ion [O:Li^+] of 5:1 as compared to 8:1 in single polymer electrolyte system. Morphology study by scanning electron microscopy (SEM) revealed homogenous surface in MG49 and MG49-PMMA films with no

http://dx.doi.org/10.5772/intechopen.69068

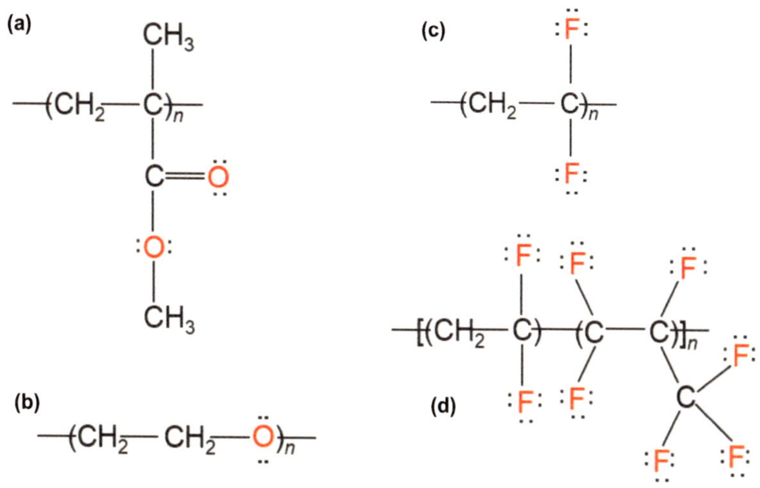

Figure 4. Chemical structure of (a) PMMA, (b) PEO, (c) PVdF, and (d) PVdF-HFP, where each atom in red can act as electron donor, and the black dots represent lone electrons. The polymers have been reported to be blended with MG polymer matrices.

phase separation. The changes in the structure of MG49 upon blending with PMMA indicated that intermolecular interactions between the two polymers have occurred. Crystals of $LiBF_4$ in MG49 sample added with 25 wt.% of the salt were reported due to salt aggregation above its optimum concentration, which explained the drop in conductivity above 20 wt.% $LiBF_4$.

A polymer blend system that comprised PEO-MG49 with varied wt.% PEO was reported by [39]. Photograph of the PEO:MG49 with wt.% ratio of 60:40 showed a more homogeneous surface and free-standing film, as compared to other wt. ratios (80:20, 20:80, 40:60) which were not homogeneous. From SEM analyses, MG49 film was smooth due to its amorphous characteristic, while PEO film displayed rough surface attributable to its crystalline nature. Upon blending between the two polymers, homogeneous surface was reported for PEO-MG49 sample, whereby the two polymers were well distributed. With increasing addition of $LiClO_4$, the co-continuous phase appeared smoother and increased in pore size. Addition of PEO into MG49 was reported to improve the conductivity by two orders of magnitude. The PEO-MG49-$LiClO_4$ samples exhibited conductivity that was also higher than PEO-$LiClO_4$ sample ($\sigma = 10^{-8}$ to 10^{-7} S cm^{-1}) [40, 41]. Maximum conductivity of 8.0×10^{-6} S cm^{-1} was obtained in the polymer blend incorporated with 15 wt.% $LiClO_4$, which was three orders higher than the sample before addition of salt ($\sigma = 4.0 \times 10^{-9}$ S cm^{-1}). The conductivity values of MG-based blend polymer electrolytes are listed in **Table 2**.

Su'ait and co-workers [24] blended MG49 with PMMA in the wt.% ratio of 70 to 30, and incorporated the polymer matrix with different lithium salts, namely $LiClO_4$ and $LiBF_4$. The best conductivity was obtained upon addition of 25 wt.% salt in both PE systems. The PE containing $LiBF_4$ exhibited higher conductivity ($\sigma = 8.6 \times 10^{-6}$ S cm^{-1}) than that of $LiClO_4$ ($\sigma = 1.5 \times 10^{-8}$ S cm^{-1}) and is attributed to the larger ionic radius of BF_4^- (0.218 pm) compared to ClO_4^- (0.215 pm), and hence to the lower lattice energy of the former salt. Infrared results showed that Li^+ ions complexed with the oxygen atoms present in both MG49 and PMMA.

Polymer electrolyte composition (wt.%)	Ambient conductivity (S cm^{-1})	Reference
MG49:PMMA (70:30)	1.1×10^{-12}	[24]
	4.1×10^{-12}	[35]
MG49:PMMA:$LiClO_4$ (52.5:22.5:25)	1.5×10^{-8}	[24]
MG49:PMMA:$LiBF_4$ (52.5:22.5:25)	8.6×10^{-6}	[24]
PEO:MG49 (60:40)	4.0×10^{-9}	[36]
PEO:MG49: $LiClO_4$ (51:34:15)	8.0×10^{-6}	[36]
PVdF-HFP: MG49: $LiBF_4$ (49:21:30)	2.8×10^{-4}	[37]
PVdF-MG49:$LiTCF_3SO_3$ (70:30)	3.3×10^{-4}	[38]

Table 2. Ionic conductivity values of MG-based polymer blend electrolytes.

Among the polymer-blend electrolyte systems listed in **Table 2**, fluorine (F)-containing polymer electrolyte systems such as PVdF-HFP-MG49 [37] and PVdF-MG49 [38] achieved the best ionic conductivity up to the order of 10^{-4} S cm^{-1} when blended with $LiBF_4$ and $LiCF_3SO_3$, respectively. The polymer blend made up of PEO-MG49 reached the maximum conductivity from 4.0×10^{-9} S cm^{-1} to 8.0×10^{-6} S cm^{-1} upon addition of 15 wt.% $LiClO_4$.

2.5. Interactions in MG-based polymer electrolytes

It is useful for researchers to investigate whether interactions have occurred between MG polymer and the salt, as well as with other components in the polymer electrolyte system, for example, plasticizer, filler, and so on, and identify the chemical group where the interactions have taken place. Interactions between components in a polymer electrolyte system will ensure miscibility and no phase separation.

As mentioned in previous sections, natural rubber does not possess polar atoms which could complex with cations from salt. In MG30 and MG49, the PMMA contains oxygen atoms in the C═O and C─O─C groups which have been reported to interact with the cations from salt through coordination bond(s) [4, 5, 20, 23, 24]. Evidence of complexation of MG30 and MG49 with various salts could be obtained from infrared (IR) spectroscopy. The wavenumber regions of interest of MG polymer matrices are ≈1720–1750 cm^{-1} and ≈950–1450 cm^{-1} which represent the C═O and C─O─C group, respectively. Shifting of the IR bands mentioned indicates complexation of MG polymer with cations from the salt.

Although interactions between MG polymer matrix and salt are usually shown from the shifting of the symmetrical C═O stretching (ν(C═O)) and IR bands corresponding to the C─O─C group of PMMA, that is, combined C─O─O stretching (ν(C─O─O)) and C─O stretching (ν(C─O)), symmetric, and asymmetric stretching of C─O─C group (ν_s(C─O─C) and ν_{as}(C─O─C)) as well as deformation of O─CH_3 (δ(O─CH_3)), some researchers did not find significant wavenumber shift that is beyond the resolution used to measure the IR spectra. Su'ait and co-workers [24] investigated on MG49-PMMA blend incorporated with different lithium salts, namely $LiBF_4$ and $LiClO_4$. The ν(C═O) band in the samples containing 10 and 20 wt.%

respective salt did not show wavenumber change but reduced in intensity. Other bands such as the combined ν(C—O—O) and ν(C—O) band at 1272 cm^{-1} and ν_s(C—O—C) at 985 cm^{-1} also did not manifest any IR wavenumber shift. The intensity of the ν_{as}(C—O—C) band was noticeably reduced upon addition of 20 wt.% $LiBF_4$. Only the δ(O—CH_3) band shifted from 1455 cm^{-1} to 1460 cm^{-1} and 1461 cm^{-1} in $LiClO_4$ and $LiBF_4$-added samples, respectively. The reduction in intensity and wavenumber shifts indicated complexation has taken place between Li^+ ions from the salts and MG49.

Several reports on plasticized MG-based polymer electrolytes found no interaction from IR studies between the polymer matrix and the plasticizer. Kumutha and co-workers [21] reported on the IR analyses of MG30-$LiCF_3SO_3$-EC system and explored the interactions between MG30 and EC and between $LiCF_3SO_3$ and EC. For the MG30-EC sample, no IR change was found, which indicated that there was no interaction between the polymer and plasticizer. However, for the $LiCF_3SO_3$-EC mixture, the ν(C=O), C=O bending, and ring breathing of EC shifted from 1803 to 1778 cm^{-1}, from 718 to 720 cm$^{-1,}$ and from 897 to 899 cm^{-1}, respectively, which suggested that interaction has occurred between the C=O group of EC and Li^+ ions from the salt to form C=O…Li^+ interaction. Hence, EC has penetrated into the MG30-$LiCF_3SO_3$ complex with interaction with the salt but no interaction with the polymer.

Ali et al. [23] investigated on MG30-$LiCF_3SO_3$-PC system. The ν(C=O) of PC displayed downshift from 1788 to 1775 cm^{-1} in the $LiCF_3SO_3$-PC mixture, which showed coordination of Li^+ ions onto the C=O group of PC. Similar to [21], no IR change was observed for MG30-PC sample, which suggested no interaction between the polymer and the plasticizer.

In filler-added polymer electrolyte system, such as MG30-$LiCF_3SO_3$-EC-Al_2SiO_5, Kumutha et al. [20] did not observe any IR change to the spectra of polymer electrolyte upon incorporation of the filler. This indicates that Al_2SiO_5 did not interact with the components in the PE complex and only existed as dispersed solid.

The formation of MG49-PMMA blend was investigated by Ahmad et al. [36] using IR spectroscopy. The δ(O—CH_3) and ν_s(C—O—C) bands of pure MG49 shifted from 984 and 1457 cm^{-1} to 992 and 1450 cm^{-1} in MG49-PMMA blend, respectively, which indicated the occurrence of intermolecular interactions between MG49 and PMMA. Upon addition of $LiBF_4$ salt, several bands of MG49 and MG49-PMMA shifted, which suggested complexation between Li^+ ions and the polymer matrix. The ν_s(C—O—C) shifted from 984 to 987 cm^{-1} in MG49-$LiBF_4$ sample, and from 992 to 985 cm^{-1} in MG49-PMMA-$LiBF_4$ sample; the band reduced in intensity in both systems. The δ(O—CH_3) also shifted from 1450 and 1457 cm^{-1} in MG49-$LiBF_4$ and MG49-PMMA-$LiBF_4$, respectively, to 1461 cm^{-1}. From the results, the authors implied that oxygen in the C—O—C group of PMMA interacted at a higher extent to Li^+ ions.

For PEO-MG49-$LiClO_4$ polymer blend electrolyte system, complexation between Li^+ ions and the polymer blend was confirmed by Ahmad and co-workers [39] from the shifting of bands belonging to PEO and PMMA in MG49. The triple band due to PEO initially located at 1146, 1099, and 1060 cm^{-1} formed a single band in the range of 1066–1094 cm^{-1} in PEO-MG49-$LiClO_4$ samples, which indicated coordination of Li^+ ions on the O atoms in PEO. The δ(O—CH_3) of PMMA in MG49 also shifted from 1466 cm^{-1} to lower wave numbers to around 1450 cm^{-1} in the $LiClO_4$-containing polymer blend electrolytes.

2.6. Determination of free ion, ion pairs and ion aggregates

Besides being able to reveal the interactions that occur in polymer electrolyte systems, IR analysis can also estimate the amount of charge carriers present in different samples. The result can be used to explain the conductivity behavior at different added contents of salt or other additives in the polymer electrolyte system.

Some IR regions representing certain bands of the salt could reveal the types of ionic species (i.e. free ions, ion pairs, ion aggregates) present and their area percentage in the polymer electrolytes. According to Huang and Frech [42], different types of triflate ($CF_3SO_3^-$) ionic species can be distinguished from the deconvolution of the IR envelope of symmetrical stretching of SO_3 ($\nu_s(SO_3)$) of the salt. Free ions, ion pairs and ion aggregates could be distinguished at 1030–1034 cm^{-1}, 1040–1045 cm^{-1}, and 1049–1053 cm^{-1}, respectively [42].

Yap et al. [5] studied MG30 PE system added with varied $LiCF_3SO_3$ contents and noted the increase of conductivity up to 30 wt.% $LiCF_3SO_3$ with maximum conductivity of 1.7 μS cm^{-1} at 30°C before dropping beyond that composition. The IR spectra of $LiCF_3SO_3$, MG30, and MG30-$LiCF_3SO_3$ samples between 1000 and 1100 cm^{-1}, where the $\nu_s(SO_3)$ of $LiCF_3SO_3$ lies, are illustrated in **Figure 5**. The band observed at 1044 cm^{-1} in $LiCF_3SO_3$ sample corresponded to the ion pairs of the salt. Upon incorporation of $LiCF_3SO_3$ into MG30, a band located at 1032 cm^{-1} attributed to free ions was found in all MG30-$LiCF_3SO_3$ samples. A shoulder appeared around 1040 cm^{-1} (as shown by the dotted line in **Figure 5**). In order to separate the IR envelope from 1000 to 1060 cm^{-1} in the varied MG30-$LiCF_3SO_3$ samples, IR deconvolution was carried out as shown in **Figure 6**.

Yap and co-workers [5] reported the presence of free ions (1032 cm^{-1}) and ion pairs (1040 cm^{-1}) in the MG30-$LiCF_3SO_3$ samples. No ion aggregate was detected in the samples. The trend of free ions obeyed the trend of conductivity, whereby the highest area % of free ions was found in the best conducting sample, which is MG30 containing 30 wt.% $LiCF_3SO_3$. Increasing free ions with $LiCF_3SO_3$ contents is due to the recombination of Li^+ and $CF_3SO_3^-$ ions. Hence, the trend of conductivity, that is, the rise of conductivity below 30 wt.% $LiCF_3SO_3$ and its subsequent drop above that composition was explained by the trend of free ions.

2.7. Determination of number density, mobility and diffusion coefficient values

By using the % area of free ions obtained from the IR deconvolution of $\nu_s(SO_3)$ band belonging to $LiCF_3SO_3$ for MG30-$LiCF_3SO_3$ samples, ion transport parameters, such as number density (n), mobility (μ), and diffusion coefficient (D), could be calculated using the equations below:

$$n = \%FI \times \frac{m}{M_W} \times \frac{N_A}{V} \tag{1}$$

$$\mu = \frac{\sigma}{n \times e} \tag{2}$$

$$D = \frac{\mu \times k \times T}{e} \tag{3}$$

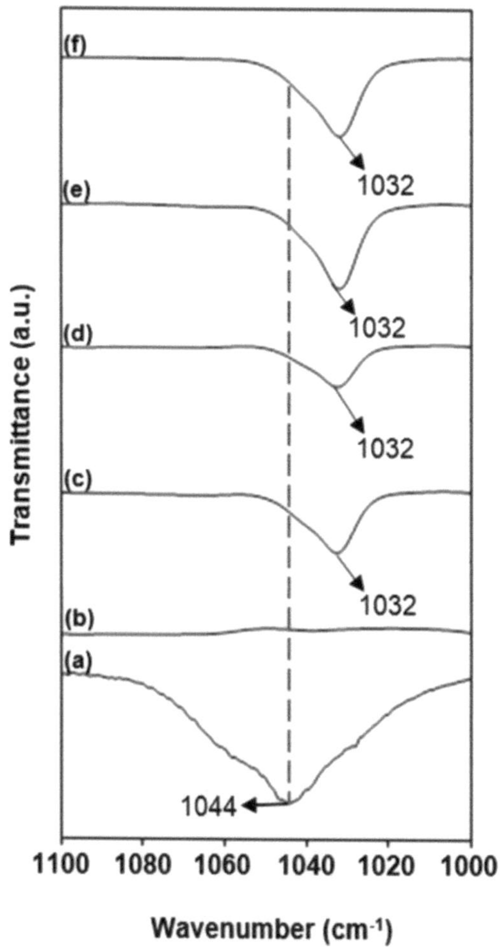

Figure 5. IR spectra between 1000 and 1100 cm^{-1} of (a) $LiCF_3SO_3$, (b) MG30, (c) MG30-$LiCF_3SO_3$ (90:10), (d) MG30-$LiCF_3SO_3$ (80:20), (e) MG30-$LiCF_3SO_3$ (70:30), and (f) MG30-$LiCF_3SO_3$ (60:40).

Where %*FI* represents the percentage of free Li^+ ions obtained from IR deconvolution, *m* refers to mass of $LiCF_3SO_3$ used, M_W is the molecular mass of $LiCF_3SO_3$ (156.01 g mol^{-1}), N_A Avogadro's number (6.02×10^{23}), *V* is the total volume of components present in the sample, σ is the conductivity of every sample at 298 K, *e* is the electron charge (1.60×10^{-19} C), *k* is the Boltzmann constant (1.38×10^{-23} J K^{-1}), and *T* is 298 K. The density of $LiCF_3SO_3$ is 1.90 g cm^{-1}, while the density of MG30 (natural rubber:PMMA = 70:30) was taken to be 1.005 g cm^{-1}. The value *n* corresponds to the number density of mobile ions, μ value refers to the mobility attained by the Li^+ ions under application of electric field, and *D* value gives the rate at which Li^+ ions are transported within a second. **Table 3** lists the conductivity, % area of free ions, *n*, μ, and *D* for MG30 added with varied wt.% $LiCF_3SO_3$. It was found that the *n* increased continuously up to 40 wt.% $LiCF_3SO_3$, while both μ and *D* were highest at 30 wt.% $LiCF_3SO_3$, which indicates that the ion transport is diffusion controlled. As the MG30 added with 30 wt.% $LiCF_3SO_3$ exhibited the best conductivity and the lowest degree of crystallinity [5], Li^+ ions are more mobile and are able to diffuse more easily through the amorphous region in the sample.

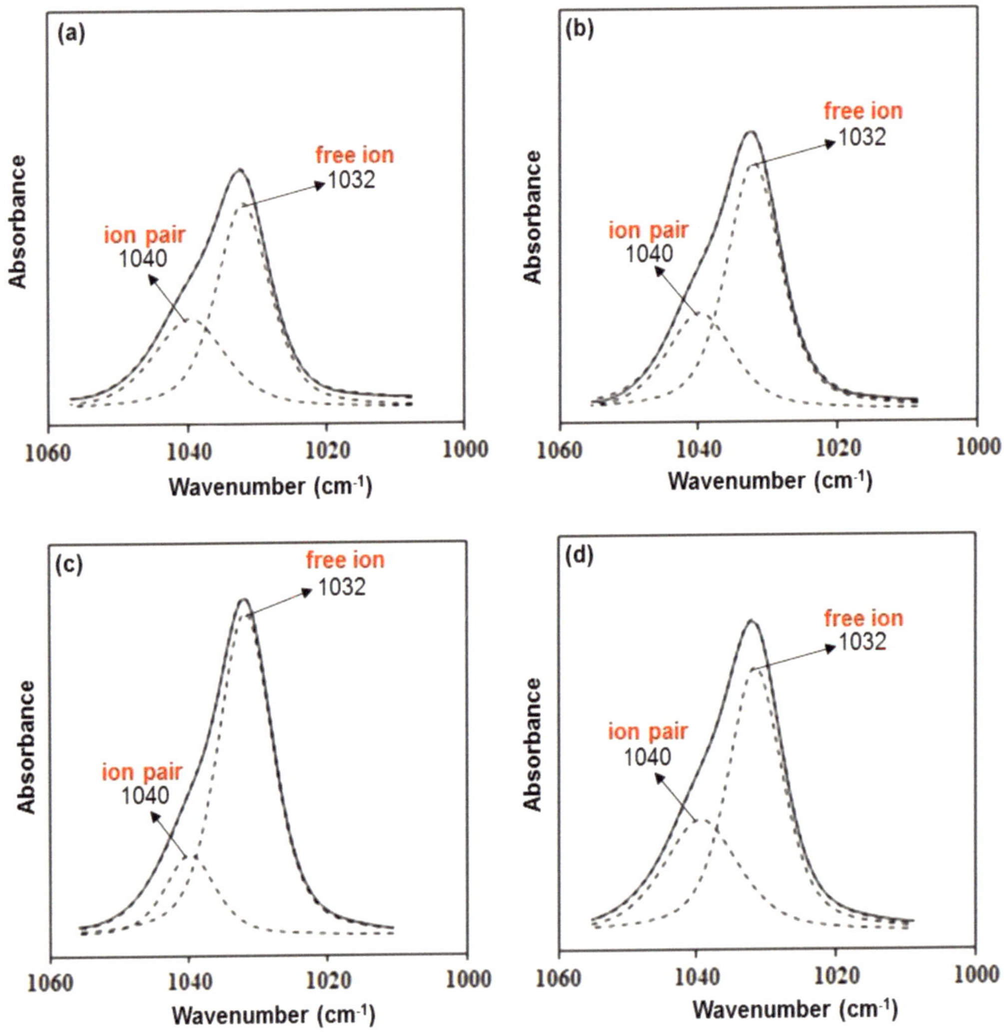

Figure 6. IR deconvolution of $v_s(SO_3)$ band belonging to $LiCF_3SO_3$ in MG30:$LiCF_3SO_3$ system with wt.% ratio of (a) 90:10, (b) 80:20, (c) 70:30, and (d) 60:40.

$LiCF_3SO_3$ content (wt.%)	σ (S cm^{-1})	FI (%)	*n* (cm^{-3})	*μ* (cm^{2}V^{-1}s^{-1})	*D* (cm^2 s^{-1})
10	1.9×10^{-9}	67.0	2.7×10^{22}	4.3×10^{-13}	1.1×10^{-14}
20	1.1×10^{-8}	70.0	6.0×10^{22}	1.1×10^{-12}	2.9×10^{-14}
30	1.7×10^{-6}	78.2	1.1×10^{23}	10.0×10^{-11}	2.6×10^{-12}
40	6.4×10^{-8}	68.6	1.3×10^{23}	3.1×10^{-12}	7.8×10^{-14}

Table 3. Conductivity (σ), % area of free ions (FI), number density (*n*), mobility (*μ*) and diffusion coefficient (*D*) of Li^+ ions of MG30 samples added with varied $LiCF_3SO_3$ contents.

3. Summary

Modified natural rubbers namely PMMA-grafted natural rubber (i.e. MG30 and MG49), are widely used as polymer matrices in polymer electrolytes. MG30 and MG49 contain polar C═O and C─O─C groups which could complex with cations from salt to produce ion transport. Plasticizers, such as EC, PC, DMC, and PEG200, were employed to enhance the conductivity of solid-based MG polymer matrices. Passive inorganic fillers, such as SiO_2, Al_2O_3, TiO_2, TiO_2-SiO_2, were used to improve the mechanical strength of the polymer electrolytes. Polymer blends of MG with PEO, PMMA, PVdF, and PVdF-HFP were able to improve the properties of MG-based polymer electrolytes in terms of conductivity and structural homogeneity. Complexation between MG polymer matrices and salt could usually be shown through IR shifts of bands corresponding to C═O and C─O─C groups, although in some cases, only changes in intensity were reported. Plasticizers and fillers added into MG polymer matrices did not manifest change on the IR spectra of the polymer, although interactions were found to occur between plasticizer and the salt (i.e., between $LiCF_3SO_3$ and EC or PC). MG30 and MG49-based polymer electrolytes could achieve high ionic conductivity up to 10^{-2} S cm^{-1} when plasticized and have potential to be used in electrochemical devices.

Acknowledgements

The authors would like to thank Ministry of Higher Education (MOHE) for grant FP053-2014A, SATU Joint Research Scheme RU022G-2014 and University of Malaya.

Author details

Li Na Sim and Abdul Kariem Arof*

*Address all correspondence to: akarof@um.edu.my

Department of Physics, Faculty of Science, Centre for Ionics University Malaya, University of Malaya, Kuala Lumpur

References

[1] Fenton DE, Parker JM, Wright PV. Complexes of alkali metal ions with poly (ethylene oxide). Polymer. 1973;**14**:589

[2] Berthier C, Gorecki W, Minier M, Armand MB, Chabagno JM, Rigaud P. Microscopic investigation of ionic conductivity in alkali metal salts–poly(ethylene oxide) adducts. Solid State Ionics. 1983;**11**:91-95

[3] Kang OL, Ahmad A, Hassan NH, Rana UA. [MG49-LiClO4]:[TiO2-SiO2] polymer electrolytes: In situ preparation and characterization. International Journal of Polymer Science. 2016;**9838067**

[4] Ali AMM, Subban RHY, Bahron H, Yahya MZA, Kamisan AS. Investigation on modified natural rubber gel polymer electrolytes for lithium polymer battery. Journal of Power Sources. 2013;**244**:636-640

[5] Yap KS, Teo LP, Sim LN, Majid SR, Arof AK. Investigation on dielectric relaxation of PMMA-grafted natural rubber incorporated with LiCF3SO3. Physica B. 2012;**407**:2421-2428

[6] Santhosh P, Gopalan A, Vasudevan T, Lee KP. Evaluation of a cross-linked polyurethane acrylate as polymer electrolyte for lithium batteries. Materials Research Bulletin. 2006;**41**:1023-1037

[7] Wu N, Cao Q, Wang X, Li S, Li X, Deng H. In situ ceramic fillers of electrospun thermoplastic polyurethane/poly(vinylidene fluoride) based gel polymer electrolytes for Li-ion batteries. Journal of Power Sources. 2011;**196**:9751-9756

[8] Zhou L, Jing B, Wang X, Tang X, Wu N. Study of a novel porous gel polymer electrolyte based on thermoplastic polyurethane/poly(vinylidene fluoride-co-hexafluoropropylene) by electrospinning technique. Journal of Power Sources. 2014;**263**:118-124

[9] Yoshimoto N, Nomura H, Shirai T, Ishikawa M, Morita M. Ionic conductance of gel electrolyte using a polyurethane matrix for rechargeable lithium batteries. Electrochimica Acta. 2004;**50**:275-279

[10] Wen TC, Chen WC. Gelled composite electrolyte comprising thermoplastic polyurethane and poly(ethylene oxide) for lithium batteries. Journal of Power Sources. 2001;**92**:139-148

[11] Wen TC, Kuo HH, Gopalan A. The influence of lithium ions on molecular interaction and conductivity of composite electrolyte consisting of TPU and PAN. Solid State Ionics. 2002;**147**:171-180

[12] Song JY, Wang YY, Wan CC. Review of gel-type polymer electrolytes for lithium-ion batteries. Journal of Power Sources. 1999;**77**:183-197

[13] Kang Y, Lee W, Suh DH, Lee C. Solid polymer electrolytes based on cross-linked polysiloxane-g-oligo(ethylene oxide): Ionic conductivity and electrochemical properties. Journal of Power Sources. 2003;**119-121**:448-453

[14] Zhang ZC, Jin JJ, Bautista F, Lyons LJ, Shariatzadeh N, Sherlock D, Amine K, West R. Ion conductive characteristics of cross-linked network polysiloxane-based solid polymer electrolytes. Solid State Ionics. 2004;**170**:233-238

[15] Kang Y, Lee J, Suh DH, Lee C. A new polysiloxane based cross-linker for solid polymer electrolyte. Journal of Power Sources. 2005;**146**:391-396

[16] Cznotka E, Jeschke S, Wiemhofer H-D. Characterization of semi-interpenetrating polymer electrolytes containing poly(vinylidene fluoride-co-hexafluoropropylene) and ether-modified polysiloxane. Solid State Ionics. 2016;**289**:35-47

[17] Seidel SM, Jeschke S, Vettikuzha P, Wiemhofer HD. PVDF-HFP/ether-modified polysiloxane membranes obtained via airbrush spraying as active separators for application in lithium ion batteries. Chemical Communications. 2015;**51**:12048-12051

[18] Rohan R, Pareek K, Chen Z, Cai W, Zhang Y, Xu G, Gao Z, Cheng H. A high performance polysiloxane-based single ion conducting polymeric electrolyte membrane for application in lithium ion batteries. Journal of Materials Chemistry A. 2015;**3**:20267-20276

[19] Iijima T, Tyoguchi Y, Eda N. Quasi-solid organic electrolytes gelatinized with PMMA and their applications for lithium batteries. 1985;**53**:619-623

[20] Brown H. A molecular interpretation of the toughness of glassy polymers. Macromolecules. 1991;**24**:2752-2756

[21] Kumutha K, Alias Y, Said R. FTIR and thermal studies of modified natural rubber based polymer electrolytes. Ionics. 2005;**11**:472-476

[22] Kumutha K, Alias Y. FTIR spectra of plasticized grafted natural rubber-LiCF3SO3 electrolytes. Spectrochimica Acta Part A. 2006;**64**:442-447

[23] Ali AMM, Subban RHY, Bahron H, Winie T, Latif F, Yahya MZA. Grafted natural rubber-based polymer electrolytes: ATR-FTIR and conductivity studies. Ionics. 2008;**14**:491-500

[24] Su'ait MS, Ahmad A, Hamzah H, Rahman MYA. Effect of lithium salt concentrations on blended 49% poly(methyl methacrylate) grafted natural rubber and poly(methyl methacrylate) based solid polymer electrolyte. Electrochimica Acta. 2011;**57**:123-131

[25] Zaki NHM, Mahmud ZS, Yahya MZA, Ali AMM. Conductivity studies on 30% PMMA grafted NR-NH4CF3SO3 Gel Polymer Electrolytes. 2012 IEEE Symposium on Humanities, Science and Engineering Research. 2012;**6268995**:797-800

[26] Ali AMM, Yahya MZA, Bahron H, Subban RHY. Electrochemical studies on polymer electrolytes based on poly(methyl methacrylate)-grafted natural rubber for lithium polymer battery. Ionics. 2006;**12**:303-307

[27] Low SP, Ahmad A, Rahman MYA. Effect of ethylene carbonate plasticizer and TiO2 nanoparticles on 49% poly(methyl methacrylate) grafted natural rubber-based polymer electrolyte. Ionics. 2010;**16**:821-826

[28] Hashim H, Adam NI, Zaki NHM, Mahmud ZS, Said CMS, Yahya MZA, Ali AMM. Natural rubber-grafted with 30% poly(methylmethacrylate) characterization for application in lithium polymer battery. 2010 International Conference on Science and Social Research. 2010;**5773825**:485-488

[29] Yap KS, Teo LP, Sim LN, Majid SR, Arof AK. Plasticised polymer electrolytes based on PMMA grafted natural rubber-LiCF3SO3-PEG200. Materials Research Innovations. 2011;**15**:34-38

[30] Kamisan AS, Kudin TIT, Ali AMM, Yahya MZA. Electrical and physical studies on 49% methyl-grafted natural rubber-based composite polymer gel electrolytes. Electrochimica Acta. 2011;**57**:207-211

[31] Adam NI, Zaki NHM, Mahmud ZS, Yahya MZA. The effect of composition nanofiller Al2O3 to the conductivity, morphology and thermal properties of MG30-LiTf polymer electrolyte. IEEE Symposium on Business, Engineering and Industrial Applications. 2012;**6422980**:701-704

[32] Aziz AFB, Ali AMM. Thermal oxidation studies on methyl grafted natural rubber polymer electrolytes with paraphenylene diamine additive. 2012 IEEE Colloquium on Humanities, Science and Engineering Research. 2012;**6504406**:719-723

[33] Li GY, Koenig JL. FTIR imaging of oxidation of polyisoprene 2. The role of N-phenyl-Nϱ-dimethyl-butyl-p-phenylenediamine antioxidant. Polymer Degradation and Stability. 2003;**81**:377-385

[34] Narathichat M, Sahakaro K, Nakason C. Assessment degradation of natural rubber by moving die processability test and FTIR spectroscopy. Journal of Applied Polymer Science. 2010;**115**:1702-1709

[35] Su'ait MS, Ahmad A, Rahman MYA. Ionic conductivity studies of 49% poly(methyl methacrylate)-grafted natural rubber-based solid polymer electrolytes. Ionics. 2009;**15**: 497-500

[36] Ahmad A, Rahman MYA, Su'ait MS, Hamzah H. Study of MG49-PMMA based polymer electrolyte. The Open Materials Science Journal. 2011;**5**:170-177

[37] Ataollahi N, Ahmad A, Hamzah H, Rahman MYA, Mohamed NS. Preparation and characterization of PVDF-HFP/MG49 based polymer blend electrolyte. International Journal of Electrochemical Science. 2012;**7**:6693-6703

[38] TianKhoon L, Ataollahi N, Hassan NH, Ahmad A. Studies of porous solid polymeric electrolytes based on poly(vinylidene fluoride) and poly(methyl methacrylate) grafted natural rubber for applications in electrochemical devices. Journal of Solid State Electrochemistry. 2016;**20**:203-213

[39] Ahmad A, Rahman MYA, Su'ait MS. Morphological, infrared, and ionic conductivity studies of poly(ethylene oxide)-49% poly(methyl methacrylate) grafted natural rubber-lithium perchlorate salt based solid polymer electrolytes. Journal of Applied Polymer Science. 2012;**124**:4222-4229

[40] Xi J, Qiu X, Zheng S, Tang X. Nanocomposite polymer electrolyte comprising PEO/ LiClO4 and solid super acid: Effect of sulphated-zirconia on the crystallization kinetics of PEO. Polymer. 2005;**46**:5702-5706

[41] Abdullah M, Lenggoro W, Okuyama. Polymer electrolyte nanocomposites. In: Nalwa HS, editor. Encyclopedia Nanoscience and Nanotechnology. American Scientific Publishers; USA: American Scientific Publishers; 2004. p. 731

[42] Huang W, Frech R. Dependence of ionic association on polymer chain length in poly (ethylene oxide)–lithium triflate complexes. Polymer. 1994;**35**:235-242

Chapter 3

Nitrile Elastomer/LDH Composites with Varying Mg/Al Ratio, Curing, Nanoparticles Dispersion and Mechanical Properties

Magdalena Lipińska

Additional information is available at the end of the chapter

http://dx.doi.org/10.5772/intechopen.68384

Abstract

Layered double hydroxides (LDHs) tend to be a promising material in the field of polymer nanocomposites because it possesses unique chemical and structural properties. The novelty of layered double hydroxides (LDHs) comparing to other conventional fillers is connected with possibility of surface functionalization by introduction of various organic species into an interlayer area. Organic modifiers could act not only as intercalation promoters but also have additional functions such as UV stabilizers, anti-aging substances, pigments, antimicrobial, and antifungal activity. Additionally, in the case of nitrile elastomers, layered double hydroxides (LDHs) could play role as curing substances, reinforcing fillers, and improve mechanical, barrier and thermal properties of rubber products. The purpose of our experimental work was to examine the effect of Mg-Al-LDHs on the crosslink density and properties of nitrile rubber. Various LDHs containing, respectively, 30, 63, 70 wt% of magnesium were applied as a curing system for carboxylated acrylonitrile-butadiene rubber (XNBR) composites prepared by melt compounding methods. The influence of ionic liquids containing saccharinate, acesulfamate ions on the curing reactions and properties of rubber was investigated.

Keywords: nanocomposites, layered double hydroxides (LDHs), elastomers, carboxylated nitrile butadiene rubber (XNBR)

1. Introduction

1.1. Layered double hydroxides synthesis, structure and modification

In commercial application, elastomers are mixed both with vulcanizing additives, sulfur, and the most popular crosslinking activators such as zinc oxide and fillers, commonly silica and carbon black are added. Reinforcing effect, improvement in mechanical properties achieved after adding these fillers depends on physicochemical properties of materials, e.g., particle shape and sizes, surface area, and surface interaction with macromolecules of polymer as well as degree of particles dispersion in the matrix. Many research works are focused on replacing conventional fillers with various layered solids, montmorillonite, hectorite, expanded graphite. The exfoliated layered structure dispersed in a nanoscale in a polymer matrix-enhanced interfacial adhesion, and thus exerts a stronger reinforcing effect. Between novel nanoscale fillers, the layered double hydroxides (LDHs) are regarded as promising materials, which are able to improve varied properties of elastomer and polymer composites [1–6]. It should be taken into account that the layered double hydroxides (LDHs) combine the unique morphology and functionality with high volume-to-surface ratio of nanofiller. As opposed to layered silicates (montmorillonites) widely used in preparing polymer nanocomposites layered double hydroxides also called hydrotalcites belong to a class of anionic minerals with general chemical structure $[M_{1-x}^{II} M_x^{III} (OH)_2]^{X+} (A^{n-})_{x/n} \cdot yH_2O$, where M^{II} is a divalent metal ion; Mg^{2+}, Ca^{2+}, Zn^{2+}, etc., M^{III} represents a trivalent metal ion; Al^{3+}, Cr^{3+}, Fe^{3+}, Co^{3+}, etc., and A^{n-} represents an interlayer anion, e.g., Cl^- and CO_3^{2-} [7, 8].

The structure of hydrotalcites depends on method of synthesis, the pure LDH phase contains narrow x values (a molar ratio of trivalent and divalent metal ions), Al/(Mg + Al) containing in the range $0.20 \leq x \leq 0.33$. The LDH structure could be imagined by analogy to brucite, which consists hexagonal packaging of hydroxyl ions with octahedral sites filled by Mg^{2+} ions. In contrast to hydrotalcites, the metal hydroxide sheets in brucite are neither positive nor negative charged, the Van der Waals forces link sheets together and in consequence the distance between layers is approximately 0.48 nm. In hydrotalcites, a part of divalent metal ions M^{II} is substituted by M^{III} and generates the excess of positive charge, which is neutralized by the exchangeable anions, as well as some water molecules. As a result of anions and water molecules, the presence the interlayer distance increases and is approximately 0.77 nm for Mg-Al-LDH [8, 9]. Their properties such as high specific surface area 100 ± 300 m^2/g, "memory effect"—an ability to reconstruct after calcination until 500°C the original structure by contact with solution containing anions and good anion exchange capacities resulted in increasing development of the research concerning industrial application of LDHs. Main industrial applications of LDHs are found in areas as a catalyst and a catalyst support (Ziegler-Natta, Ce02, hydrogenation, polymerization, steam reforming), adsorbent and flame retardant (wastewater, PVC stabilizer, molecular sieve), pharmacy and medicine (ion exchanger, stabilizer, drug delivery systems, molecular containers) [10]. Opposite to layered silicates, cationic clays that occur abundantly in nature, natural anionic clays, are rare (e.g., hydrotalcite $Mg_6Al_2(OH)_{16}CO_3 \cdot 4H_2O$ [11] and mostly obtained by synthesis in the laboratory [12]. LDHs with different structures can be synthesized using various precipitation methods, among them are the most common, co-precipitation [13], and homogenous precipitation [14]. It has been reported that the sol-gel [15] techniques and hydrothermal reactions are also useful for synthesis of LDHs with desired composition. **Table 1** summarizes brief description of the most common LDH synthesis and modification methods.

Synthesis methods	Brief description	Reference
Co-precipitation	A solution containing M^{II} and M^{III} metal cations in precise concentration reacts with an alkaline solution	[13, 16–23]
Co-precipitation together with *"in-situ"* synthesis of surfactant-modified LDH	Co-precipitation of magnesium and aluminum hydroxide layers carried out in the presence of selected anions resulting modified LDHs	[19]
Co-precipitation together with *"in-situ"* synthesis of silane-modified LDH	A reacting solutions containing dissolved anionic surfactants and silanes, the synthesis procedure allows obtain *"in situ"* anionic surfactant-modified and silane-grafted hydrotalcites	[18]
Homogenous precipitation, extended homogenous precipitation method	Precipitation reactions took place in the presence of urea selected as ammonia releasing reagent. Urea hydrolysis controls the dynamic of the homogenous precipitation	[14, 19]
Hydrothermal reactions	Mg-Al-CO_3-HT was prepared via hydrothermal synthesis using a hydrothermal reaction kettle from precursors' solution (120°C, 16 h) and then was organic modified.	[24]
Sol-gel techniques	Procedure of synthesis based on hydrolysis of a mixture of metal alcoholates	[15, 25]
Regeneration method (memory effect)	Rehydration of mixed oxides obtained after calcination (450–500°C) of LDH precursor containing volatile interlayer anions (e.g., CO_3^{2-} or NO_3^-) then regeneration in solution containing new anions	[4, 26–28]
Ion exchange reaction	Original LDH is dispersed in an aqueous solution of the desired anionic species and stirred at ambient temperature for several hours	[17, 22, 23, 27, 29],

Table 1. Methods of LDH synthesis.

In the case of LDHs obtained from natural sources, the easily exchangeable interlayer anion species provide opportunity of an ion exchange with different organic compounds such as carboxylates, sulfonates, and phosphates. Various reports [9, 30, 31] claim that the incorporation of organic, anionic species between LDH layers can be done by method in which the LDH is dispersed in an aqueous solution of anionic species and ion exchange occurs in room temperature during stirring the dispersion for couple of hours. The goal of modification process is to facilitate the intercalation and exfoliation of the layers; however, the effect of d-spacing increases and an interchange reaction strongly depends on the affinity of the anions, their charge, and size. In general, the purpose of organic modification is to enlarge the basal spacing, change the surface energy, and make particles' surface more hydrophobic by replacing of interlayer anions such as Cl^-, NO^{3-}, and CO_3^{2-} with organic anions such as

- dodecyl sulfate DS anions [32–35]
- 1-decanesulfonate anions [4, 28]
- 1-hexadecanesulfonate [28]
- dodecylbenzenesulfonate [36, 37]
- stearate anions [38–40]

The interlayer distance after LDH organic modification is usually determined by wide-angle X-ray diffraction (XRD, WAXS). **Table 2** summarizes the basal spacing of functionalized, modified LDH comparing to pristine, before modification LDHs.

Intercalated anion	Basal spacing of pristine LDH [nm]	Basal spacing of modified LDH [nm]	Reference
Stearate	0.76	2.75	[40]
	0.90	3.10	[39]
Laurate	0.76	2.45	[26]
Decanesulfonate	0.76	2.27	[4]
	0.76	2.23	[28]
Hexadecanesulfonate	0.76	2.99	[28]
Dodecylsulfate	0.77	2.56	[32]
	0.76	2.68	[26]
	0.76	2.76	[41]
	0.76	2.59	[34]
	0.76	2.77	[35]
	0.76	2.48	[33]
Dodecylbenzenesulfonate	0.76	2.96	[36]
	0.76	2.95	[26]
	0.77	2.90	[42]
	0.78	2.80	[24]
Bis (2-ethylhexyl) hydrogenphosphate	0.76	1.52	[26]
Bis (2-ethylhexyl)phosphate	0.77	2.34	[42]
2-Ethylhexylsulfate	0.77	2.14	[7]

Table 2. Basal spacing of pristine and modified LDHs.

After modification LDHs, hybrid filler containing varied functional groups, having specific properties like light/UV/color/thermal stability can be dispersed in a nanoscale in a matrix leading to improve properties of final materials.

1.2. Application of ionic liquids as surface functionalizing agents for layered fillers/ rubber composites

Ionic liquids are described as salts, which are composed solely of cations and anions with melting below 100°C. The most common organic cations are imidazolium and pyridinium derivatives or compounds based on phosphonium or tetra-alkyl ammonium cations, commonly used anions are $[BF_4]^-$, $[PF_6]^-$, $[CF_3COO]^-$, and $[CF_3SO_3]^-$. ILs had been used in a variety of applications including electrochemistry, chemical synthesis, and catalysis. It is possible to immobilize this type of compound onto solid supports. Ionic liquids immobilized onto solids exhibit high selectivity and activity. One of the important ILs properties is their high thermal stability comparing to conventional organic modifiers used to enhance dispersion and compatibility of nanoparticles in organic media. Treatment of different nanofillers with ILs with proper selection of constituent ions able to react with filler and improve their compatibility to polymers may open new routes for the preparation of composite materials. The adsorption of imidazolium and pyridinium ionic liquids onto layered fillers (montmorillonites) was studied. X-ray diffraction studies confirmed the intercalation of the alkyl imidazolium and pyridinium cations into the interlayer space [42–46]. Takahashi et al. [47] reported the application of ionic liquids in synthesis of anion exchangeable layered silicates. The cation exchangeable sites in the interlayer space of silicate were converted to anion exchangeable sites by a two-step reaction. The first step was the intercalation of hexadecyltrimethylammonium ions to widen the interlayer distance. The second step was immobilization of triethoxysilyl-terminated imidazolium salts with butyl or octyl group. This way obtained materials showed the affinity for Cl^-, Br^-, I^-, and NO_3^- but different from those of conventional anion exchangeable layered double hydroxides (LDHs). According to the literature [48], ionic liquids such as didecyldimethylammonium sachcarinate [DDA][Sac], didecyldimethylammonium acesulfamate [DDA][Ace], benzalkonium saccharinate [BA][Sac], and benzalkonium acesulfamate [BA][Ace] exhibited antimicrobial, antibacterial, and antifungal activities. For two of ionic liquids [DDA][Sac] and [DDA][Ace], oral toxicity, skin irritation, and deterrent activity were also established well then to prevent harmfulness the application after immobilization on the solid surface is strongly recommended. In our previous work, we modified layered double hydroxide (LDH) and montmorillonite MMT surface with ionic liquids composed of antibacterial quaternary ammonium compounds with artificial sweetener anions saccharinate and acesulfamate, as well as derivatives of imidazolium [9, 49]. The modified fillers were used in two matrices with different polarity: the ethylene-propylene copolymer (EPM) and hydrogenated nitrile rubber (HNBR). Elastomers were crosslinked with dicumyl peroxide (DCP). The influence of ionic liquids modification on the crosslinking density of the vulcanisates, rheometric, and mechanical properties of filled elastomers was found. The phosphonium-based ionic liquids were applied as an intercalation promoter and an antibacterial substance for the LDH-polylactic acid composites. The hydrophobic modification of LDH was conducted by direct ion exchange reaction with ionic liquids having phosphonium-based cation and different anions [27]. Successful anionic intercalation took place when calcined LDH reacted with a phosphinate anion containing ionic liquids in excess of the stechiometric ratio. In this way, modified LDHs could be applied as a reservoir with a controlled amount and release of phosphorous containing antibacterial substance. The ionic liquid immobilized on LDH surface minimized the plasticizing/degradative effect on polylactic PLA composites. Addition of imidazolium salts to the ionic conductivity and curing of carboxylated nitrile rubber (XNBR) were analyzed [50]. It was found that imidazolium salts containing tetrachloroaluminate ions participate in the crosslinking process of XNBR rubber. Additionally, ionic liquids showed a plasticizing effect on a rubber matrix, shifting glass temperature *Tg* to lower values with increasing ILs concentration. Other studies showed that directly incorporation of imidazolium ionic liquids into rubber mixtures increased the conductivity of XNBR and NBR vulcanizates [51].

1.3. Preparation of rubber/LDH nanocomposites

Many research works demonstrated that LDH anionic clay materials compounded with different elastomers significantly improve mechanical, thermal (flame retardancy), and barrier properties of polymeric materials. Various methods have been applied to obtain nanocomposites and microcomposites containing anionic clay, between them were:

- "*in-situ*" polymerization of previously intercalated by monomers clay [33],
- solution supported blending [32, 52],
- melt blending, diverse variants, e.g., two-step melt compounding method [4], melting directly carried out in a laboratory two-roll mill [36, 38, 39], compounding by a twin-screw extruder [53]. **Table 3** summarizes diverse techniques used to obtain rubber/LDH nanocomposites.

A commonly applied method for rubber/LDH composite preparation is a two-step melt compounding procedure. First, the raw rubber is milled using two-roll mill with curing agents, then the mixture is melted at elevated temperature in an external mixer with high speed rotors, during mixing a layered filler is added [37]. A variant of the method was proposed by Costa et al. [28], first the raw rubber is melted using a temperature controlled mixer, then at elevated temperature at high-rotor speed clay is added and further mixed. Curing additives and rest of ingredients are mixed at high speed and room temperature. For easy dissolved in water or organic solvents (toluene, chloroform, THF), polymers' solution blending method provides opportunity to facilitate the intercalation of LDH because of polymer viscosity decrease. Many research studies have been devoted to develop and produce LDH/polymer composites with thermoplastic matrices such as polyethylene (PE), polypropylene (PP), poly(vinyl alcohol) (PVA), poly(vinyl chloride) (PVC), polystyrene (PS), poly(methyl methacrylate) (PMMA), poly(ethylene-co-vinyl acetate), poly(ethylene terephthalate) (PET) as well as epoxy resins and coatings [2, 7, 22, 27, 42, 56–58], while it has been only several reports published in the last years which referred to as properties of LDH/elastomers composites (**Table 4**).

Rubber	Method of rubber/LDH compounding	Reference
Polychloroprene CR	Melt compounding in laboratory size open two-roll mixing mill	[36]
Nitrile butadiene rubber NBR	Melt compounding in laboratory size open two-roll mixing mill	[38]
Hydrogenated nitrile rubber HNBR	Melt compounding in laboratory size open two-roll mixing mill	[54]
Carboxylated nitrile rubber XNBR	Internal mixer with temperature and rotors speed control	[8, 55]
Ethylene propylene diene rubber EPDM	Melt compounding using twin-screw extruder	[53]
Ethylene propylene diene rubber EPDM	Two-step melt compounding method	[4]
Nitrile butadiene rubber NBR	Two-step melt compounding method	[51]
Carboxylated nitrile rubber XNBR	Two-step melt compounding method	[4]
Carboxylated nitrile rubber XNBR	Two-step melt compounding method	[37]
Carboxylated nitrile rubber XNBR	Two-step melt compounding method	[28]
Ethylene propylene diene rubber EPDM	Solution blending in toluene	[32, 52]
Polyurethane/nitrile rubber blends	Solution blending in THF	[34]
Silicon rubber SR	Solution blending in CCl_4	[35]
Polyurethane	*In situ* polymerization	[33]

Table 3. Methods of rubber/LDH composites preparation.

Type of rubber	LDH loadings	Improvement of rubber properties	Structure of composite	Reference
PUR/NBR blends	1, 3, 5, 8 wt%	Mechanical, thermal properties, flame retardancy, solvent resistance	partially exfoliated	[34]
PUR	1, 3, 5, 8 wt%	Mechanical, adhesive properties, thermal stability	partially exfoliated	[59]
PUR	0.5, 1, 2, 5 wt%	Mechanical properties, gases permeability	exfoliated < 2wt% LDH, intercalated > 2 wt% LDH	[33]
EPDM	0, 5, 7.5, 10 phr	Thermal stability	Not exfoliated	[4]
EPDM	0, 2, 3, 4, 8 wt%	Mechanical properties, thermal stability	Partially exfoliated	[32, 52]
CR	5 phr	Mechanical properties, thermal stability, improvement of compression set properties,	Partially intercalated	[36]

Type of rubber	LDH loadings	Improvement of rubber properties	Structure of composite	Reference
SiR	1, 3, 5, 8 wt%	Mechanical properties, thermal stability, solvent resistance	Exfoliated	[35]
XNBR	5, 7.5, 10 phr	Significant improvement in mechanical properties because of polar/ionic interaction	Exfoliated	[4]
XNBR	5, 7.5, 10 phr	Improvement in mechanical properties, influence on the strain-induced crystallization process	Exfoliated, partially exfoliated	[28]
XNBR	2.5, 5, 10, 20, 30 phr	Improvement in the tensile strength, air permeability, enhanced UV stability	Not studied	[55]
XNBR	30 phr, various ionic liquids 5, 10, 15 phr	Increased conductivity (result of ILs addition)	Not studied	[52, 53]
XNBR	10, 20, 30, 40 phr	Mechanical properties	Not studied	[8]
XNBR	2.5, 5, 10, 20, 30 phr	Mechanical, barrier properties, cure degree,	Not studied	[60]

Table 4. Properties of LDH/rubber composites according to various authors' studies.

1.4. Application of LDHs as a part of curing system and reinforcing filler for various rubbers

This section discusses the effectiveness of LDHs in various elastomeric systems. Currently, the possibilities of LDH application both as a curing substances and reinforcing filler in elastomer composite synthesis seem particularly interesting and have attracted research and industrial interest. Emphasis has been given to use LDH as ZnO substitute during rubber mixture compounding, which could lead to decrease in zinc oxide amount used in rubber crosslinking and offer the environmentally friendly rubber products. In sulfur vulcanization, the mixture of stearic acid with zinc oxide is used. According to European Union regulation, the use of zinc oxide in rubber compounds should be restricted because of its toxicity to aquatic species. Compounds containing zinc in form such as zinc stearate, zinc-2-ethylhexanoate, and zinc borate were studied as a possible substitute of zinc oxide (ZnO) for sulfur crosslinking of SBR and EPDM rubbers. Applied substances characterized the differences in crosslinking activity related to the stability of the zinc complexes [61]. It was found that zinc-stearate-modified LDH could be applied as a part of NR and NBR rubber crosslinking system able to cure elastomer in shorter time. Recently, Das et al. [38] designed and applied LDH mineral filler in rubber formulation preparing environmental friendly rubber composites. Stearic acid and zinc are essential activators in sulfur vulcanization to promote crosslinking of rubber chains. Synthesized zinc-stearate-modified LDH was a source of both zinc and stearate ions, and it served as a triple multifunctional additive: (1) accelerator able to deliver zinc ions in the vulcanization reactions, (2) activator delivering stearate ions, (3) acting as highly anisotropic nanofillers able to reinforce a rubber matrix. Additionally, nitrile rubber NBR vulcanizates crosslinked by sulfur and modified LDH zinc oxide in the absence of ZnO and stearic acid were transparent (showing a high degree of optical transparency) what was not possible to obtain in the case of traditional ZnO, sulfur, and stearic acid curing system. WAXS and TEM studies confirmed intercalation and exfoliation of LDHs particles in rubber but the rubber matrix still contained ordered structures to some extent. The important advantage of this approach is the reduction of components added during rubber mixture processing as well as decrease in necessary ZnO content what has potentially great importance for the rubber industry.

Another possibility of LDH application is its activity as a coagent in peroxide crosslinking. The activity of hydrotalcite as coagents in the crosslinking of hydrogenated acrylonitrile-butadiene rubber was studied [54]. The layered material was surface modified with unsaturated itaconic acid. Application of layered material LDH together with itaconic acid increased the efficiency of dicumyl peroxide DCP crosslinking without scorching. Owing to the presence of unsaturated bonds, LDH/itaconic acid coagents reacted with peroxide or polymer radicals and as a consequence it was grafted to the polymer chains. The decrease of the optimal vulcanization time and improved crosslink density as well as tensile strength was observed. The formation of additional ionic crosslinks in the elastomer network was concluded. It was found that multifunctional ionic clusters showed ability to slip on the surface of the solid LDH coagent particles resulted in a high ability of vulcanisates for stress relaxation. Modified LDHs allow the replacement of traditional coagents, zinc salts. Layered double hydroxides and modified multifunctional LDHs have been applied not only to natural and nitrile rubbers but also elastomers with different polarity and chemical structure [62]. Polar particles such as LDH solids are badly dispersible in the nonpolar EPDM. This elastomer due to unsaturated bonds present in structure offers possibility of sulfur vulcanization. The presence of a saturated backbone causes a substantial anti-aging property comparing to other rubbers. The improvement of LDHs dispersion in nonpolar matrices has been of increasing research interest [4, 32, 52, 53]. Pradhan et al. [4] studied Zn-Al LDHs potential as a reinforcing filler in ethylene-propylene diene terpolymer (EPDM). Their studies showed that conventional melt compounding of EPDM with unmodified LDHs is not causing the exfoliation of LDHs layers. Analysis of WAXS patterns confirmed that there is no intercalation of the modified LDH by elastomer chains. Some improvement of mechanical properties was observed with the increase of LDH loadings. The influence of LDH on the curing behavior (higher curing level, faster curing time) of EPDM was observed for higher amount (≥ 10 parts per hundred rubber, phr) of Zn-Al LDH comparing to EPDM mixtures containing 4 phr of ZnO. The effect on the curing level, higher crosslinking state was confirmed by DSC study, the glass transition temperature T_g of EPDM/Zn-Al-LDH composites shifted to higher values. The XRD patterns and TEM micrographs confirmed that the application of solution intercalation method instead of conventional melt compounding for LDH/EPDM composites led to exfoliation of the dodecylsulfate-modified Mg-Al LDH in EPDM rubber caused improving of mechanical properties for 2–8 wt% clay loadings [32, 52]. The effect of Mg-Al-LDH addition on the EPDM photo-oxidation stability was reported by Kumar et al. [53]. The photostabilizing efficiency of LDH on EPDM was monitored after accelerated UV irradiation at 60°C for different time intervals by the measurement of the carbonyl and hydroxyl group changes using FTIR spectroscopy. For the samples containing LDH, it was confirmed that the presence of filler enhanced the degradation of EPDM. EPDM/LDH composites showed higher degradation rate than pristine EPDM. The photodegradation of filled rubber was studied under acidic conditions, the samples before accelerated aging were kept in vapor phase containing 5% nitric acid concentration for 3 h. Acidic-treated EPDM/LDH showed almost equal degradation as under isolated conditions. The results underlined the advantages of LDHs as an acid killer. Wang et al. [63] prepared a high performance elastomeric flame retardant nanocomposite based on maleic anhydride grafted ethylene-propylene-diene terpolymer MA-EPDM and synthesized organic-modified LDH. Dodecyl benzene sulfonate-modified LDH was applied as a separate filler (40 phr) and in the presence of flame retardant FR composed pentaerythritol, ammonium polyphosphate, and methyl cyanoacetate (2 phr LDH/38 phr FR). The rubber compounds were prepared using a melt compounding method by a two-roll mixing mill. SAXS and TEM studies confirmed an exfoliated structure of LDH in composite containing 2 phr LDH and 38 phr FR. The introduction of a small amount of LDH to the flame retardant MA-EPDM led to the significant decrease in the heat release rate, mass loss, fire growth rate of the nanocomposite measured by microscale combustion colorimeter and cone calorimeter test. The addition of LDH did not affect the mechanical properties of MA-EPDM.

The effect of two opposite nanoclays, montmorillonite MMT and layered double hydroxide LDH on the properties of chloroprene rubber was discussed [36]. Das et al. [36] reported that both filler improved mechanical properties of CR rubber, though reinforcing effect was stronger for organic-modified OMMT. Dynamic mechanical analysis (DMA) confirmed stronger MMT tendency to induce the crystallization of CR rubber comparing to LDH, which resisted crystallization. OMMT restricted the mobility of elastomer chains stronger than LDH resulting in the reduced height of tgδ peak. Storage modulus data obtained for organic-modified MMT and LDH-filled CR samples at room showed a significant increase of this parameter for OMMT/CR than for OLDH/CR which indicated stronger reinforcing nature of MMT. Additionally, a decrease in the value of storage modulus at higher temperature (35–45°C) in the case OMMT/CR composites was observed, this effect was attributed to the melting of the polychloroprene crystal domain and did not occurred for LDH-CR composites. Authors suggested that reduced crystallization and restricting effect of OLDH addition on crystals domain formation might be due to the presence of different heterogeneous microstructure on the LDH and MMT surface, which prevents immobilizing and ordering of elastomer chains.

Using a solution styrene-butadiene elastomer SSBR smart, thermotropic materials based on Zn-Al-LDH filler were obtained [64]. Thermotropic materials are defined as materials able to change reversibly from being transparent to a milky white (refractive) state as a function of temperature. Material was fabricated using sulfur-cured solutions of styrene-butadiene rubber (SSBR) containing up to 100 parts per hundred rubber (phr) of incorporated Zn-Al-LDH.

Zn-Al-LDH/SSBR composite becomes opaque when heated and restores the transparency after cooling to room temperature. This behavior was reversible and the material showed a gradual change in transparent/opaque characteristics upon heating, without showing any critical transition temperature. Effect was found to be increased as the amount of LDH was increased indicating that the amount of LDH loaded inside elastomer matrix qualitatively controls the thermotropic properties of the material. The addition of LDH gradually improved the mechanical, dynamic mechanical performance and thermal stability of the base elastomer.

Pradhan et al. [35] investigated dodecylsulfonate-modified LDH as a promising filler for the preparation of inorganic elastomer composites based on silicone rubber. Organic DS-LDHp/silicon rubber composites with significant degree of exfoliation were prepared by the solution intercalation in CCl_4 method. A vinyl-terminated, linear polydimethylsiloxane polymer, and the crosslinking agent as a polysiloxane containing silicone-bonded hydrogen atoms were base materials for nanocomposite preparation with different filler loadings. A platinum-catalyzed hydrosilylation reaction forming the crosslinked silicone rubber SR. The results of X-ray diffraction and transmission electron microscopy analysis demonstrated the formation of exfoliated structures of organo-modified LDH layers in the SR matrix. The interactions between Si-O polar functionality and –OH group of DS-LDH caused improvement of mechanical properties, especially the tensile strength. Thermogravimetric analysis indicated that the thermal degradation temperature (at 50% weight loss) of the exfoliated SR/DS-LDH (1 wt%) nanocomposites was higher than that of pure SR [35].

A similar solution intercalation method was used for the fabrication of a stearate-modified layered double hydroxide silicone rubber (ST-LDH/SR) composites [65]. The characterization of the material confirmed the formation of a predominantly exfoliated dispersion of St-LDH layers of 75–100 nm in width and about 1–2 nm in thickness. As it took place for the DS-LDH/SR composites, thermogravimetric analysis indicated the improvement in thermal properties of final materials, thermal degradation temperature of the exfoliated ST-LDH/SR (1 wt%) was about 80°C higher than that of pristine SR. After the addition of ST-LDH, a significant improvement in tensile strength and storage modulus was attained. A reduction of oxygen permeability was observed for ST-LDH/SR containing 3 wt% of layered filler [65].

1.5. Ionic elastomers based on carboxylated nitrile rubber (XNBR)

Ionomers (ionic elastomers) belong to a group of polymers containing a low amount of ionic functional groups (e.g., COO^-, or SO_3^{2-}). The presence of ionic functionality in the hydrocarbon backbone provides special properties, ability to form hierarchal microstructure of ionic clusters. Carboxylated acrylonitrile rubber (XNBR) has excellent abrasion resistance and chemical resistance (especially to aliphatic hydrocarbons) as compared to other elastomers. XNBR is a terpolymer of acrylonitrile, butadiene, and monomers containing carboxyl groups, such as acrylic and metacrylic acids. The presence of carboxyl groups provides additional curing sites and allows to form ionic clusters.

It was reported for ionic elastomers that two principal types of ionomer salts such as multiplets and clusters are formed [66]:

1. Small aggregates with a low number of ion pairs (approx. 6–8), this type of cluster act a similar way as physical crosslinks. The structure of ionic aggregate depends on the size of ion pair in multiplet as well as is limited by steric effects of the polymer chain segments surrounding the ion functional groups. The multiplets are statistically dispersed in a hydrocarbon matrix and do not form any separate phase.
2. Large aggregates, clusters, the ordered ionic structure containing a higher amount of ion pairs as well as nonionic hydrocarbon content. Clusters behave as a separate phase, showing special relaxation behavior at high temperature and acting not only as physical crosslinks but also as a reinforcing filler. Therefore, the properties of ionic elastomers are enhanced as compared with those containing only conventional crosslinks.

The crosslinking reaction and its kinetic depend on the type of crosslinked substance used. Different functional sites present in carboxylated nitrile rubber, unsaturated bonds, nitrile groups (–CN), and carboxylic groups (–COOH) provide an opportunity of the formation of covalent, ionic, and physical crosslinks. The functional –COOH groups of XNBR rubber cause the reaction with many compounds that may occur allowing the formation of carboxylic salts. Among them, zinc oxide remains the most popular substance used as an XNBR ionic activator.

Ibarra et al. [67–69] examined various XNBR crosslinking agents, the studies compared conventional sulfur/accelerator systems from which ZnO was excluded and peroxide system with magnesium oxide, zinc peroxide system Struktol ZP 1014 (50 wt% ZnO and ZnO_2, 30 wt% inorganic dispersing agent, 20 wt% organic dispersing agent) and anhydrous copper sulfate (II). Significant differences were found in vulcanization rheometric curves for applied curing agents. The different viscous component curves S″ (dNm) with a time-indicated type of bonds obtained for each crosslinking agent. The descendant curve was observed from the start of the reaction. During the formation of ionic bonds (MgO), the curve increased until it reached a plateau [67]. The formation of two types crosslinks ionic and covalent (ZnO) or ionic and coordination ($CuSO_4$) entailed changed in curve shapes, graphs showed a maximum followed by a decrease in normalized viscous component value S′. The presence of ionic associations influenced rubber properties increasing stress at 100%, 300% strain, tensile, and tear strength, Shore A hardness, and compression set and room temperature [67]. The crosslinking of XNBR with magnesium oxide led to obtain an ionic elastomer with thermoplastic nature [70]. Ionic associations were thermally labile and formed a segregated microphase showing a relaxation at high temperatures called ionic transition. Ionic elastomers XNBR crosslinked with magnesium oxide and filled with organic-modified layered montmorillonite (OMMT) have been prepared [71]. Vulcanization parameters and rheological properties of MgO cured XNBR were affected depending on a filler type. The glass transition temperature and ionic transition temperature were shifted in comparison with unfilled MgO cured XNBR which suggested that the addition of OMMT influenced the formation of ionic clusters during crosslinking. The XNBR rubber was crosslinked through ionic bonds with calcium oxide CaO [72]. Two reaction temperatures 100 and 150°C have been studied. The changes in curing behavior, reduction in the reaction time, and mechanical properties corresponded with the rise in temperature and were directly proportional to the CaO amount used.

As it takes place for zinc, magnesium, and calcium oxides, the layered double hydroxides are able to participate in crosslinking of carboxylated nitrile rubber XNBR. LDHs as a filler for ionic elastomers can act as:

(a) reinforcing additives dispersed in intercalated or exfoliated state,

(b) activators promoting sulfur crosslinking replacing traditionally used zinc oxide,

(c) substance providing metal cations for metal-carboxylate crosslinking.

Additionally, the basic hydroxyl groups present on LDHs' surfaces enhance their potential to form filler/filler interactions and this way strengthen the reinforcing effect. It was observed that for the polar functionalized elastomer XNBR containing carboxylic groups in the presence of LDHs, the secondary interactions between surface of the filler and elastomer matrix occurred [4]. For XNBR/LDHs composites, strong and stable filler/polymer interface is formed and tensile failure and fracture do not take place through the interface but rather through a matrix. Another important aspect is that LDHs can influence the strain-induced crystallization of XNBR rubber [4, 28]. Pradhan's research group studies showed that Mg-Al-LDHs can significantly improve the mechanical properties of XNBR rubber as a result of strong interfacial interactions between filler and elastomer chains [4]. Storage modulus of LDH/XNBR composites measured using dynamic mechanical analysis increased with increasing LDH loading, on the other hand, tanδ maximum values decreased with increasing filler concentration, confirming the strong interactions at the filler/rubber interface. The ability of LDHs to interact with XNBR has been proven by FTIR analysis [28] and FTIR-ATR analysis [55, 60] with increasing content of LDH the intensity of the band corresponding to –COOH decreased remarkably. The Mg-Al-LDH filler may provide metal ions for reaction with the –COOH, thereby leading to the formation of ionic clusters. The effect of various LDHs' structures on the curing behavior was investigated [8, 9, 60, 73]. The Mg/Al ratio strongly affected the crosslink density of Mg-Al-LDH/XNBR composites resulted in mechanic-dynamical and barrier properties of vulcanizates [55, 60].

The next section will discuss the influence of different Mg-Al-LDH structures (Mg/Al ratio, surface area, particles sizes) on the curing behavior, mechanical, thermal, and barrier properties of nitrile rubber. It is well known that the modification with organic compounds containing an anionic end group and long hydrophobic chains leads to better filler dispersion in polymeric materials as a result of it, a decrease in surface energy and increase in the thermodynamic compatibility with polymer as compared with materials containing unmodified LDH. The study will be focused on the possibility of LDHs' modification using different ionic liquids to improve the properties nitrile rubbers composites, NBR, HNBR, and XNBR.

2. Materials and characterization

2.1. Preparation of ionic liquid-modified LDH fillers

The ionic liquids such as didecyldimethylammonium saccharinate [DDA][Sac], didecyldimethylammonium acesulfamate [DDA][Ace], benzalkonium saccharinate [BA][Sac], and benzalkonium acesulfamate [BA][Ace] were synthesized using a procedure described by Hough-Troutman et al. [48]. Hydrotalcite HT with 14.1/4.5/42.2/39.2 atomic% Mg/Al/O/C ratio (indicated using XPS analysis) supplied by Sigma-Aldrich after calcination at 490°C for 3 h was dispersed in the solution of a modifying agent containing 10 wt% excess of ionic liquid comparing to stoichiometric ratio. The dispersion was mixed for 30 min using ultrasonic bath with a frequency of 35 kHz. Then, the mixture was stirred for the following 24 h, and the solvent was evaporated using a vacuum evaporator at 50°C. Surface modified LDHs were dried in a vacuum drier at 60°C for 48 h.

2.2. LDH properties analysis

2.2.1. *Aggregates size analysis*

The size of the metal oxides and LDHs was determined using a Zetasizer Nano Serie S90 (Malvern Instrument). The size of particles in a water dispersion was measured based on the dynamic light scattering (DLS) method. The concentration of the dispersion was 0.2 g/dm^3. Additionally, the size of LDH agglomerates in paraffin oil (model of elastomer matrix) was determined to estimate the tendency of LDH agglomeration in a nonpolar medium. Before the measurement, dispersions were stabilized using ultrasonic treatment (Bandelin Sonorex DT 255).

2.2.2. *Surface area measurement*

The specific surface area was measured based on nitrogen adsorption using a Gemini 2360 V2.01 porosimeter (Micrometrics, USA). The samples were degassed for 20 h at 120°C under vacuum. Measurements were performed according to the Brunauer-Emmett-Teller (BET) nitrogen adsorption method, full adsorption isotherms were collected.

2.2.3. *Oil absorption measurement*

The oil absorption parameter OAN of layered double hydroxides (LDHs) was measured according to ASTM D2414 using Absorptometer C equipment connected with Brabender station. The process parameters were sample weight 20 g, titration rate 4.0 ml/min. The oil used in this study was dibutyl phthalate (DBP).

2.2.4. *X-ray diffraction analysis (XRD)*

Room-temperature powder X-ray diffraction patterns were collected using a DRON 2EJ apparatus with (CuKα) radiation. Data were collected in the 2θ range of 1–21° with 0.05° step and 5 s exposition per step. The d-spacing within clay galleries was calculated according to Bragg equation ($n\lambda = 2d\sin\theta$).

2.2.5. *Analysis of filler particles morphology*

The morphology of filler particles was evaluated by scanning electron microscopy (SEM) using an LEO 1530 Gemini equipment (Zeiss/Leo, Germany). Samples with a graphite-coated structure were used in these investigations.

2.2.6. *Thermogravimetric analysis*

The thermal decomposition of layered double hydroxides was performed using TGA/DSC1 (Mettler Toledo) analyzer. Samples were heating from 25 to 800°C in an argon atmosphere with the heating rate of 10°C/min.

2.3. Preparation of rubber mixtures

The rubber-layered double hydroxide composites were prepared using:

- hydrogenated acrylonitrile-butadiene rubber HNBR (Therban 3407, Mooney viscosity ML (1 + 4) 100°C 70 ± 7 MU—ISO 289/ASTM D 1646, specific gravity 0.95, total ash ≤0.49 wt.%—ISO 247/ASTM D 5668) containing 34 ± 1.0 wt% acrylonitrile (ISO 24698-1) and ≤0.9% of residual double bonds (FTIR spectroscopy) after hydrogenation was obtained from Arlanxeo. It was crosslinked with dicumyl peroxide DCP (Sigma-Aldrich).
- Carboxylated acrylonitrile-butadiene rubber XNBR (Krynac X750, Mooney viscosity ML (1 + 4) 100°C MU—47 ± 5 MU—ISO 289/ASTM D 1646, specific gravity 0.99, total ash ≤1.0 wt.%—ISO 247/ASTM D 5668) containing 27 ± 1.5 wt% acrylonitrile (ISO 24698-1) and ≤7% of carboxylic group –COOH (FTIR spectroscopy) was obtained from Arlanxeo. Elastomer was crosslinked using nanometric ZnO and MgO (Sigma-Aldrich).

Various layered double hydroxides were applied as fillers as well as curing substances. Among them were magnesium aluminum layered double hydroxides Mg-Al-CO_3-LDHs Pural® MG 30 HT, 63 HT, 70 HT with different MgO/Al_2O_3 ratio, products of Sasol Germany GmbH (thereafter designated as LDH30, LDH63, LDH70). These materials were used without modification as curing substances for XNBR rubber at different filler loadings. The composites contained 10, 20, 30 phr (parts per hundred rubber). The LDHs/XNBR rubber mixtures were prepared using a melt compounding method. The rubber and filler were homogenized using Brabender Measuring Mixer N50. The rubber compounds were processed using 50 rpm (revolutions per minute) rotors speed at the initial temperature 60°C. After 5 min of rubber mastication LDH was added and mixed for 15 min.

Synthetic hydrotalcite HT (Sigma-Aldrich) was used to prepare HNBR composites cured with dicumyl peroxide (DCP) and after modification with various ionic liquids, IL-HT was applied as an interface strengthening functional filler for HNBR rubber. The ionic liquid-modified HT was used as a reinforcing filler for metal oxide cured carboxylated nitrile rubber XNBR as well. HT/HNBR and HT/XNBR rubber compounds with the formulation given below were prepared using a laboratory two-roll mill in the temperature about 40°C.

- HNBR 100 phr, DCP 3 phr, various HT loadings, IL-HT 10 phr
- XNBR 100 phr, ZnO 3 phr, IL-HT 10 phr.

Additionally, two-step melt compounding method was applied for HNBR rubber composites. First, hydrotalcite HT was grinded with 10wt% of various ionic liquids till homogenous paste was obtained. The rubber compounds were processed in an internal mixer Brabender N50 at 50 rpm speed and 50°C initial temperature. After 5 min of rubber mastication, the IL/HT paste was added and homogenized during 15 min. Subsequently, the compounded rubbers were homogenized with a curing system in a laboratory two-roll mill (temperature 40°C, friction ratio 1:1.2, dimension of rolls—diameter 200 mm, length 450 mm).

2.4. Characterization techniques

2.4.1. Vulcanization studies

The curing characteristics of the LDH/rubber composites were determined using MonTech RPA 3000 dynamic rheometer, a sinusoidal strain of 0.5 and 7% was applied at a frequency of 1.67 Hz. The curing curves were studied at different temperatures. The optimum cure time (t_{90}), scorch time ($t_{\Delta 2}$), minimum torque (S'_L), maximum torque (S'_H), and delta torque ($\Delta S' = S'_H - S'_L$) were calculated. The rheometric studies were done according to ISO 13145, ISO 6502, ASTM D5289, and ASTM D6204 standards.

2.4.2. Crosslink density

The crosslink density (ν_t) was calculated from measurement of equilibrium swelling in toluene for 48 h at 25°C using the Flory-Rehner equation [74].

$$\nu_t = \frac{\ln(1 - V_p) + V_p + \chi \cdot V_p^2}{V_s\left(V_p - \frac{V_p}{2}\right)} \tag{1}$$

where ν_t is crosslink density, V_p is the volume fraction of elastomer in the swollen gel, and V_s is the molar volume of solvent (mol/cm³).

The Huggins parameters of the elastomer-solvent interaction χ were calculated for XNBR and HNBR rubber from the equations:

- $\chi = 0.501 + 0.273\ V_r$ (HNBR) [54],
- $\chi = 0.487 + 0.228\ V_r$ (XNBR) [75],

where V_r is the volume fraction of elastomer in the swollen gel.

The amount of ionic crosslinks in the elastomer network was determined from the equilibrium swelling in toluene placed in a desiccator with saturated ammonia vapor (25% aqueous solution).

The content of ionic crosslinks $\Delta\nu$ [76] was calculated according to equation:

$$\Delta\nu = \frac{\nu_t - \nu_a}{\nu_t} \times 100\% \qquad (2)$$

where ν_a is the crosslink density determined for samples treated with ammonia vapor.

2.4.3. *Stress-strain behavior*

Tensile properties were characterized on a Zwick1435 tensile testing machine. The moduli at 100, 200, and 300% elongation, tensile strength, and elongation at break were measured at 25°C with crosshead speed of 500 mm/min for testing type 2 dumb-bell specimens prepared according to the ISO-37-2005 standard. Five different specimens were tested and the average value for each formulation was reported.

2.4.4. *Thermogravimetric analysis (TGA) and DSC analysis*

Thermal decomposition of the rubber compounds was performed using TGA/DSC1 Metler Toledo equipment. Samples were analyzed in the presence of air from 25 to 800°C using a heating rate of 10°C/min.

2.4.5. *Analysis of filler dispersion*

The LDH dispersion in the elastomer matrix was evaluated by scanning electron microscopy (SEM) (Leo153 Gemini). The rubber composites were broken down in liquid nitrogen and the fracture surface of the material was examined. Prior to analysis, the samples were coated with graphite.

2.4.6. *Aging studies*

Accelerated aging studies under different conditions were carried out. After aging, stress-strain tests were performed. The aging coefficients ($K_{\text{type of aging}}$) were calculated according to the equation:

$$K = \frac{(TS \cdot EB)\text{ after aging}}{(TS \cdot EB)\text{ before aging}} \qquad (3)$$

where TS is tensile strength (MPa) and EB is elongation at break (%). In addition, the changes in crosslink density were estimated by the swelling method using toluene as a solvent as described above.

The conditions of UV aging studies were as follows: 120 h during which alternating day (radiation intensity 0.7 W/m^2, temperature 60°C, duration 8 h) and night segments (without UV radiation, temperature 50°C, duration 4 h) were repeated.

Weathering aging studies were carried out using Atlas Weather Ometer Ci 4000 equipment, in 100 h using selected day and night panels with parameters such as: day panel: time 240 min, energy of radiation 0.7 W/m^2, humidity 60%, night panel: time 120 min, humidity 50%.

3. Results and discussion

3.1. Characterization of LDHs

It is well known that the particle size and the distribution of the filler aggregates in the elastomer matrix can be the key factors to achieve the significant improvement of mechanical properties. An uneven dispersion and the occurrence of single larger agglomerates acted as the "weak center" initiating breaking during deformation lead to deterioration of the material properties. Moreover, reduction of aggregates size and increase the specific surface area resulting in enhancement of the interphase contact between the solid surface and elastomers and thus has strong impact on the reinforcing effect. Therefore, the aggregates size in water and paraffin oil for various Mg/Al ratio layered double hydroxides was studied (**Table 5**).

Layered double hydroxides, LDH-30 and LDH-70, characterized a strong tendency to agglomerate due to electrostatic interactions and hydrogen bonds between the layers and as a result formed aggregates in range larger than 2 μm in water and larger than 4 μm in nonpolar medium paraffin oil. Size of the main fraction by a number increased with the increasing ratio of Mg/Al. The LDHs with the lowest Mg/Al ratio LDH-30 characterized the highest value of surface area after activation 3 h in 550°C what is an advantage considering its application as potentially reinforcing filler but unfortunately it also showed strong tendency to agglomeration. One of the filler characteristic contributed to reinforcing effect is the tendency to create "own structure" in an elastomer matrix as a result of filler/filler interactions. This tendency can be estimated on the basis of oil absorption parameters (OAPs). The empty space (void volume) between aggregates and agglomerates linked together through physical interactions and creating network can be expressed as the volume of dibutyl phthalate (DBP) absorbed by a given amount of filler (g DBP/100 g filler). Based on the OAP parameter measurement, the ability of LDHs with different Mg/Al ratios to form "own structure" was estimated. The data

	LDH 30	**LDH 63**	**LDH 70**
MgO [%]	28.8	63.4	69.7
Al_2O_3 [%]	71.2	36.6	30.3
Surface area, m^2/g BET, after activation 3 h at 550°C	276	200	196
Range of aggregates size [nm] (water medium)	165–2253	526–1056	940–2118
Size of the main fraction [nm] (water medium)	307	745	1332
Percentage by number [%]	36	38	38
Range of aggregates size, nm (paraffin oil)	≤4698	≤2125	≤4863
Oil absorption values OAV, g/100 g (dibutyl phthalate)	172.9	104.2	86.5
d-Spacing, nm	0.77	0.77	0.78

Table 5. Characteristics of LDH containing different Mg/Al ratios.

from **Table 5** confirmed that all LDH-30 show the tendency to create the structure similar to conventional silica with the value of surface area (140 m^2/g) [77].

From the X-ray diffraction analysis patterns, the interlayer distances were calculated using Bragg's equation and are presented in **Table 5**. All Mg-Al-CO_3-LDH types filler featured a similar d-value of approximately 0.77 nm, what should be noted the basal reflections of LDH-30 were reduced compared with two other types of LDHs indicating a less-ordered, less-lamellar structure of particles.

Synthetic hydrotalcite (HT) with 14.1/4.5/42.2/39.2 atomic% Mg/Al/O/C ratio (indicated using XPS analysis) was modified using four different ionic liquids: didecyldimethylammonium saccharinate [DDA][Sac], didecyldimethylammonium acesulfamate [DDA][Ace], benzalkonium saccharinate [BA][Sac], and benzalkonium acesulfamate [BA][Ace], according to the method described in Section 2.1. The surface properties of pristine HT are compiled in **Table 6**.

As compared to previously described LDH, hydrotalcite (HT) characterizes lower value of surface area and porosity. The pore size is in range of 0.65–80.51 nm, with a maximum of pore radius distribution curve occurred at 1.08 nm. The studies indicated lower tendency to agglomeration comparing to other fillers used, especially in polar medium water. Scanning electron microscopy (SEM) images at various magnifications for pristine HT are shown in **Figure 1**.

The size and shape of layered double hydroxide particles are usually depended on the synthesis method. Images of hydrotalcite particles revealed that HT-like fillers exhibit platy-like shape in the varied dimensions of sheets with lateral dimension ≤500 nm, and occasionally up to 1.5 μm. Hexagonal platy particles stacked one on the top of each other and formed more aggregated structures due to the physical electrostatic interactions and hydrogen bonds. The modification with ionic liquids did not influence the particle tendency to agglomeration. The aggregated structures formed by various ionic liquid-modified HT in water were in range of 315–600 nm, with the size of the main fraction ± 450 nm (percentage by number 40%), in paraffin oil IL-HT formed larger aggregates up to 1.5 μm.

Surface area BET [m^2/g]	**68**
Volume of monolayer [cm^3/g]*	14
Specific volume of pores [cm^3/g]**	0.103
Total amount of adsorbed nitrogen N_2 [cm^3/g]	89.6
Range of pores size [nm]	0.65–80.51
Range of aggregate size [nm] (water medium)	295–526
Size of the main fraction [nm] (water medium)	417
Percentage by number [%] (water medium)	39
Size of the main fraction [nm] (paraffin oil)	2671
Oil adsorption parameter OAP [g/100 g]	57.4

Table 6. Characteristic of pristine, synthetic hydrotalcite HT.

The structure of synthetic hydrotalcite HT and ionic liquid-modified IL-HT was characterized by X-ray diffraction patterns as it is shown for pristine HT in **Figure 2**.

XRD analysis of the unmodified HT structure revealed basal reflections (003) corresponding to a Bragg diffraction peak at $2\theta = 11.60°$ and d-spacing of 0.76 nm (**Figure 2**). The modification with ionic liquids did not cause the intercalation of the layer, modified IL-HT showed d-spacing = 0.76 nm. Probably, the immobilization of modifying compound occurred only on the outer surface of the filler. Particles were modified using the regeneration method, HT was previously calcined. The rehydration with solution of ionic liquids was done in an air atmosphere what probably caused recarbonation with small size anions as CO_3^{2-} instead of reconstruction of structure with anions as saccharinate and acesulfamate. Modified IL-HT formed slightly larger aggregates as is shown on SEM images (**Figure 3**) for didecyldimethylammonium acesulfamate-modified hydrotalcite [DDA][Ace]-HT and didecyldimethylammonium saccharinate-modified hydrotalcite [DDA][Sac].

The behavior during heating and thermal stability of layered double hydroxides was determined using thermogravimetric analysis. The thermal stability of LDHs is influenced by crystallinity, the composition of the brucite-like layer, the interlayer anions.

For the LDH-30, the first weight loss (13%) occurred in the temperature range of 180–230°C and was attributed to dehydration, the removal of the intercalated water molecules. The second transition peak at 306°C in the temperature range of 255–355°C (DTG analysis) was due to dehydroxylation, the removal of hydroxyl groups from the brucite layers. The third peak at 407°C in the temperature range of 390–450°C was due to the completion of dehydroxylation and removal of interlayer carbonate anions CO_2. Similar TGA/DTG curves as for LDH-30 were revealed for hydrotalcite HT [DDT][Sac]-HT and [DDT][Ace]-HT (**Figures 4** and **5**, **Table 7**) compiles the weight loss (wt%), char residue (wt%), and the temperature ranges attributed to DTG peaks.

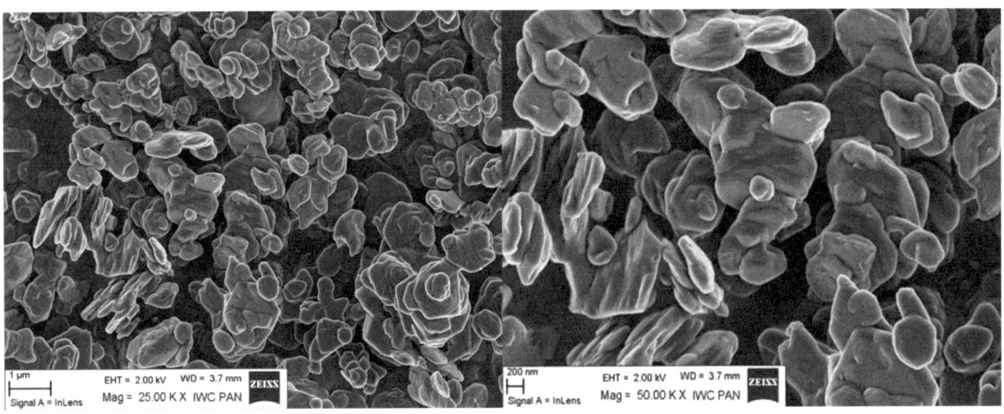

Figure 1. Characterization of the morphological structures. SEM images of hydrotlcite particles at various magnifications.

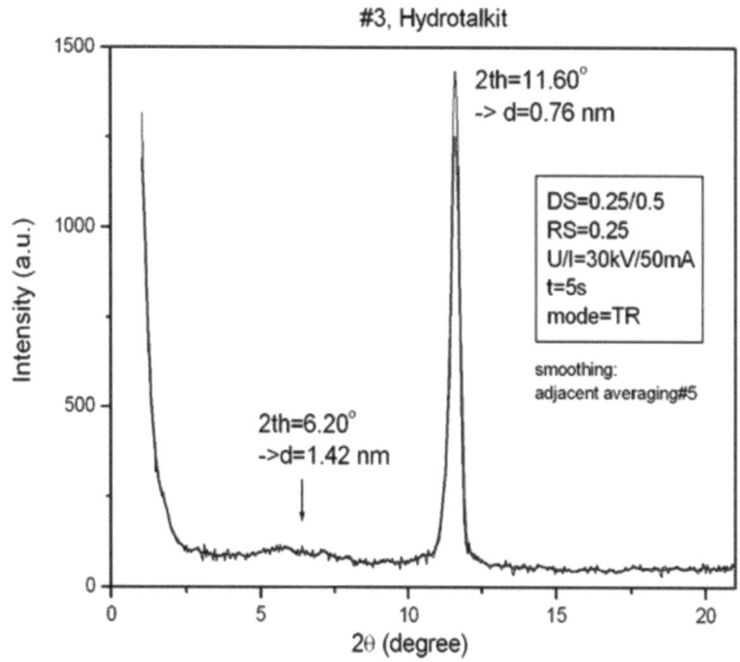

Figure 2. X-ray diffraction pattern of synthetic hydrotalcite HT.

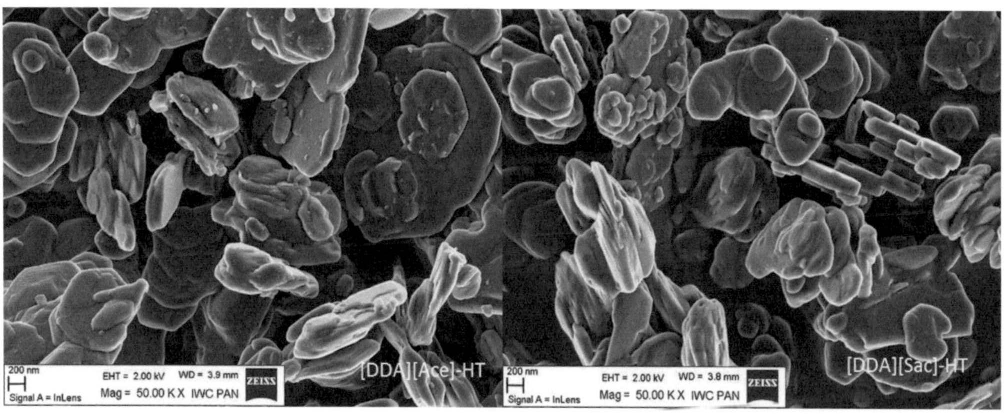

Figure 3. SEM images of [DDA][Ace]-HT and [DDA][Sac]-HT.

3.2. Ionic liquids modified hydrotalcite IL-HT as a filler for hydrogenated nitrile rubber HNBR

Table 8 shows the properties of hydrogenated nitrile rubber HNBR containing various HT loading. A significant improvement of mechanical properties of the polymer composite was related to filler loading. Effect occurred up to 10 parts of HT, higher amounts of hydrotalcite

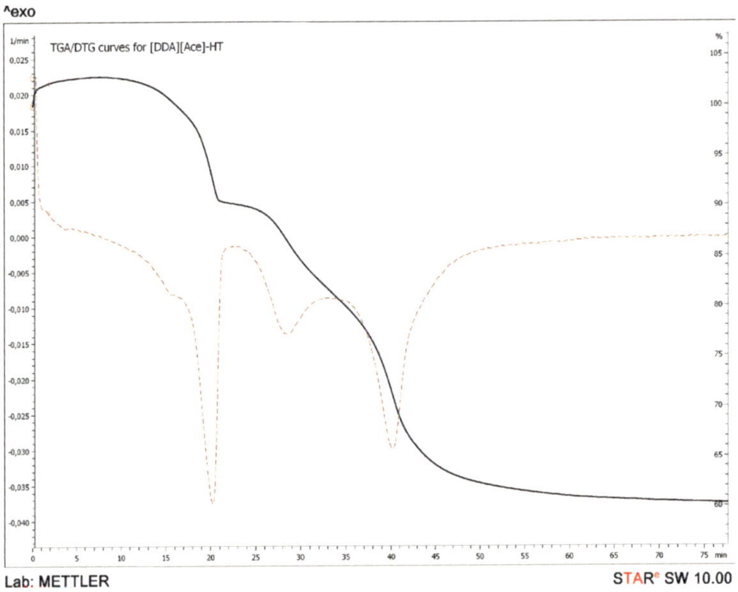

Figure 4. TGA/DTG curve for [DDA][Ace]-HT.

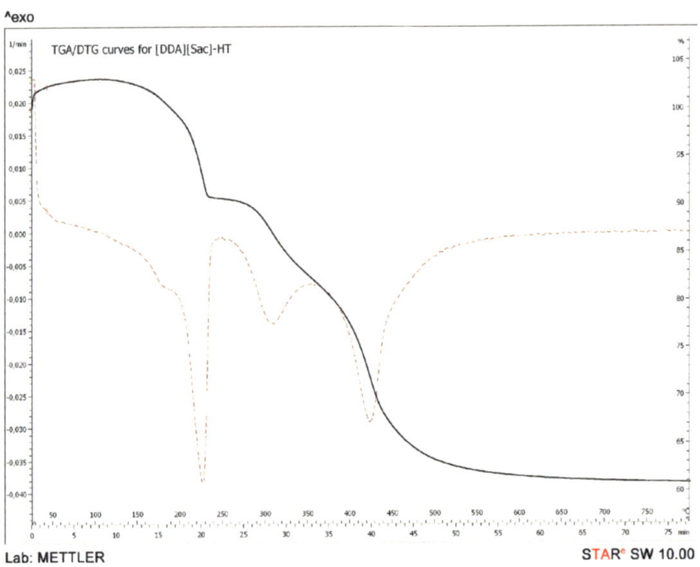

Figure 5. TGA/DTG curve for [DDA][Sac]-HT.

led to deterioration of mechanical properties as a result of uneven dispersion of clay in the matrix. Increasing clay loading influenced crosslink density of the vulcanizates, as a result of physical interactions between clay surface and polar functional group –CN in elastomer chains. Direct application of didecyldimethyl ammonium saccharinate and acesulfamate as

	LDH-30	HT	[DDA][Ace]-HT	[DDA][Sac]-HT
Temperature range of DTG peak [°C]	189–231	204–231	199–236	200–236
Temperature of peak maximum [°C]	220	243	226	226
Weight loss [wt%]	13.1	13.4	12.7	12.7
Temperature range of DTG peak [°C]	258–355	259–365	283–363	259–363
Temperature of peak maximum [°C]	306	314	307	307
Weight loss [wt%]	7.4	7.6	7.9	9.6
Temperature range of DTG peak [°C]	353–564	392–454	397–451	392–448
Temperature of peak maximum [°C]	407	429	428	425
Weight loss [wt%]	19.4	22.0	22.0	20.0
Char residue [wt%]	60.1	57.0	57.4	57.7

Table 7. Thermal stability of layered fillers determined by TGA/DTG.

a previously prepared pastes led to the enhancement of mechanical properties as a result of filler dispersion improvement. The long hydrocarbon chain presents in ionic liquid cations facilitated the compounding of rubber composites acting as plasticizing agents. Compared to vulcanizates containing unmodified hydrotalcite, a decrease of modulus at 100% SE_{100} with a simultaneous increase in crosslink density ν_t, tensile strength TS, and elongation at break E_b for IL-HT/HNBR composites was observed what confirm that this way modified filler characterizes lower surface polarity, better dispersion in rubber matrix; on the other hand, the interactions with polar elastomer functional groups are still possible what resulted in the presence of physical crosslinks in the matrix. Application of HT influenced resistance of the composites during accelerated aging. Inorganic compounds such as metal oxides are known for their effective UV screening. The analysis of the aging coefficients indicated that a higher HT amount resulted in the UV and weathering stability. Effect was strongly evident for the IL-HT. This behavior may be explained by more uniform dispersion of the filler in the elastomer which resulted more effective UV screening.

Analysis of SEM images for vulcanizates containing HT and [DDA][Sac]-HT concluded that filler particles were distributed uniformly (**Figure 6**). The melt compounding method using a laboratory two-roll mill in the temperature about 40°C did not favor any preferential orientation of the layer and still some randomly distributed agglomerates with an average size of 3 μm for pristine HT were detected.

3.3. Curing studies, mechanical properties of IL-HT/XNBR composites

Mg-Al-LDHs are active fillers able to interact with –COOH functional groups. Addition of large quantities of LDH might cause difficulties in obtaining a homogenous dispersion. The melt-mixing method in industry scale usually is supported by the addition during processing plasticizers improving and able to facilitate the dispersion. It was anticipated that the incorporation of ILs may help decrease the size of aggregates formed in the XNBR matrix.

	t_{90} [min]	$\Delta S'$ [dNm]	$\upsilon_t \times 10^5$ [mol/dm^3]	SE_{100} [MPa]	TS [MPa]	E_B [%]	K_{UV}	K_W
–	20	21.8	2.5	0.85	10.3	808	0.24	0.22
HT, 5 phr	19,7	31.5	7.7	1.26	14.0	570	0.28	0.26
HT, 10 phr	16,9	32.8	9.6	1.37	14.6	606	0.28	0.27
HT, 20 phr	15,0	31.9	13.7	1.47	15.8	578	0.32	0.36
[DDA][Sac]-HT, 10 phr	21,5	32.8	10.4	0.97	12.8	732	0.38	0.36
[DDA][Ace]-HT, 10 phr	21,5	32.3	11.4	0.98	13.3	697	0.33	0.53
*[DDA][Sac]/HT, 10 phr	16,0	36.3	13.3	1.00	16.9	728	0.47	0.46
*[DDA][Ace]/HT, 10 phr	17,5	31.1	14.4	0.97	15.2	720	0.49	0.42
[BA][Sac]-HT, 10 phr	20.0	32.3	13.9	0.92	16.8	678	0.37	0.32
[BA][Ace]-HT, 10 phr	20.5	35.3	15.8	0.94	16.5	700	0.54	0.55

t_{90}, optimum cure time, $\Delta S'$, torque increment; υ_t, crosslink density; SE_{100}, modulus at 100% strain; TS, tensile strength; K, aging coefficient; K_{UV}, UV treatment; W, weathering aging.

*Application of ionic liquids, directly as a mixture with filler.

Table 8. Properties of HT/HNBR and IL-HT/HNBR composites.

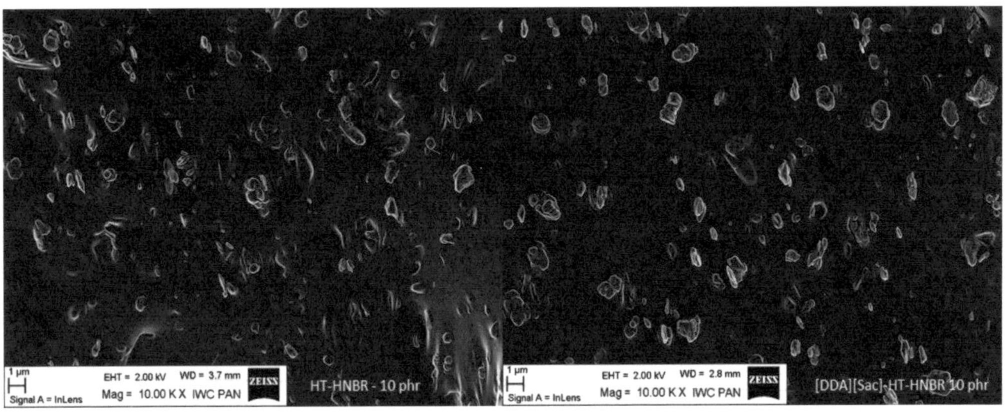

Figure 6. SEM images of HT/HNBR and [DDA][Sac]-HT/HNBR composites.

The IL-HT/XNBR composites were cured using nanoparticles of zinc oxide ZnO. The properties of vulcanizates are compiled in **Table 9**. A strong influence of ionic liquids on the crosslink network density of XNBR rubber was observed. Ionic liquids act here as activators, increased crosslinking density. Probably in the presence of ionic liquids, the structure of formed clusters was more ordered containing a higher amount of ion pairs as well as nonionic hydrocarbon content. The influence of the ionic liquids, type of anion and cation on the crosslink density was observed. Hydrotalcites modified with ionic liquids containing saccharinate anion [DDA]

[Sac]-HT and[BA][Sac]-HT increased the percentage of ionic crosslinks. As a result, the tensile strength of vulcanizates was almost decreased double. At the same time, such a modification did not affect the increase in stiffness. Additionally, as shown in **Figures 7–9**, the presence of long hydrocarbons segments in ILs cations exerted the plasticizing effect of IL-HT-based composites improving dispersion and caused the orientation of the layers of filler in the direction of the shear force during compounding.

Selected ionic liquids can be advantageously used to modify the hydrotalcite in order to obtain rubber vulcanizates XNBR characterized by excellent mechanical properties.

	t_{90} [min]	$\Delta S'$ [dNm]	$v_t \times 10^5$ [mol/dm^3]	Δv [%]	SE_{100} [MPa]	TS [MPa]	E_B [%]
–	31	6.50	9.6	16.8	1.27	11.8	693
HT, 10 phr	30	7.20	21.8	28.2	3.12	13.8	642
[DDA][Sac]/ HT, 10 phr	24	9.48	39.48	86.6	2.19	23.9	823
[DDA][Ace]/ HT, 10 phr	24	7.10	28.77	39.2	2.17	21.6	706
[BA] [Sac]-HT, 10phr	27	8.83	39.4	72.5	2.13	23.8	777
[BA][Ace]- HT,10phr	26	7.20	45.01	42.06	2.07	20.3	817

t_{90}, optimum cure time; $\Delta S'$, torque increment; v_t, crosslink density; Δv, ionic crosslinks number; SE_{100}, modulus at 100% strain; TS, tensile strength.

Table 9. Properties of HT/XNBR and IL-HT/XNBR composites.

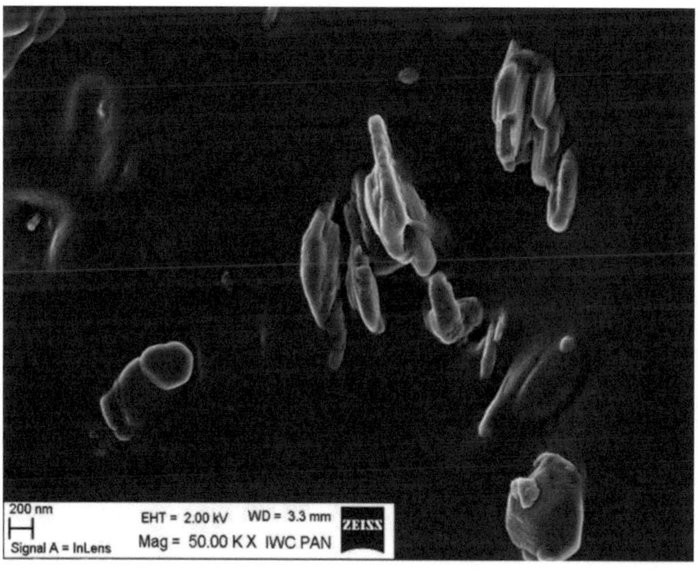

Figure 7. SEM image of HT-XNBR composite.

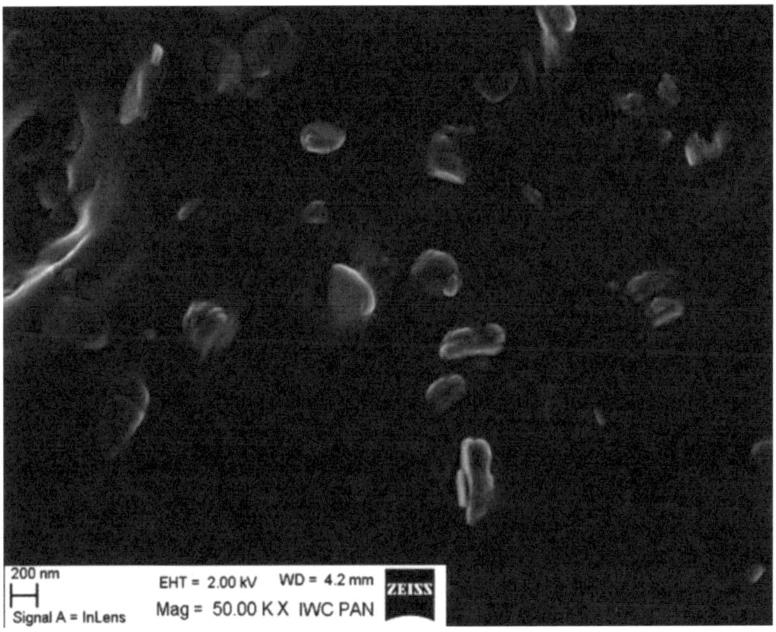

Figure 8. SEM image of [BA][Sac]-HT/XNBR composite.

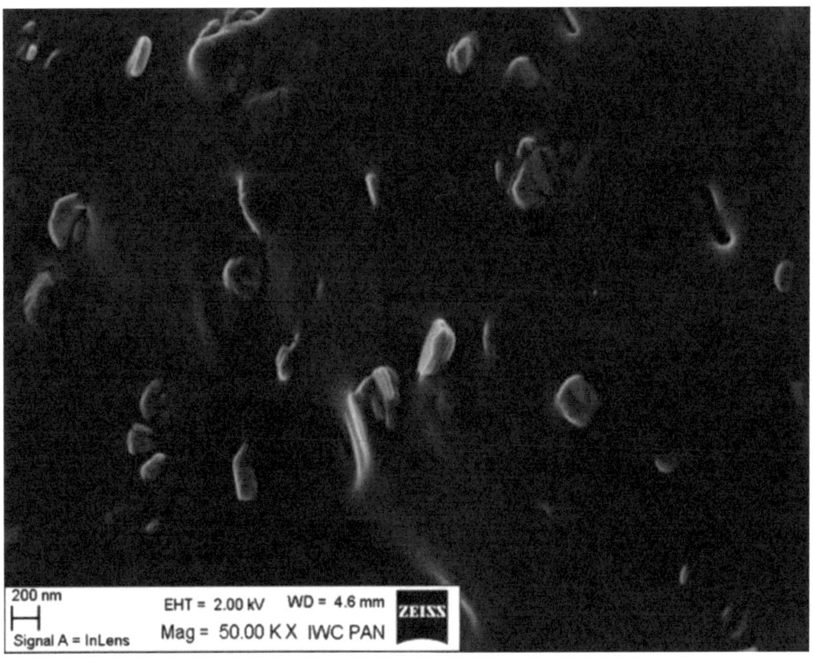

Figure 9. SEM image of [BA][Ace]-HT/XNBR composite.

3.4. Ionic elastomers based on carboxylated nitrile rubber XNBR and LDHs with various Mg/Al ratios

Carboxylated nitrile rubber composites containing LDH-30, LDH-63, and LDH-70 at various clay loadings (10, 20, and 30 parts per hundred rubber) were prepared according to method described in Section 2.3. The curing at various temperatures was investigated. **Table 10** shows the parameters that describe the vulcanization reaction at three temperatures 150, 160, and 170°C for XNBR containing 30 phr of various LDH.

Every LDH used was able to cure carboxylated nitrile rubber when it was added to rubber mixture in content of 30 parts per hundred rubber. As the temperature was increased, the cure index rate CRI raised. The effect was stronger for the LDH-30 with a lower Mg/Al ratio indicating that ionic cluster formation is strongly induced by temperature.

Regarding the normalized elastic component S′, which is the one that is usually taken as the vulcanization curve (**Figure 10**), it can be observed that at any temperature, the curve rises with reaction time, no reaching plateau. A reaction rate, in terms of the slope, in the first stage of the curve increases with temperature for LDH-63 and the maximum torque achieved higher value at 170°C than it took place for the temperature of 160°C. The opposite effect was observed for LDH-70 the filler with a higher Mg/Al ratio. At lower temperatures 150°C, the reaction is slower and the curves cannot achieve a plateau during the test time (120 min). Higher vulcanization temperature 170°C enhanced the reaction at the first stage, during first 30 min, indicating a higher value of the elastic component S′ (**Figure 10**) at this period as compared with values achieved at 170°C. However, the maximum torque decreases with the increment in reaction temperature for LDH-70/XNBR composites. It should be taken into consideration that ionic bonds formed during reaction are thermally labile. If the curve that corresponds to the normalized viscous component S″ (**Figure 11**) is analyzed a different behavior occurred. For each temperature

	LDH-30, 30 phr			LDH-63, 30 phr		LDH-70, 30 phr		
	150°C	160°C	170°C	160°C	170°C	150°C	160°C	170°C
S'_H, dNm	7.84	7.34	6.23	4.36	4.67	8.21	7.78	6.42
S'_L, dNm	1.87	1.63	1.47	0.75	0.66	1.90	1.64	1.43
$\Delta S'$	5.97	5.71	4.76	3.61	4.01	6.31	6.14	4.99
$t_{\Delta 2}$, min	10.21	6.35	4.81	34.16	16.58	11.66	1.17	0.62
t_{50}, min	18.68	11.12	6.34	28.92	16.63	23.89	13.95	6.58
t_{90}, min	80.18	68.07	47.96	116.88	101.29	115.90	101.79	55.56
$(t_{90}-t_{\Delta 2})$, min	69.97	61.72	43.15	82.72	84.71	104.24	100.62	55.04
CRI, $100/(t_{90}-t_{\Delta 2})$	1.43	1.62	2.32	1.21	1.18	0.96	0.99	1.82

Table 10. Vulcanization parameters as function of various LDHs and temperature of curing.

growing curves, characteristics for ionic vulcanization are observed. It confirms that ionic bonds that are formed during reaction show thermal liability. Thermal liability and the breakage of the ionic associations cause the loss in elasticity of the sample (S′ values) at higher temperature and increment of the viscous component. Studies of crosslink density and mechanical properties of XNBR composites were performed for the vulcanizates cured at 160°C.

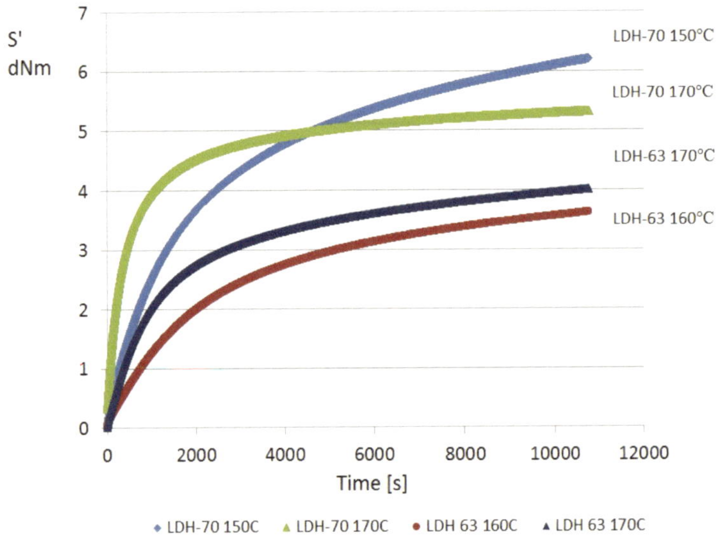

Figure 10. Normalized elastic component S′ (dNm) of the torque as a function of time for XNBR compounds crosslinked at various temperatures.

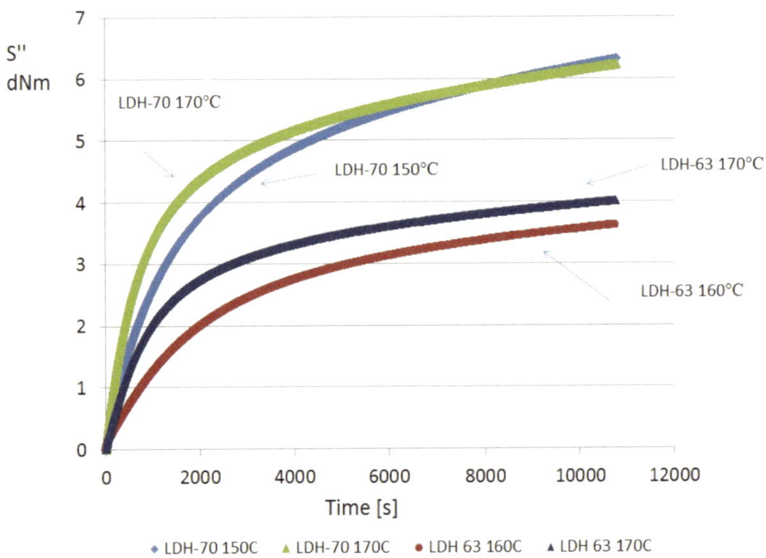

Figure 11. Normalized viscous component S″ (dNm) of the torque as a function of time for XNBR compounds crosslinked at various temperatures.

http://dx.doi.org/10.5772/intechopen.68384

The influence of various LDH contents on the rubber properties was analyzed (**Table 11**).

The tensile measurement shown in **Table 11** clearly indicates the reinforcement effect as a filler in carboxylate nitrile rubber. Both the high Mg/Al ratio and filler concentration lead to improvement in tensile strength. The increase in tensile strength with the content of various LDH fillers is likely due to a change in crosslink density. It was also noted that the modulus at 100% increased with higher LDH content. The highest tensile strength was achieved in the case of an XNBR composite filled with LDH-70 containing the highest amount of magnesium.

	$v_t \times 10^5$ [mol/dm^3]	SE_{100} [MPa]	TS [MPa]	E_B [%]
Cured with 3 phr ZnO	9.6	1.27	11.8	693
LDH-30, 10 phr	2.2	1.21	3.9	974
LDH 30, 20 phr	6.8	1.81	4.5	898
LDH 30,30 phr	8.6	2.08	12.1	831
LDH-63, 10 phr	5.9	1.32	8.92	812
LDH-63, 20 phr	7.5	1.56	10.9	793
LDH-63, 30 phr	8.1	2.06	12.2	739
LDH-63, 30 phr**	**6.8**	**1.71**	**8.67**	**735**
LDH-70, 10 phr	6.1	1.69	14.4	901
LDH -70, 20 phr	8.4	1.86	15.6	912
LDH -70, 30 phr	12.9	2.00	16.2	901
LDH-70, 30 phr**	**7.1**	**1.82**	**9.81**	**743**

v_t, crosslink density; SE_{100}, modulus at 100% strain; TS, tensile strength.
**Cured sample was cut and vulcanized again.

Table 11. Properties of LDH-30/XNBR, LDH-63/XNBR, and LDH-70/XNBR composites.

4. Conclusions

Layered double hydroxides are promising materials in the rubber industry. The presence of both unsaturated bonds and two functional groups nitrile and carboxyl in carboxylated nitrile rubber allows it to be crosslinked via conventional agents (such as sulfur and accelerators, metallic oxides (e.g., zinc oxide), organic and metallic peroxides.

This investigation shows that magnesium-rich LDH can be applied at once as a reinforcing filler and as a curing agent in a carboxylated elastomer. An interesting conclusion is that the application of LDH Mg-Al as a filler may allow the amount of metal oxide curative to be reduced in XNBR compounds. This revelation is very essential from an ecological aspect because the European Union requires that the concentration of zinc oxide should be reduced as much as possible. When layered double hydroxides are used as an alternative to conventional curing systems, the production of mechanically strong and transparent elastomer composites without any supplementary additives is enabled that is an important ecological and economical advantage. It is huge benefit for layered double hydroxides (LDHs) to serve multifunctional suitability in elastomers, such as crosslinking substance, UV screener, thermal, mechanical, and barrier property enhancer. Simplicity of synthesis and easy way of surface modification with different anions give this filler highly tunable characteristics, the possibility to control filler structure and interlayer spacing. Study has demonstrated that the incorporation of layered double hydroxides can remarkably change elastomer properties. Layered double hydroxides were able to cure carboxylated nitrile rubber. The tensile measurements indicated the enhancement of mechanical properties. Magnesium-rich layered double hydroxides (LDHs) act as a reinforcing filler and as a curing agent in carboxylated elastomers. Application of ionic liquids consisting of saccharinate and acesulfamate ions as surface functionalizing agents for layered double hydroxide enhanced dispersion and compatibility of nanoparticles in an elastomer matrix. Ionic liquids facilitated the compounding of nitrile rubber acting as plasticizing agents and influenced ionic clusters formation, resulting in increase of crosslink density.

Author details

Magdalena Lipińska

Address all correspondence to: magdalena.lipinska@p.lodz.pl

Institute of Polymer and Dye Technology, Lodz University of Technology, Stefanowskiego, Łódź, Poland

References

[1] Leroux F, Besse JP. Polymer interleaved layered double hydroxide: A new emerging class of nanocomposites. Chemistry of Materials. 2001;**13**:3507-3515. DOI: 10.1021/cm0110268

[2] Wang GA, Wang CC, Chen CY. The disorderly exfoliated LDHs/PMMA nanocomposite synthesized by in situ bulk polymerization. Polymer. 2005;**46**:5065-5074. DOI: 10.1016/j.polymer.2005.04.054

[3] Hsueh HB, Chen CY. Preparation and properties of LDHs/epoxy nanocomposites. Polymer.2003;**44**:5275-5283. DOI: 10.1016/S0032-3861(03)00579-2

[4] Pradhan S, Costa FR, Wagenknecht U, Jehnichen D, Bhowmick AK, Heinrich G. Elastomer/LDH nanocomposites: Synthesis and studies on nanoparticle dispersion, mechanical properties and interfacial adhesion. European Polymer Journal. 2008;**44**:3122-3132. DOI: 10.1016/j.eurpolymj.2008.07.025

[5] Zammarano M, Bellayer S, Gilman JW, Franceschi M, Beyer FL, Harris RH, Meriani S. Delamination of organo-modified layered double hydroxides in polyamide 6 by melt processing. Polymer. 2006;**47**:652-662. DOI: 10.1016/j.polymer.2005.11.080

[6] Li YC, Yang YH, Shields JR, Davis RD. Layered double hydroxide-based fire resistant coatings for flexible polyurethane foam. Polymer. 2015;**56**:284-292. DOI: 10.1016/j.polymer.2014.11.023

[7] Wang L, Shengpei S, Chen D, Wilkie CA. Variation of anions in layered double hydroxides: Effects on dispersion and fire properties. Polymer Degradation and Stability. 2009;**94**:770-781. DOI: 10.1016/j.polymdegradstab.2009.02.003

[8] Laskowska A, Lipińska M, Zaborski M. Hydrotalcite as a filler for carboxylated nitrile rubber. Przemysł Chemiczny. 2011;**90**(5):878-882

[9] Laskowska A, Lipińska M, Zaborski M. Properties of carboxylated nitrile elastomer containing hydrotalcites with varying Mg/Al ratio. In: Boudenne A, editor. Proceedings of Nanocomposite 2011, Multiphase Polymers and Polymer Composites, Conference and Training School; 07-10 June 2011; Paris-Est University, France. Paris, France: Paris-Est University; 2011

[10] Del Hoyo C. Layered double hydroxides and human health: An overview. Applied Clay Science. 2007;**36**:103-121. DOI: 10.1016/j.clay.2006.06.010

[11] Mills SJ, Christy AG, Genin JMR, Kameda T, Colombo F. Nomenclature of the hydrotalcite supergroup: Natural layered double hydroxides. Mineralogical Magazine. 2012;**76**(5):1289-1336

[12] Xu PZ, Zhang J, Adebajo MO, Zhang H, Zhou C. Catalytic applications of layered double hydroxides and derivatives. Applied Clay Science. 2011;**53**:139-150. DOI: 10.1016/j.clay.2011.02.007

[13] Leroux F, Adachi-Pagano M, Intissar M, Chauviere S, Forano C, Besse JP. Delamination and restacking of layered double hydroxides. Journal of Materials Chemistry. 2001;**11**:105-112. DOI: 10.1039/b002955f

[14] Ogawa M, Kaiho H. Homogeneous precipitation of uniform hydrotalcite particles. Langmuir. 2002;**18**:4240-4242. DOI: 10.1021/la0117045

[15] Ramos E, Lopez T, Bosch P, Asmoza M, Gomez R. Thermal stability of sol-gel hydrotalcites. Journal of Sol-Gel Science and Technology. 1997;**8**:437-442. DOI: 10.1007/BF02436879

[16] Kovanda F, Jindova E, Lang K, Kubat P, Sedlakowa Z. Preparation of layered double hydroxides intercalated with organic anions and their application in LDH/poly(butyl methacrylate) nanocomposites. Applied Clay Science. 2010;**48**(1-2):260-270. DOI: 10.1016/j.clay.2009.11.012

[17] Depege C, El Metoui FZ, Forano C, de Roy A, Dupuis J. Polymerization of silicates in layered double hydroxides. Chemistry of Materials. 1996;**8**(4):952-960. DOI: 10.1021/cm950533k

[18] Zhu J, Yuan P, He H, Frost R, Tao Q, Shen W, Bostrom T. In-situ synthesis of surfactant/silane modified hydrotalcites. Journal of Colloid and Interface Science. 2008;**319**:498-504. DOI: 10.1016/j.jcis.2007.11.037

[19] Kutlu B, Leuteritz A, Boldt R, Jehnichen D, Heinrich G. Effect of LDH synthesis and modification on the exfoliation and introduction of a robust anion-exchange procedure. Chemical Engineering Journal. 2014;**243**:394-404. DOI: 10.1016/j.cej.2014.01.026

[20] Hibino T. A new method for preparation of nanoplates of Zn-Al layered double hydroxides. Applied Clay Science. 2011;**54**:83-89. DOI: 10.1016/j.clay.2011.07.016

[21] Brei VV, Melezhyk OV, Starukh GM, Oranskaya EI, Mutovkin PA. Organic precursor synthesis of Al-Mg mixed oxides and hydrotalcites. Microporous and Mesoporous Materials. 2008;**113**:411-417. DOI: 10.1016/j.micromeso.2007.11.040

[22] Zhelud kevich ML, Poznyak SK, Rodrigues LM. Raps D, Hack T, Dick LF, et al. Active protection coatings with layered double hydroxide nanocontainers. Corrosion Science. 2010;**52**:602-611. DOI: 10.1016/j.corsci.2009.10.020

[23] Arizaga GGC, Mangrich AS, da Costa Gardolinski JEF, Wypych F. Chemical modification of zinc hydroxide with dicarboxylic acids. Journal of Colloid and Interface Science. 2008;**320**:168-176. DOI: 10.1016/j.jcis.2007.12.038

[24] Li M, Guo Q. The preparation of the hydrotalcite-based composite phase change material. Applied Energy. 2015;**156**:207-212. DOI: 10.1016/j.apenergy.2015.07.040

[25] Longo P, Guerra G. Layered double hydroxides with low Al content and new intercalate structures. Applied Clay Science. 2013;**71**:27-31. DOI: 10.1016/j.clay.2012.10.016

[26] Costa FR, Leuteritz A, Wagenknecht U, Jehnichen D, Häußler L, Heinrich G. . Intercalation of Mg-Al double hydroxide by anionic surfactants: Preparation and characterization. Applied Clay Science. 2008;**38**(3):153-164. DOI: 10.1016/j.clay.2007.03.006

[27] Ha JU, Xanthos M. Novel modifiers for layered double hydroxides and their effects on the properties of polylactic acid composites. Applied Clay Science. 2010;**47**(3-4):303-310. DOI: 10.1016/j.clay.2009.11.033

[28] Costa FR, Pradhan S, Wagenknecht U, Bhowmick AK, Heinrich G. XNBR/LDH nanocomposites: Effect of vulcanization and organic modifier on nanofiller dispersion and strain-induced crystallization. Journal of Polymer Science: Part B: Polymer Physics. 2010;**48**:2302-2311. DOI: 10.1002/polb.22116

[29] Zhou CH. Emerging trends and challenges in synthetic clay-based materials and layered double hydroxides. Applied Clay Science. 2010;**48**(1-2):1-4. DOI: 10.1016/j.clay.2009.12.018

[30] Khan AI, O'Hare D. Intercalation chemistry of layered double hydroxides: Recent developments and application. Journal of Materials Chemistry. 2002;**12**:3191-3198. DOI: 10.1039/B204076

[31] Lipińska M, Zaborski M. The properties of butadiene-acrylonitrile rubber NBR filled with modified layered fillers. In: Book of abstracts 15th European Conference on Composite Materials; 24-28 June 2012; Venice, Italy. 2012. p. abstract ID 1355

[32] Acharya H, Srivastava SK, Bhowmick AK. Synthesis of partially exfoliated EPDM/LDH nanocomposites by solution intercalation: Structural characterization and properties. Computer Science and Technology. 2007;**67**(13):2807-2816. DOI: 10.1016/j.compscitech.2007.01.030

[33] Guo S, Zhang C, Peng H, Wang W, Liu T. Structural characterization, thermal and mechanical properties of polyurethane/CoAl layered double hydroxide nanocomposites prepared via in-situ polymerization. Computer Science and Technology. 2011;**71**(6):791-796. DOI: 10.1016/j.compscitech.2010.12.001

[34] Kotal M, Srivastava SK, Bhowmick AK. Thermoplastic polyurethane and nitrile butadiene rubber blends with layered double hydroxide nanocomposites by solution blending. Polymer International. 2010;**59**(1):2-10. DOI: 10.1002/pi.2686

[35] Pradhan B, Srivastava SK, Ananthakrishnan R, Saxena A. Preparation and characterization of exfoliated layered double hydroxide/silicone rubber composites. Journal of Applied Polymer Science. 2011;**119**(1):343-351. DOI: 10.1002/app.32614

[36] Das A, Costa FR, Wagenknecht U, Heinrich G. Nanocomposites based on chloroprene rubber: Effect of chemical nature and organic modification of nanoclay on the vulcanizate properties. Journal of European Polymer. 2008;**44**(11):3456-3465. DOI: 10.1016/j.eurpolymj.2008.08.025

[37] Thakur V, Das A, Mahaling RN, Rooj S, Gohs U, Wagenknecht U.et al. Influence of layered double hydroxides on the curing of carboxylated nitrile rubber with zinc oxide. Macromolecular Materials and Engineering. 2009;**294**(9):561-569. DOI: 10.1002/mame.200900083

[38] Das A, Wang DY, Leuteritz A, Subramaniam K, Greenwell HC, Wagenknecht U, et al. Preparation of zinc oxide free, transparent rubber nanocomposites using a layered double hydroxide filler. Journal of Materials Chemistry. 2011;**21**:7194-7200. DOI: 10.1039/c0jm03784b

[39] Das A, Wang DY, Stoeckelhubner KW, Jurk R, Fritzche J, Kluppel M, et al. Rubber-clay nanocomposites: Some recent results. In: Heinrich G, editor. Advanced Rubber Composition Advances in Polymer Science. Vol. 239. Berlin Heidelberg: Springer-Verlag; 2010. pp. 85-166. DOI: 10.1007/12_2010_96

[40] Kumar S, Das CK. Stearate modification of layered double hydroxide (LDH) for polyurethane elastomeric nanocomposites. Elastomery. 2010;**14**(4):15-22.

[41] Kulia T, Srivastava SK, Bhowmick AK. Rubber/LDH nanocomposites by solution blending. Journal of Applied Polymer Science. 2009;**111**:635-641. DOI: 10.1002/app.29117

[42] Reinert L, Batouche K, Lévêque J-M, Muller F, Bény J-M, Kebabi B, et al. Adsorption of imidazolium and pyridinium ionic liquids onto montmorillonite: Characterization and thermodynamic calculation. Chemical Engineering Journal. 2012;**209**:13-19. DOI: 10.1016/j.cej.2012.07.128

[43] Lv G, Li Z, Jiang WT, Chang PH, Liao L. Interlayer configuration of ionic liquids in Ca-montmorillonite as evidenced by FTIR, TG-DT, and XRD analyses. Material Chemistry Physics. 2015;**162**:417-424. DOI: 10.1016/j.matchemphys.2015.06.008

[44] Livi S, Duchet-Rumeau J, Gerard JF. Supercritical CO_2-ionic liquid mixture for modification of organoclays. Journal of Colloid and Interface Science. 2011;**353**:225-230. DOI: 10.1016/j.jcis.2010.09.049

[45] Kim NH, Malhotra SV, Xanthos M. Modification of cationic nanoclay with ionic liquids. Microporous and Mesoporous Materials. 2006;**96**:29-35. DOI: 10.1016/j.micromeso.2006.06.017

[46] Lawal IA, Moodley B. Synthesis, characterization and application of imidazolium based ionic liquid modified montmorillonite sorbents for the removal of amaranth dye. RSC Advances. 2015;**5**:61913-61924. DOI: 10.1039/c5ra09483f

[47] Takahashi N, Hata H, Kuroda K. Anion exchangeable layered silicates modified with ionic liquids on the interlayer surface. Chemistry of Materials. 2010;**22**:3340-3348. DOI: 10.1021/cm9037439

[48] Hough-Troutman WL, Smiglak M, Griffin S, Reichert WM, Mirska I, Jodynis-Liebert J, et al. Ionic liquids with dual biological function: Sweet and anti-microbial, hydrophobic quaternary ammonium-based salts. New Journal of Chemistry. 2009;**33**:26-33. DOI: 10.1039/b813213p

[49] Lipińska M, Laskowska A, Zaborski M. Elastomer composites containing layered fillers modified with ionic liquids. Materials Science Forum. 2012;**714**:73-78. DOI: 10.4028/www.scientific.net/MSF.714.73

[50] Marzec A, Laskowska A, Boiteux G, Zaborski M, Gain O, Serghei A. Properties of carboxylated nitrile rubber/hydrotalcite composites containing imidazolium ionic liquids. Macromolecular Symposia. 2014;**341**:7-17. DOI: 10.1002/masy.201300149

[51] Laskowska A, Marzec A, Boiteux G, Zaborski M, Gain O, Serghei A. Effect of imidazolium ionic liquid type on the properties of nitrile rubber composites. Polymer International. 2013;**62**:1575-1582. DOI: 10.1002/pi.4550

[52] Acharya H, Srivastava SK, Bhowmick AK. A solution blending route to ethylene propylene diene terpolymer/layered double hydroxide nanocomposites. Nanoscale Research Letters. 2007;**2**(1):1-5. DOI: 10.1007/s11671-006-9020-x

[53] Kumar B, Rana S, Singh RP. Photo-oxidation of EPDM/layered double hydroxides composites: Influence of layered double hydroxides and stabilizers. eXPRESS Polymer Letters. 2007;**1**(11):748-754. DOI: 10.3144/expresspolymlett.2007.103

[54] Maciejewska M, Zaborski M, Krzywania-Kaliszewska A. Mineral oxides and layered minerals in combination with itaconic acid as coagents for peroxide crosslinking of hydrogenated acrylonitrie-butadiene elastomer. Comptes Rendus Chimie. 2012;**15**:414-423. DOI: 10.1016/j.crci.2012.01.001

[55] Laskowska A, Zaborski M, Boiteux G, Gain O, Marzec A, Maniukiewicz W. Ionic elastomers based on carboxylated nitrile rubber (XNBR) and magnesium aluminium layered double hydroxide (hydrotalcite). eXPRESS Polymer Letters. 2014;**8**(6):374-386. DOI: 10.3144/expresspolymlett.2014.42

[56] Zhang M, Ding P, Du L, Qu B. Structural characterization and related properties of EVA/ZnAl-LDH nanocomposites prepared by melt and solution intercalation. Materials Chemistry and Physics. 2008;**109**:206-211. DOI: 10.1016/j.matchemphys.2007.11.013

[57] Leggat RB, Taylor SA, Taylor SR. Adhesion of epoxy to hydrotalcite conversion coatings: I. Correlation with wettability and electrokinetic measurements. Colloids Surface A: Physicochemical Engineering Aspects. 2002;**210**:69-81.

[58] Leggat RB, Taylor SA, Taylor SR. Adhesion of epoxy to hydrotalcite conversion coatings: II. Surface modification with ionic surfactants. Colloids and Surface A: Physicochemical and Engineering Aspects. 2002;**210**:83-94.

[59] Kotal M, Srivastava SK. Structure-property relationship of polyurethane/modified magnesium aluminium layered double hydroxide nanocomposites. Journal of Plastics Technology. 2011;**15**:61-68. DOI: 10.1007/s12588-011-9006-0

[60] Laskowska A, Zaborski M, Boiteux G, Gain O, Marzec A. Effect of unmodified layered double hydroxides MgAl-LDHs with various structures on the properties of filled carboxylated acrylonitrile-butadiene rubber XNBR. European Polymer Journal. 2014;**60**:172-185. DOI: 10.1016/j.eurpolymj.2014.09.013

[61] Heideman G, Noordermeer JWM, Datta RN, Van Baarle B. Effect of zinc complexes as activator for sulfur vulcanization in various rubbers. Rubber Chemistry and Technology. 2005;**78**:245-257. DOI: 10.5254/1.3547881

[62] Basu D, Das A, Stoeckelhuber KW, Wagenknecht U, Heinrich G. Advances in layered double hydroxide (LDH)-based elastomer composites. Progress in Polymer Science. 2014;**39**:594-626. DOI: 10.1016/j.progpolymsci.2013.07.011

[63] Wang DY, Das A, Leuteritz A, Mahaling RN, Jehnichen D, Wagenknecht U, et al. Structural characteristics and flammability of fire retarding EPDM/layered double hydroxide (LDH) nanocomposites. RSC Advances. 2012;**2**:3927-3933. DOI: 10.1039/C2RA20189E

[64] Das A, George JJ, Kutlu B, Leuteritz A, Wang DY, Rooj S, et al. A novel thermotropic elastomer based on highly filled LDH-SSB composites. Macromolecular Rapid Communication. 2012;**33**:337-342. DOI: 10.1002/marc.201100735

[65] Pradhan B, Srivastava SK, Bhowmick AK, Saxena A. Effect of bilayered stearate ion-modified Mg-Al layered double hydroxide on the thermal and mechanical properties of silicone rubber nanocomposites. Polymer International. 2012;**61**:458-465. DOI: 10.1002/pi.3218

[66] Basu D, Das A, Stoeckelhuber KW, Jehnichen D, Formanek P, Sarlin E, et al. Evidence for an is situ developed polymer phase in ionic elastomers. Macromolecules. 2014;**47**:3436-3450. DOI: 10.1021/ma500240v

[67] Ibarra L, Rodriguez A, Mora-Barrantes I. Crosslinking of unfilled carboxylated nitrile rubber with different systems: Influence on properties. Journal of Applied Polymer Science. 2008;**108**:2197-2205. DOI: 10.1002/app.27893

[68] Ibarra L, Alzorriz M. Ionic elastomers based on carboxylated nitrile rubber (XNBR) and zinc peroxide: Influence of carboxylic group content on properties. Journal of Applied Polymer Science. 2002;**84**:605-615.

[69] Ibarra L, Rodriguez A, Mora-Barrantes I. Crosslinking of carboxylated nitrile rubber (XNBR) induced by coordination with anhydrous copper sulfate. Polymer International. 2009;**58**:218-226. DOI: 10.1002/pi.2519

[70] Ibarra L, Alzorriz M. Ionic elastomers based on carboxylated nitrile rubber and magnesium oxide. Journal of Applied Polymer Science. 2006;**103**:1894-1899. DOI: 10.1002/app.25411

[71] Ibarra L, Rodriguez A, Mora I. Ionic nanocomposites based on XNBR-OMg filled with layered nanoclay. European Polymer Journal. 2007;(43):753-761. DOI: 10.1016/j.eurpolymj.2006.12.007

[72] Ibarra L, Alzorriz M. Ionic elastomers based oncarboxylated nitrile rubber and calcium oxide. Journal of Applied Polymer Science. 2003;**87**:805-813.

[73] Laskowska A, Lipińska M, Zaborski M. Properties of carboxylated nitrile elastomer contained hydrotalcites with varying Mg/Al ratio. Journal of Engineering Technology. 2011;**2**(1):24-33.

[74] Flory PJ, Rehner J. Statistical mechanics of cross-linked polymer networks. II. Swelling. Journal of Chemistry and Physics. 1943;**11**:521-526. DOI: 10.1063/1.1723792

[75] Przybyszewska M, Zaborski M. The effect of zinc oxide nanoparticle morphology on activity in crosslinking of carboxylated nitrile elastomer. eXPRESS Polymer Letters. 2009;**3**:542-552. DOI: 10.3144/expresspolymlett.2009.68

[76] Mandal UK, Tripathy DK, De SK. Dynamic mechanical spectroscopic studies on plasticization of an ionic elastomer based on carboxylated nitrile rubber by ammonia. Polymer. 1996;**37**:5739-5742. DOI: 10.1016/S0032-3861(96)00545-9

[77] Blume A. Analytical properties of silica-a key for understanding silica reinforcement. Kautschuk Gummi Kunstoffe. 2000;**53**(6):338-345.

Chapter 4

Nanocomposites Based on Thermoplastic Polyester Elastomers

Sandra Paszkiewicz, Iman Taraghi, Anna Szymczyk,
Elżbieta Piesowicz and Zbigniew Rosłaniec

Additional information is available at the end of the chapter

http://dx.doi.org/10.5772/intechopen.68216

Abstract

The use of fillers in order to enhance the properties of polymers has been already well documented. Fundamentally, traditional fillers were applied to reduce the cost of the final polymeric products. Moreover, most micron-sized fillers required high loading for slight properties enhancement, thus causing problems in processing and melt flow due to the high viscosity of the obtained composite. Nanofillers might constitute the answer to the requirements made to the modern polymer materials. Nanofillers in the range of 3–5 wt% achieve the same reinforcement as 20–30 wt% of micron-sized fillers. Therefore, this study presents the influence of three different types of nanofillers that differ in shape (aspect ratio) on the morphology, electrical conductivity, and thermal stability of polyester thermoplastic elastomer (TPE) matrix, by means of poly(trimethylene)-*block*-poly(tetramethylene oxide) copolymer (PTT-PTMO). The morphology in this copolymer consisted of semicrystalline PTT domains dispersed in the soft phase of amorphous, noncrystallisable PTMO. The PTT-PTMO copolymer has been combined with 0.5 wt% of 1D (single-walled carbon nanotubes (SWCNTs), silicon carbide (SiC) nanofibers), 2D (graphene oxide (GO), graphene nanoplatelets (GNPs)), and 3D (polyhedral oligomeric silsesquioxane (POSS)) through *in situ* synthesis to obtain nanocomposites (NCs) samples.

Keywords: polymer nanocomposites, thermoplastic elastomer, graphene derivatives, carbon nanotubes, SiC, POSS particles, *in situ* synthesis, electrical conductivity, morphology

1. Introduction

Thermoplastic elastomers (TPEs) belong to the widely studied group of polymers, which express characteristics of both thermoplastics and rubbery materials based on weight concentrations of each part [1–5]. Recently, the incorporation of nanofillers into the TPEs has been converted to a challenging issue for many researchers to obtain unique functional materials with superior mechanical, thermal, and electrical properties [6–16]. Organic and inorganic nanoadditives, such as three-dimensional (3D) fullerenes, polyhedral oligomeric silsesquioxanes (POSSs), carbon black (CB); two-dimensional (2D) graphene nanoplatelets (GNPs), montmorillonite (MMT); and one-dimensional (1D) carbon nanotubes and nanofibers (CNTs and CNFs), are widely used as fillers in order to obtain polymer composites with enhanced physical properties at a very low content of nanoparticles. However, the shapes and aspect ratios, amount of concentrations, and large scale of aggregation, dispersion, and orientation degree of the nanofillers are the main factors for the overall characteristics of the nanocomposites (NCs) [17–20]. Furthermore, the dispersion state of nanofillers in the NCs could be influenced by viscosity of the matrix [21]. Hence, providing the excellent dispersion state of nanofillers into the matrix is one of the factors that the scientists design and produce novel functional materials based on the TPEs.

Poly(ether-ester)s (PEEs), which are segmented blocks of copolymers consisting of alternating sequences of flexible polyether and rigid polyester segments, exhibit a TPE behavior. They are of special interest due to their excellent mechanical properties, such as strength and elasticity in a wide range of temperature. PEEs based on poly(butylene terephthalate) (PBT) as rigid segments, and poly(tetramethylene oxide) (PTMO) as soft segments (PBT-PTMO) are available as commercial products (Elitel™, Arnitel, Hytrel®, DSM, etc.). Recently developed polyester thermoplastic based on poly(trimethylene terephthalate) (PTT) as the rigid segment and PTMO as flexible ones (PTT-PTMO) were first synthesized and characterized by Szymczyk et al. [5]. PTT is a recently commercialized aromatic polyester, which has become one of the most important polymer materials, since DuPont obtained PTT (Sorona® EP), which contains 20–37% renewable material from nonfood biomass, and has performance similar to conventional PBT plastics [22]. Additionally, the incorporation of nanofillers with different shapes and aspect ratios into the PTT-PTMO matrix has been already studied [7, 10–16]. The main objective of those studies was to investigate the effect of the addition of various types of nanofillers on the mechanical, thermal, electrical, and viscoelastic properties of the PTT-PTMO. For instance, the incorporation of SWCNTs into the PTT-PTMO caused the significant increase in the values of the Young's modulus, yield stress, and elongation at break, while, the GNPs indicated the opposite effect on the mechanical properties of neat PTT-PTMO [11]. Moreover, the presence of SWCNTs and GNPs in the polymer speeded up the crystallization process as it was evidenced by a shift of the crystallization peak to up to 45°C, which was recorded for the hybrid nanocomposite that contained 0.5 wt% of SWCNTs and 0.1 wt% of GNPs. Furthermore, the addition of SWCNTs and GNPs individually, as well as the mixture of both, caused substantial enhancement of thermooxidative stability, shifting the beginning of the chemical decomposition temperature by 20–25°C [11]. At the same time, a remarkable synergistic effect between GNPs and SWCNTs leading to an improvement of the electrical and thermal conductivities of the PTT-PTMO/GNPs/SWCNTs hybrid NCs was observed [10, 12]. Despite the fact that PTT-PTMO/GNPs nanocomposites were found to be nonconductive at a total content of 0.1 and 0.3 wt%, which was perhaps due to large number of defects, free radicals, and other irregularities on the surface of nanoplatelets [12], a significant synergistic effect between SWCNTs and GNPs on improving electrical conductivity of nanocomposites based on segmented block copolymers has been observed. Moreover, ceramic silicon carbide (SiC) nanofibers, which improve electron mobility and thermal conductivity, and at the same time, ensure chemical and high temperature resistivity [23–25], can be a good candidate as a 1D nanostructure to combine with novel polymeric materials to be used in photovoltaic utilizations. For instance, Mdletsche et al. [16] has investigated the efficiency of the addition of SiC nanoparticles on the mechanical, thermal, and biodegradation properties of polycaprolactone. Moreover, the effect of GO as 2D nanofillers on the phase structure, melt viscosity, and the mechanical properties of PTT-PTMO was studied by the means of DSC, ARES rheometer, and tensile tests [11]. An improvement of the Young's modulus and yield stress was observed in the PTT-PTMO/GO NCs with the increase of GO content from 0 to 0.5 wt% [11]. Such an improvement was observed most probably due to the large interfacial area and high aspect ratio of GO. Furthermore, the effects of POSS particles on the phase separation and glass transition temperature (T_g) of the PTT-PTMO has been investigated using dynamic mechanical thermal analysis (DMTA) [15]. POSS affected the phase separation of the polymer matrix, and it was shown that the glass transition temperature of PTMO-rich soft phase, melting temperature of PTT hard phase, and the degree of crystallinity of the nanocomposites were not affected by the presence of POSS cages in PTT-PTMO matrix. The crystallization temperatures shifted from 154°C for the neat copolymer to 130–133°C for nanocomposites. This was indicated on the anti-nucleating behavior of POSS particles for crystallization of PTT hard phase [15].

The main objective of this chapter is to compare the influence of nanofillers that differ in shape (aspect ratio) on the supramolecular structure, phase separation, thermal stability, and electrical conductivity of the PTT-PTMO-based nanocomposites. The 1D (SWCNT, SiC), 2D (GO, GNP), and 3D (POSS) type nanofillers were selected to be mixed with the polymer matrix. The next sections exhibit complete discussion of the present study.

2. Materials and methods

2.1. Materials and synthesis

For the manufacturing process of the PTT-PTMO copolymers, the dimethyl terephtalate (DMT, Sigma Aldrich), 1,3-propanenediol (PDO, Sigma Aldrich), and poly(tetramethylene oxide) glycol (PTMG) with the molecular mass of 1000 g/mol (DuPont, United States) were used. Tetrabutyl orthotitaniate (TBT, Fluka) was applied as a catalyst in transesterification and polycondensation steps. Irganox 1098 (Ciba-Geigy, Switzerland) was used as an antioxidant.

Moreover, the selected nanofillers, which were described in details in Refs. [10–15, 26] have been listed as follows: SWCNTs with a diameter of <2 nm, length of 5–30 μm, purity higher than 95%, and surface area of 380 m^2/g were bought from Grafen Chemical Industries, Grafen Co., Ankara, Turkey [10–12]; SiC nanofibers were produced via self-propagating high-temperature synthesis (SHS) from elemental Si and poly(tetrafluoroethylene) (PTFE) powder mixtures and provided by the group of Prof. A. Huczko [26]; GNPs in the form of a powder with less than three graphene layers, x–y dimensions of up to 10 μm, carbon content of ~97.0%, and the oxygen content of ~2.10% were bought from ANGSTRON Materials, Dayton, Ohio, USA [10–12]; GO sheets with average particle size of 50 μm obtained from expanded graphite (SLG Technologies GMbH, Germany) by Brodie oxidation method [27] were provided by the Polymer Institute of Slovak Academy of Science) [10, 11]; octakis[(n-octyl)dimethylsiloxy]octasilsesquioxane (POSS) was obtained according to a sequential methodology presented in scheme in Ref. [12] and provided by the Centre for Advanced Technologies, Poznan, Poland). The neat PTT-PTMO copolymer and PTT-PTMO-based NCs were produced by *in situ* polymerization technique. The procedure details were already published elsewhere [7–12]; however, the scheme of the synthesis process is presented in **Figure 1**. First, the nanofillers were dispersed for 30 min using high-speed stirrer device (Ultra-Turax T25) and ultrasonicator apparatus (Homogenizer HD 2200, Sonoplus, with frequency of 20 kHz and power 200 W) in PDO. Then, the polymerization process conducted in two stages has been applied. In the first stage, the dispersion of the selected nanofiller in PDO, DMT, and TBT catalyst were charged into 1 dm^3 steel reactor (Autoclave Engineers Inc, USA), where DMT was transesterified with PDO in the presence of catalyst under nitrogen flow at 165°C and atmospheric pressure under nitrogen flow. PDO was used in a sixfold molar excess over DMT. During the reaction, methanol was distilled off. After ca. 1.5 h to the reaction mixture, comprises mostly of *bis*-(3-hydroxypropyl) terephthalate, the PTMG, Irganox 1010, and second portion of catalyst were added. Then, the temperature was slowly lifted up to 210°C and

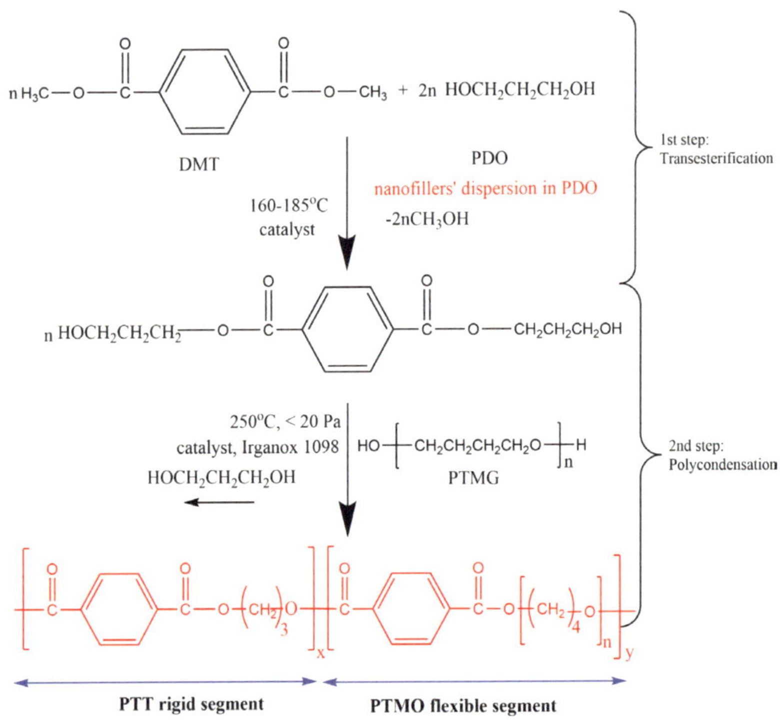

Figure 1. Scheme of the preparation process of PTT-PTMO and PTT-PTMO-based nanocomposites.

held to reach the endpoint of transesterification. Subsequently, the excess of PDO, used in the first stage, was distilled off during increasing the temperature and reducing the pressure. The second step, melt polycondensation was carried out at 250°C under reduced pressure of ~20 Pa. During polycondensation, the torque was monitored in order to detect changes in viscosity. All syntheses were finished when the melt reached the same value of viscosity at 250°C. The obtained PTT-PTMO and PTT-PTMO NCs were extruded from the reactor under nitrogen flow. The content of rigid PTT and flexible PTMO segments was approximately the same (i.e., 50 wt% of each). Finally, the extruded neat PTT-PTMO and NCs were granulated, injection molded, and compressed to prepare the specimens in accordance to the standard tests.

2.2. Characterization methods

The SEM analyses have been applied using two different scanning electron microscopes (SEM, JEOL JSM 6100; SEM ULTRA-55 Zeiss and SEM SUPRA-55 VP Zeiss). Before SEM evaluation, the extruded specimens were cryo-fractured in liquid nitrogen and then fractured surfaces were coated with thin gold layers.

In order to evaluate DMTA analysis, a Polymer Laboratories MK II apparatus has been applied in the bending mode at the constant frequency of 1 Hz and a heating rate of 3°C/min from −100°C to the sample softening temperature.

The thermal behavior was observed using differential scanning calorimetry (DSC, TA Instrument Q-100) at the heating and cooling rates of 10°C/min in the heating-cooling-heating cycle. Cooling and second heating were used in order to determine melting and crystallization temperatures. The heat of fusion was determined by integration of the normalized area of melting endotherm. The procedure of calculation degree of crystallinity and determination of glass transition temperature was already published in Refs. [5, 7, 10–12].

The electrical conductivity measurements were conducted by a Novocontrol broadband dielectric spectrometer. The complex permittivity $\varepsilon^* = \varepsilon' - i\varepsilon''$, where ε' indicates the permittivity and ε'' the dielectric loss, which was operated as a function of frequency ranging from 10^{-2} Hz < F < 10^6 Hz. The broadband electrical conductivity is usually represented as $\sigma(F) = \sigma_{DC} + \sigma_{AC} = \sigma_{DC} + A\cdot(F)^S$, where σ_{DC} is the frequency independent direct current (DC) conductivity, caused by the movement of electrons in phase with applied electric field. $A\cdot(F)^S = \sigma_{AC}$ is the component of conductivity associated with alternating current (AC), where F is frequency. The electrical conductivity of the samples was derived by $\sigma(F) = \varepsilon_0 2\pi F \varepsilon''$, where ε_0 is the vacuum permittivity [12]. Among all of the NCs specimens, the PTT-PTMO/POSS NC was not considered for the dielectric measurements, since the electrical conductivity of PTT-PTMO matrix as a host polymer cannot be influenced by the POSS nanoparticles.

Thermal and thermooxidative stability of the prepared polymer nanocomposites were evaluated by thermogravimetry (TGA 92-16.18 Setaram) at simultaneous TG-DSC system. Measurements were carried out in inert atmosphere (argon) and an oxidizing atmosphere, that is, dry, synthetic air (N_2:O_2 = 80:20 vol.%). The study was conducted at the heating rate of 10°C/min in the temperature range of 20–700°C. Measurements were conducted in accordance with the PN-EN ISO 11358:2004 standard.

3. Results and discussion

3.1. Nanocomposite morphologies

In this part, the morphologies of the produced nanocomposites were investigated by means of scanning electron microscopy (SEM). SEM is a useful tool to give a distinct insight into the nanocomposite morphology and for assessing the dispersion of nanoparticles. Especially for conductive nanoparticles, such as SWCNTs, SiC, GNPs, and contrast imaging techniques, one can use this to visualize the nanoparticles networks due to a different state of charge of the matrix and nanoparticles [28–30].

Figure 2a–e has indicated the SEM images of PTT-PTMO-based nanocomposites with 0.5 wt% of SWCNTs, SiC, GO, GNPs, and POSS nanofillers, respectively. SEM analysis presents the well-dispersed nanoparticles in the whole volume of polymer matrix, and thus the micrographs confirm that *in situ* polymerization is a highly efficient method for preparing NCs. In **Figure 2a**, it is obvious that the SWCNTs were uniformly distributed in the PTT-PTMO matrix. Moreover, the fractured surface of PTT-PTMO/0.5SWCNT indicates that the SWCNTs were pulled-out from the matrix and still embedded at both ends in the matrix. It reveals the high potential of the SWCNTs to enhance the mechanical properties of the polymer nanocomposite [11]. Similarly, in the case of the SiC nanofibers, well-distributed nanoparticles with no obvious agglomerations were observed (**Figure 2b**). Moreover, most of the SiC NFs were embedded in the polymer matrix with both ends, just like the previously reported SWCNTs. Thus, the SEM

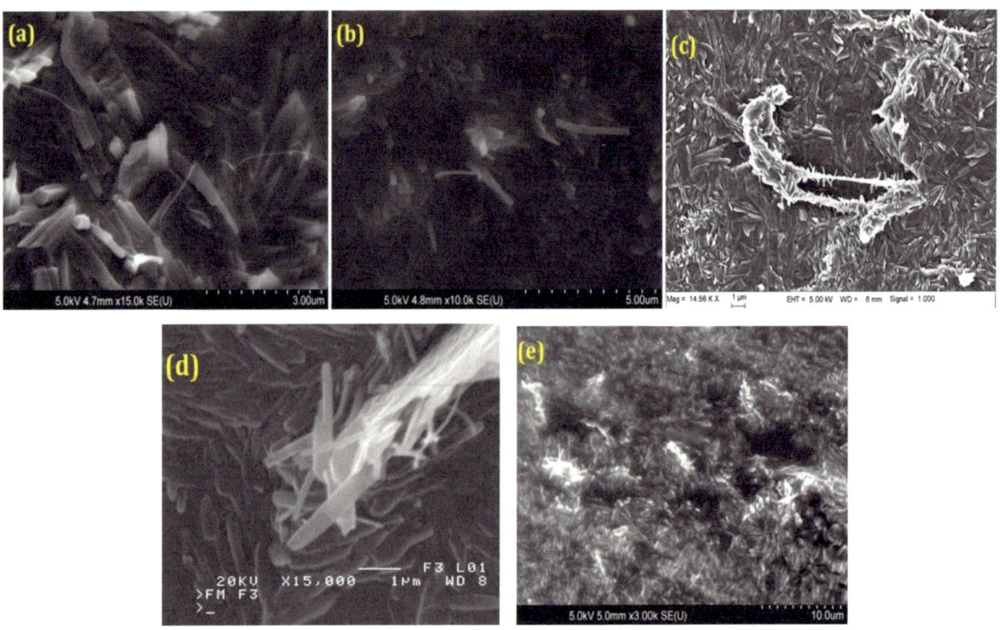

Figure 2. The SEM micrographs of PTT-PTMO copolymer reinforced with 0.5 wt% of (a) SWCNTs, (b) SiC, (c) GO, (d) GNP, and (e) POSS nanofillers.

images of NCs surface suggest the "pulling out" SiC NFs from the polymer matrix during the brittle fracture of the sample [10]. Thus, it can be concluded that 1D type nanofillers, that is, SWCNTs, SiC NFs, with high aspect ratio distinctly represent their efficiency for avoiding the existence of agglomerates due to the high shear mixing with high-frequency vibration [10–12, 26], thus confirming the legitimacy of the use of *in situ* method to obtain nanocomposites.

In the case of PTT-PTMO nanocomposites containing 2D-type nanofillers, that is, GNP and GO (**Figure 2c** and **d**, respectively), equally good dispersion was obtained. The PTT-PTMO/0.5GNPs nanocomposite (**Figure 2c**) indicates the presence of lower interfacial interactions between the GNPs and polymer matrix that is due to the existence of defects, free radicals, and other regularities on the surface of graphene nanosheets [12]. Additionally, the observations on the dispersion of GNPs were expanded upon Raman spectroscopy [12] and transmission electron microscopy (TEM) [13]. Since in our case, the Raman spectrum [12] of GNPs resembled the spectrum of reduced graphene oxide [31, 32], one could have expected strong interfacial interactions between GNPs and polymer matrix. Furthermore, in the case of PTT-PTMO/0.5GNPs [13] nanocomposite, the bent or crumpled/wrinkled platelets were visible. Exfoliated graphene-based materials are often compliant, and when dispersed in a polymer matrix are typically not observed as rigid disks (flat platelets), but rather as wrinkled ones. Moreover, randomly oriented, exfoliated platelets were observed, possibly due to restacking of the platelets. Ipso facto, the good dispersion of GO was also visible in **Figure 2d**, that it can result from the excellent dispersion of GO in PDO at the level of individual sheets. During *in situ* polymerization, a stable interphase interaction between GO nanofillers and PTT-PTMO copolymer was created due to the strong connections among the oxygen-containing functional groups [33–35] through the polymer [14]. Therefore, extreme reinforcement effects [14] were observed, while GO nanofillers were added in the polymer matrix without exhibition of agglomerates.

Finally, from the SEM image of PTT-PTMO/0.5POSS (**Figure 2e**), one can clearly see that the (n-octyl)dimethylsiloxy groups of POSS can react with functional groups of the matrix during *in situ* polymerization process. Therefore, POSS nanofillers can be completely connected to the PTT-PTMO due to excellent interfacial adhesion [15]. Such a study expanded upon SEM-EDX analysis [15] and provided the detailed insight into the distribution of POSS cages in the PTT-PTMO matrix along with the confirmation on the homogeneity in the silica distribution in the whole volume of polymer matrix.

Moreover, in our previous works [11, 13], the relation between dispersion of the nanofillers within the matrix and the values of melt viscosity has been verified and well discussed through the morphological studies. One can find that various shapes and aspect ratios could affect the values of melt viscosity, and consequently, the dispersion quality and orientation of the nanofillers in the polymer matrix.

3.2. Phase structure of PTT-PTMO-based nanocomposites

The phase structure of PTT-PTMO-based nanocomposites resulting from the addition of different types of nanofillers at the same content of 0.5 wt% was investigated by dynamic mechanical thermal analysis (DMTA). Temperatures of β_2-relaxation ($T_{\beta 2}$) and β_1-relaxation

($T_{\beta 1}$) were designated from the maximum of the loss modulus change curve and the loss angle tangent of temperature curve. Herein, the storage modulus (E'), loss modulus (E''), and tan δ for PTT-PTMO and PTT-PTMO-based nanocomposites are shown in **Figure 3a–c**, respectively. At a low temperature, an obvious inflection on the E' curves (related with the β_1 in the E'' plot) can be observed, that is associated with the glass transition of the soft polyether-rich phase. At the temperature range between 20 and 120°C, a decrease and a subsequent increase in the quantities of E' was perceived which is due to the changes associated with the glass transition temperature of amorphous polyester (PTT) phase which is followed by cold crystallization. Moreover, the addition of carbon nanofillers into the multiphase block copolymer initiated the heterophase structure with one crystalline and two amorphous phases [11]. The elastic features of the segmented block copolymer result from the aforementioned microseparated phase structure, which is a result of the chemical nature and incompatibility between the rigid and flexible blocks/segments build into the polymer chains.

As a result, at a lower temperature, the increase in the values of E' was only observed for the PTT-PTMO/0.5GO; however, the E' decreased with the addition of SWCNT, SiC, GNP, and POSS. Above 20°C, the storage modulus depends on the type and amount of nanofillers, in

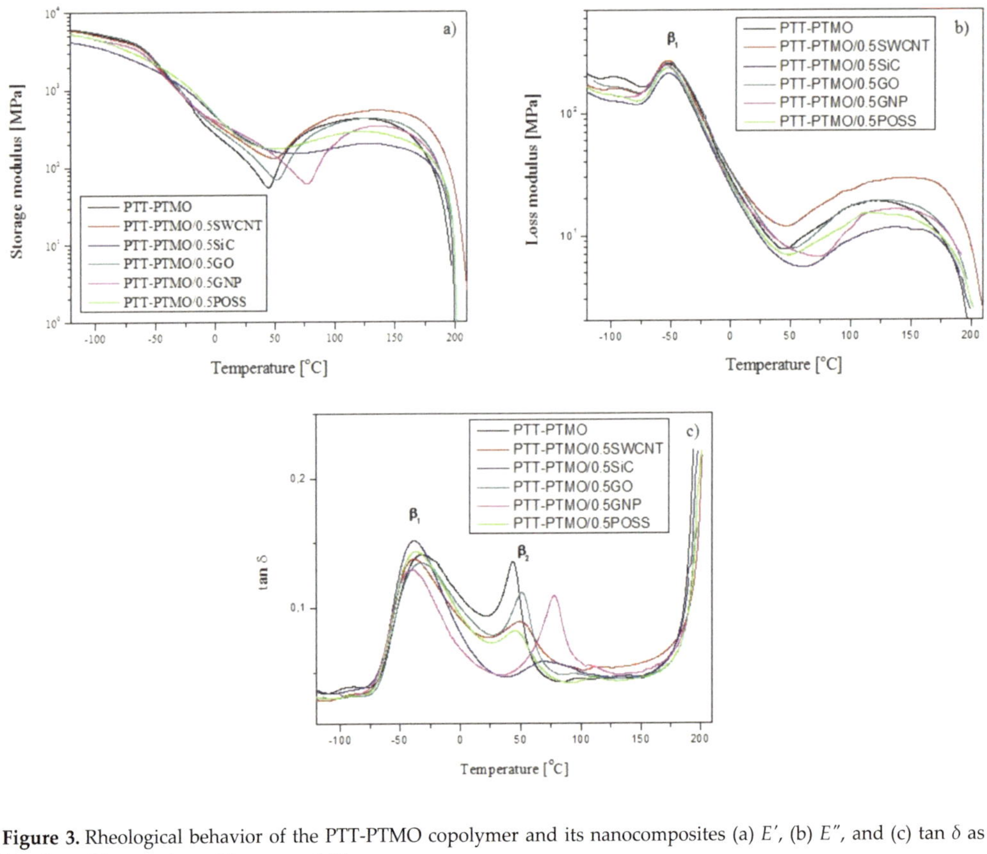

Figure 3. Rheological behavior of the PTT-PTMO copolymer and its nanocomposites (a) E', (b) E'', and (c) tan δ as a function of temperature at frequency 1 Hz.

which the highest quantities of solid-like behavior for the nanocomposites were obtained with introduction of SWCNTs into the PTT-PTMO, which was probably due to the higher stiffening effect of the rigid SWCNTs and their extremely high modulus [36–39]. At peak β_1, the difference in the values of the peaks height is due to the variety effects of nanofillers on the mobility of the chains in the polymer matrix. SWCNTs caused the greatest molecular motions in comparison to the other nanofillers. This may come from the strong connection of SWCNTs with PTT-PTMO matrix (confirmed previously by TEM and Raman spectroscopy [10, 12]). In **Figure 3c** (tan δ curves), the β_1 and β_2 relaxation peaks are related to the glass transition of amorphous polyether phase and amorphous polyester phase, respectively [11]. The incorporation of carbon nanofillers into this polymer caused stronger phase separation as evidenced by a shift in the β_1 relaxation peak toward lower temperatures. In this case, at higher temperature, the GO nanoparticles exhibited the highest impact for the efficiency on the phase separation of the PTT-PTMO copolymer, comparing with other nanofillers such as SWCNT, SiC, GNPs, and POSS. And since, for PTT-PTMO/GO nanocomposites, an increase of long period (L) was observed [14], which resulted from an increase of amorphous layer thickness that was caused by the restriction of mobility of polymer chain, one can confirm that the addition of GO-induced interfacial interactions between polymer and nanosheets in the stronger manner. Moreover, in all of nanocomposites, the composition of each phase has taken place in the matrix, since the β_2 peak showed the smaller and wider peaks.

Additionally, the DSC analyses performed for the series of PTT-PTMO-based nanocomposites (**Table 1**) confirmed that the glass transition temperature of amorphous part in semicrystalline PTT hard phase (T_{g2}) of nanocomposites remained unaffected by incorporation of SWCNT, GNP, GO, and POSS nanoparticles. At the same time, regardless of the effect of the addition of nanoparticles on the melting temperature (T_m), a slight increase (2–6°C) was observed, while the degrees of crystallinity of the prepared nanocomposites were comparable to the neat PTT-PTMO block copolymer. At the loading of 0.5 wt% SiC, GO, GNP, and POSS in PTT-PTMO matrix, the values of crystallization temperatures (T_c) of nanocomposites were close to the T_c of neat PTT-PTMO [11, 14, 15]. The only exception was the nanocomposite that contains 0.5 wt% of SWCNTs that accelerated the rate of crystallization probably due to affinity of nanotubes to act as crystallization agents. These states in agreement with many studies [40–45] proved that SWCNTs are the strongest nucleation agents.

3.3. Electrical conductivity of nanocomposites

Polymer nanocomposites based on carbon nanoparticles have enjoyed a big interest due to their enhanced mechanical and thermal properties along with electrical and thermal conductivity. In particular, nanocomposites based on carbon nanotubes (CNT), both SWCNTs and MWCNTs, carbon nanofibers, like SiC NFs [24, 26, 46], as well as graphene derivatives like GNP or expanded graphite (EG) have shown to exhibit exceptional mechanical strength and electrical conductivity [47–55]. In turn, in order to restore the electrical conductivity of

Sample	T_{g1}	T_{g2}	T_m	ΔH_m	T_c	ΔH_c	x_c	Reference
	°C	°C	°C	J/g	°C	J/g	%	
PTT-PTMO	–58	53	206	30.9	126	28.3	21.2	[69]
PTT-PTMO/0.5SWCNT	–63	54	206	34.6	171	34.5	24.9	[69]
PTT-PTMO/0.5SiC	–64	–	206	30.1	153	29.7	21.7	–
PTT-PTMO/0.5GO	–67	54	204	31.5	123	32.0	22.7	[14]
PTT-PTMO/0.5GNP	–67	53	203	34.3	148	34.2	24,7	[69]
PTT-PTMO/0.5POSS	–68	54	207	34.2	131	30.3	24.6	[15]

T_{g1}: glass transition temperature of soft phase; Tg_2: glass transition temperature of hard phase; T_m: melting temperature of polyester crystalline phase; T_c: crystallization temperature of polyester crystalline phase; ΔH_m, ΔH_c: enthalpy of melting and crystallization of polyester crystals, respectively; x_c: mass fraction of crystallinity.

Table 1. Thermal properties of neat PTT-PTMO and PTT-PTMO-based nanocomposites determined by DSC.

GO nanosheets, at least partial reduction is needed [18], since it is electrically insulating and thermally unstable. Therefore, in this work, the electrical conductivity of PTT-PTMO/0.5GO nanocomposite was also investigated, since during the synthesis process some partial reduction (second step of the synthesis takes place at 250°C) could have appeared. The use of *in situ* polycondensation method allows to obtain conductive nanocomposites filled with a relatively low loading (less than 0.5 wt%), which has been confirmed for the many types of polymer matrices [56–59]. **Figure 4** shows the alternating current electrical conductivity (σ) as a function of frequency (F) for PTT-PTMO nanocomposites and neat PTT-PTMO copolymer (reference sample).

PTT-PTMO block copolymers show at low frequencies characteristic conducting behavior (the presence of a frequency independent component, σ_{dc}) associated with the presence of PTMO. Such behavior may be due to ionic conductivity [60]. According to the results, the creation of the conducting paths in the insulating PTT-PTMO copolymer was strongly influenced by the presence of carbon nanofillers. The electrical conductivity of the nanofillers plays a main role to produce conductive polymer nanocomposites. POSS nanoparticles can improve the dielectric properties of the materials; however, this improvement is attributed to various factors, such as large particle-polymer interfacial area, particle-polymer nanoscopic structure, change in internal electric field (polarity) due to the presence of nanoparticles, functionality [61–63]), herein only the effect of carbon nanofillers on the electrical conductivity of the PTT-PTMO copolymer was investigated. In order to obtain higher electrical conductivity at lower cost, several factors need to be

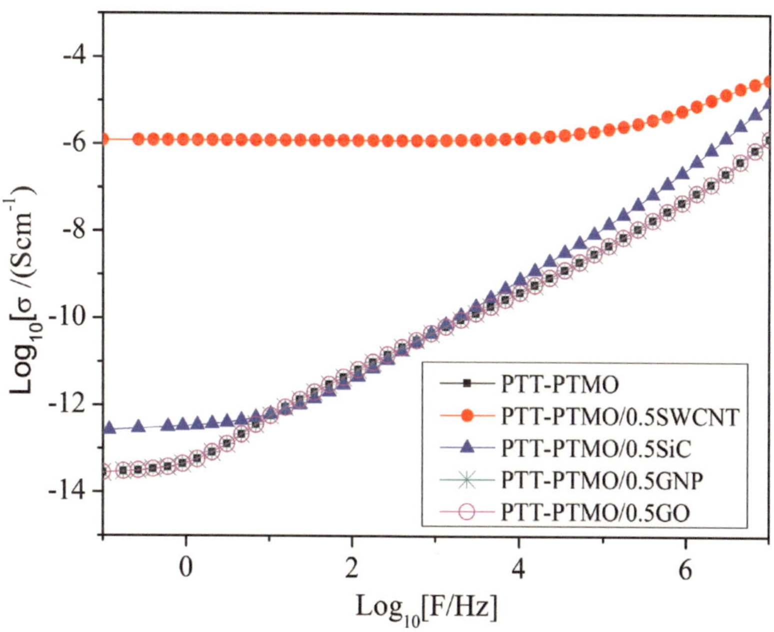

Figure 4. Broadband electrical conductivity ($\sigma(F)$) as a function of frequency (F) at room temperature for neat PTT-PTMO and its nanocomposites.

taken into consideration, that is, proper dispersion, orientation degree, and the amounts of agglomerates within the polymer-based nanocomposites [64–66]. The study of the broadband electrical conductivity confirms that the incorporation of the SWCNTs into the PTT-PTMO can covert the insolating polymers with two rigid and soft phases to the conductive specimens. The incorporation of 0.5 wt% of SWCNTs was high enough to provide conductive paths in the polymer matrix. However, even with the smaller loading of nanoparticles (0.1 wt% of SWCNTs), a slight increase in conductivity was observed, similarly as in the case of 0.3 wt% of SWCNTs [12]. Unlike SWCNTs, the GNPs and GO could not affect the electrical conductivity of the PTT-PTMO copolymer in the same manner. Due to the existence of rich defects, free radicals, residual functional groups, and impurities on the surface of GNPs, the percolation threshold has not taken place in PTT-PTMO/GNPs nanocomposites [10, 12]. However, by the way, it is also worth mentioning that conducting networks in PTT-PTMO block copolymer was already formed at 0.3 wt% of SWCNT alone [12], and further increase of 0.1 wt% of GNPs provided more electron pathways by synergy between the SWCNTs and GNPs. Once the filler content exceeded the percolation threshold, agglomerates could have even improved the electrical conductivity than well-dispersed CNTs. Along with an increase of the SWCNTs to GNPs content ratio to 5:1 (PTT-PTMO/0.5SWCNTs + 0.1GNPs), one observed the characteristic flat plot, where an extended frequency region of constant σ was detected [12]. The hybrid PTT-PTMO/0.5SWCNTs + 0.1GNPs exhibited the typical behavior for semiconducting samples with conductivity of about 10^{-6} S/cm. These observations made for the improvement in electrical conductivity of PTT-PTMO/SWCNT by GNPs were supported by the fact that the GNPs' surface rich of defects, free radicals, and other irregularities ensured strong nanotube-to-nanoplatelet interactions. These results highlighted the remarkable potential in industrial application of a new group of polymer hybrid nanocomposites, based on nanofillers with different shapes (SWCNTs and GNPs).

In addition, as mentioned above, GO nanosheets cannot create electric paths through the polymer due to the presence of functional groups such as –COOH and –OH on its surface [33–35]. Even the high temperature during *in situ* synthesis did not restore the conductivity of GO, thus the nanocomposite with 0.5 wt% of GO was found to be not conductive. Moreover, as shown in **Figure 4**, for nanocomposites with SiC nanofibers, the values of electrical conductivity in low frequency range exhibit higher values than those of pristine matrix. This fact suggests that there are many connections between nanofibers, with small gaps of polymer between them, which promotes polarization phenomena [26]. Our further study on the electrical properties of PTT-PTMO/SiC nanocomposites with higher content of SiC confirmed the electrical percolation threshold (ϕ_c) equals to 1.7 [26]. Additionally, the calculated critical exponent "*t*" [67] (equaled to 1.7) proved that a three-dimensional system was obtained [68]. Despite the fact that this study deals with nanocomposites based on PTT-PTMO block copolymer, with the total concentration of 0.5 wt%, it should be mentioned that the rather low percolation threshold, the low cost of preparation of SiC nanofibers compared to nanocomposites containing SWCNTs (ca. 1000 USD/kg for SiC nanofibers and about hundreds USD/g for SWCNT) [69–71], and their low viscosity values, which makes it easier for extrusion process after *in situ* method, give rise to a new class of interesting materials with potential use in a wide range of applications.

3.4. Thermal stability of nanocomposites

The incorporation of nanofillers with high thermal conductivity into polymer matrix may facilitate heat distribution in the material and thereby improve its heat resistance. Most carbon nanostructures, such as CNTs, SiC, and graphene derivatives, exhibit electron affinities similar to those of fullerenes, and they are therefore capable of acting as radical scavengers in free radical chain reactions, including polymerization and the thermo-oxidative degradation of polymers [69, 72]. Similarly, several studies confirmed that the incorporation of POSS can enhance the thermal stability of nanocomposites [61, 62]. The influence of the presence of SWCNT, SiC, GNP, GO, and POSS particles on the thermal and thermo-oxidative decomposition of the nanocomposites based on PTT-PTMO block copolymer has been investigated during heating in air and argon atmosphere. The weight loss (TG) and its derivative of weight loss (DTG) curves are shown in **Figure 5a** and **b**. The mechanisms of thermal and thermo-oxidative degradation of copoly(ether-ester) have been already widely discussed [73, 74]. Decomposition of copoly(ether-ester) begins with the flexible segment PTMO. Oxygen mainly affects the carbon atom located in the α position relative to the ether oxygen atom in ether [73]. Detailed studies have been done for PBT-PTMO copolymers, however, for the PTT-PTMO, the mechanism is identical, differing only in decay fragments. The thermal decomposition process of poly(1,4-tetraoksymethylene) (PTMO) chains has a radical nature, and in the initial stage of PTMO chain decomposition is observed the secretion of tetrahydrofurane (THF) aldehydes and low-boiling and volatile alkenes. At the temperature of 200°C occurs the thermal oxidation of PTMO segment with releasing volatile substances [69]. An analysis of the values of the characteristic temperatures of decomposition, including the temperature of 5, 10, and and 50% weight loss and the temperature at the maximum weight loss rate (T_{DTG}) of neat PTT-PTMO and PTT-PTMO nanocomposites (**Table 2**) showed that the presence of different types of nanofillers at the concentration of 0.5 wt% does not affect the thermal stability in an inert atmosphere (**Figure 5b**), regardless of nanofillers' content, whereas in oxidizing atmosphere, an effect on thermo-oxidative stability of the polymer matrix has been observed. The studies on thermal decomposition of PTT-PTMO nanocomposites proved that in an oxidized atmosphere, the thermal degradation process proceeds in two steps (**Figure 5a**), whereas in inert atmosphere, proceeds in only one step (**Figure 5b**). Thermal degradation profiles of PTT-PTMO nanocomposites displayed that thermal stability of the nanocomposites based on SWCNT and graphene derivatives (GO and GNP) was improved at the concentration of 0.5 wt%, while the incorporation of 0.5 wt% of SiC NFs and POSS particles caused a decrease in thermal stability of PTT-PTMO nanocomposites at the first stage of decomposition process (values of $T_{5\%}$ and $T_{10\%}$). In turn, the values of $T_{50\%}$ for the whole series of PTT-PTMO-based nanocomposites were comparable to one another.

Only in the case of PTT-PTMO/SWCNT composite, do the thermal degradation temperatures increase to about several degrees. Furthermore, the values of the temperature at the maximum weight loss rate (T_{DTG2}) of neat PTT-PTMO and PTT-PTMO nanocomposites containing SWCNT and graphene derivatives suggested the strongest effect on the thermal stabilization behavior. Since all nanofillers are found to be well-dispersed in PTT-PTMO matrix (**Figure 2**), carbon nanotubes and graphene derivatives caused the annihilation of free radicals gener-

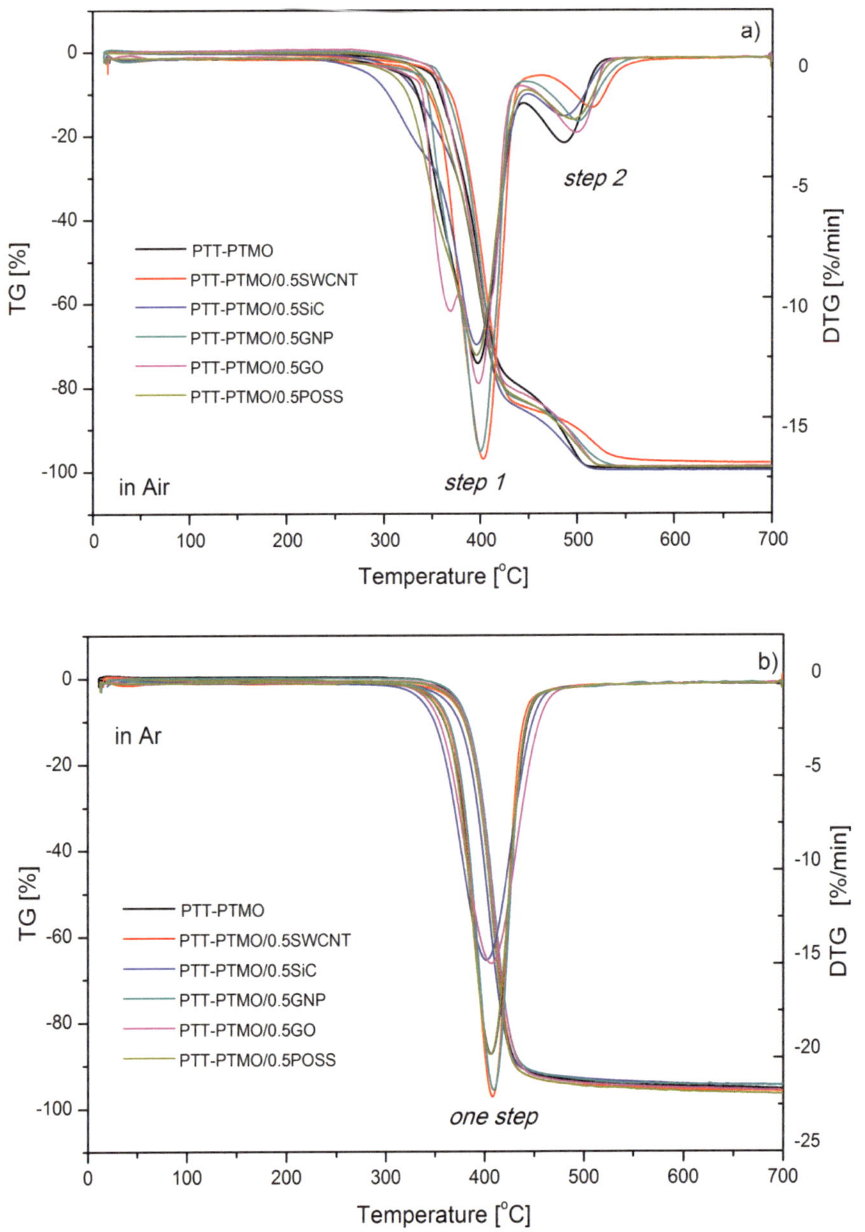

Figure 5. Weight loss and derivative weight loss versus temperature for neat PTT-PTMO and PTT-PTMO-based nanocomposites in air (a) and in argon (b) at a heating rate of 10°C/min.

ated during thermal decomposition of polymer matrix in nanocomposite in the strongest manner, and thus retarding thermal degradation of the nanocomposites.

Symbol	$T_{5\%}$, °C	$T_{10\%}$, °C	$T_{50\%}$, °C	T_{DTG1}, °C	T_{DTG2}, °C
Measurement carried out in an oxidizing atmosphere					
PTT-PTMO	347	358	398	397	485
PTT-PTMO/0.5SWCNT	350	369	402	403	516
PTT-PTMO/0.5SiC	315	333	394	395	486
PTT-PTMO/0.5GNP	356	365	399	400	502
PTT-PTMO/0.5GO	353	360	396	368/397	501
PTT-PTMO/0.5POSS	336	350	395	395	496
Measurement carried out in argon					
PTT-PTMO	371	382	407	406	–
PTT-PTMO/0.5SWCNT	367	380	406	408	–
PTT-PTMO/0.5SiC	360	374	403	401	–
PTT-PTMO/0.5GNP	373	383	409	410	–
PTT-PTMO/0.5GO	371	381	408	406	–
PTT-PTMO/0.5POSS	369	380	406	406	–

Table 2. Temperatures corresponding to 5, 10, and 50% weight loss and the temperature at maximum of weight loss rate for PTT-PTMO and PTT-PTMO-based nanocomposites obtained in an air and argon atmosphere.

4. Conclusions

PTT-PTMO block copolymer-based nanocomposites were prepared by *in situ* polymerization with an addition of 0.5 wt% of several types of nanofillers, that is, SWCNTs, SiC, GO, GNP, and POSS nanoparticles. SEM micrographs verified that the dispersion of the nanofillers in the PTT-PTMO matrix was rather homogeneous suggesting that *in situ* polymerization is a highly efficient method for preparing nanocomposites with low loading of nanofillers. DMTA has been used in order to investigate the effect of nanofillers on the phase separation and phase transition temperatures ($T_{\beta1}$, $T_{\beta2}$) of the thermoplastic elastomer matrix. One can conclude that both, organic and inorganic nanofillers affected the phase separation of PTT-PTMO block copolymer. Additionally, it was observed that PTT-PTMO/0.5GO and PTT-PTMO/0.5GNP were found to be nonconductive. In turn, a conducting network has been formed by 0.3 wt% of SWCNT. While in the case of PTT-PTMO nanocomposites containing SiC, the percolation threshold equals to 1.7 wt%. Moreover, SWCNT and graphene derivatives caused the annihilation of free radicals generated during thermal decomposition in oxidizing atmosphere of polymer matrix in nanocomposites in the strongest manner, and thus retarding thermal degradation of the nanocomposites, in comparison to the effect of SiC NFs and POSS particles, while in an inert atmosphere no significant influence due to the addition of nanofillers that differ in shape was observed.

Acknowledgements

This work is the result of the research project GEKON2/O5/266860/24/2016 funded by The National Centre for Research and Development and National Fund for Environmental Protection and Water Management, Poland.

Author details

Sandra Paszkiewicz[1*], Iman Taraghi[1,2], Anna Szymczyk[3], Elżbieta Piesowicz[1] and Zbigniew Rosłaniec[1]

*Address all correspondence to: spaszkiewicz@zut.edu.pl

1 Institute of Materials Science and Engineering, Faculty of Mechanical Engineering and Mechatronics, West Pomeranian University of Technology, Szczecin, Poland

2 Department of Mechanical Engineering, Semnan University, Semnan, Iran

3 Institute of Physics, Faculty of Mechanical Engineering and Mechatronics, West Pomeranian University of Technology, Szczecin , Poland

References

[1] Adams RK, Hoeschele GK, Witsiepe WK. Thermoplastic polyether-ester elastomers. In: Holden G, Kricheldorf HR, Quirck RP. Thermoplastic Elastomers. 2nd ed. Munich: Hanser; 2004. p. 183-216

[2] Gabriëlse W, Soliman M, Dijkstra K. Microstructure and phase behavior of block copoly(ether-ester) thermoplastic elastomers. Macromolecules. 2001;**34**:1681693. DOI: 10.1021/ma0012696

[3] Schmalz H, Abetz V, Lange R, Soliman M. New thermoplastic elastomers by incorporation of non-polar soft segments in PBT based copolyesters. Macromolecules. 2001;**34**:795-800. DOI: 10.1021/ma001226p

[4] Roslaniec Z. Polyester thermoplastic elastomers: Synthesis, properties, and some applications. In: Fakirov S, editor. Handbook of Condensation Elastomers. Weinheim: Wiley-VCH; 2005. p. 77-116

[5] Szymczyk A, Senderek E, Nastalczyk J, Roslaniec Z. New multiblock poly(ether-ester)s based on poly(trimethylene terephthalate) as rigid segments. European Polymer Journal. 2008;**44**:436-443. DOI: 10.1016/j.eurpolymj.2007.11.005

[6] Koerner H, Price G, Pearce NA, Alexander M, Vaia RA. Remotely actuated polymer nanocomposites-stress-recovery of carbon-nanotube-filled thermoplastic elastomers. Nature Materials. 2004;**3**:115-120. DOI: 10.1038/nmat1059

[7] Szymczyk A. Poly(trimethylene terephthalate-block-tetramethylene oxide) elastomer/single-walled carbon nanotubes nanocomposites: Synthesis, structure, and properties. Journal of Applied Polymer Science. 2012;**126**:796-807. DOI: 10.1002/app.36961

[8] Taraghi I, Fereidoon A, Paszkiewicz S, Roslaniec Z. Electrically conductive polycarbonate/ethylene-propylene copolymer/multi-walled carbon nanotubes nanocomposites with improved mechanical properties. Journal of Applied Polymer Science. 2017;**134**:44661. DOI: 10.1002/app.44661

[9] Guskos N, Maryniak M, Typek J, Guskos A, Szymczak R, Senderek E, Roslaniec Z, Petridis D, Aidinis K, Influence of maghemite concentration on magnetic interactions in maghemite/PTT-block-PTMO nanocomposite. Journal of Non-Crystalline Solids. 2008;**354**: 4401-4406. DOI: 10.1016/j.jnoncrysol.2008.06.059

[10] Paszkiewicz S, Szymczyk A, Pilawka R, Przybyszewski B, Czulak A, Rosłaniec Z. Improved thermal conductivity of poly(trimethylene terephthalate-block-poly(tetramethylene oxide) based nanocomposites containing hybrid single-walled carbon nanotubes/graphene nanoplatelets fillers. Advances in Polymer Technology. DOI: 10.1002/adv.21611

[11] Paszkiewicz S, Szymczyk A, Livanov K, Wagner HD, Rosłaniec Z. Enhanced thermal and mechanical properties of poly(trimethylene terephthalate-block-poly(tetramethylene oxide) segmented copolymer based hybrid nanocomposites prepared by in situ polymerization via synergy effect between SWCNTs and graphene nanoplatelets. eXPRESS Polymer Letters. 2015;**9**:509-524. DOI: 10.3144/expresspolymlett.2015.49

[12] Paszkiewicz S, Szymczyk A, Sui XM, Wagner HD, Linares A, Ezquerra TA, Rosłaniec Z, Synergetic effect of single-walled carbon nanotubes (SWCNT) and graphene nanoplatelets (GNP) in electrically conductive PTT-block-PTMO hybrid nanocomposites prepared by in situ polymerization. Composites Science and Technology. 2015;**118**:72-77. DOI: 10.1016/j.compscitech.2015.08.011

[13] Paszkiewicz S, Pawelec I, Szymczyk A, Rosłaniec Z. Thermoplastic elastomers containing 2D nanofillers: montmorillonite, graphene nanoplatelets and oxidized graphene platelets. Polish Journal of Chemical Technology. 2015;**17**:74-81. DOI: 10.1515/pjct-2015-0071

[14] Paszkiewicz S, Szymczyk A, Špitalsky Z, Mosnáček J, Kwiatkowski K, Rosłaniec Z. Structure and properties of nanocomposites based on PTT-block-PTMO copolymer and graphene oxide prepared by in situ polymerization. European Polymer Journal. 2014;**50**:69-77. DOI: 10.1016/j.eurpolymj.2013.10.031

[15] Paszkiewicz S, Pilawka R, Dudziec B, Dutkiewicz M, Marciniec B, Kochmańska A, Jedrzejewski R, Roslaniec Z. Morphology and phase separation in PTT-block-PTMO nanocomposites containing POSS particles. European Polymer Journal. 2015;**70**:37-44. DOI: 10.1016/j.eurpolymj.2015.07.004

[16] Szymczyk A, Paszkiewicz S, Rosłaniec Z. Influence of intercalated organoclay on the phase structure and physical properties of PTT-PTMO block copolymers. Polymer Bulletin. 2013;**70**:1575-1590. DOI: 10.1007/s00289-012-0859-y

[17] Kovacs JZ, Velegala BS, Schulte K, Bauhofer W. Two percolation thresholds in carbon nanotube epoxy composites. Composites Science and Technology. 2007;**67**:922-928. DOI: 10.1016/j.compscitech.2006.02.037

[18] Kim H, Abdala AA, Macosko CW. Graphene/polymer nanocomposites. Macromolecules. 2010;**43**:6515-6530. DOI: 10.1021/ma100572e

[19] Coleman JN, Khan U, Blau WJ, Gun'ko YK. Small but strong: a review of the mechanical properties of carbon nanotube polymer composites. Carbon. 2006;**44**:1624-1652. DOI: 10.1016/j.carbon.2006.02.038

[20] Moniruzzaman M, Winey KI. Polymer nanocomposites containing carbon nanotubes. Macromolecules. 2006;**39**:5194-5205. DOI: 10.1021/ma060733p

[21] Naderi G, Lafleur PG, Dubois C. Microstructure-properties correlations in dynamically vulcanized nanocomposite thermoplastic elastomers based on PP/EPDM. Polymer Engineering and Science. 2007;**47**:207-217. DOI: 10.1002/pen.20673

[22] www.dupont.com [Internet]. [Accessed: 1.02.2017]

[23] Guo Z, Kim TY, Lei K, Pereira T, Sugar JG, Hahn HT. Strengthening and thermal stabilization of polyurethane nanocomposites with silicon carbide nanoparticles by a surface initiated-polymerization approach. Composite Science and Technology. 2008;**68**:164-170. DOI: 10.1016/j.compscitech.2007.05.031

[24] Mavinakuli P, Wei S, Wang Q, Karki AB, Dhage S, Wang Z, Young DP, Guo Z. Polypyrrole/silicon carbide nanocomposites with tunable electrical conductivity. Journal of Physical Chemistry C. 2010;**114**:3874-3882. DOI: 10.1021/jp911766y

[25] Kueseng K, Jacob KI. Natural rubber nanocomposites with SiC nanoparticles and carbon nanotubes. European Polymer Journal. 2006;**42**:220-227. DOI: 10.1016/j.eurpolymj.2005.05.011

[26] Paszkiewicz S, Taraghi I, Szymczyk A, Huczko A, Kurcz M, Przybyszewski B, Stanik R, Linares A, Ezquerra TA, Rosłaniec Z, Electrically and thermally conductive thin elastic polymer foils containing SiC nanofibers. Composite Science and Technology. 2016. Forthcoming

[27] Špitalský Z, Danko M, Mosnáček J. Preparation of functionalized graphene sheets. Current Organic Chemistry. 2011;**15**:1133-1150. DOI: 10.2174/138527211795202988

[28] Loos J, Alexeev A, Grossiord N, Koning CE, Regev O. Visualization of single-wall carbon nanotube (SWNT) networks in conductive polystyrene nanocomposites by charge contrast imaging. Ultramicroscopy. 2005;**104**:160-167. DOI: 10.1016/j.ultramic.2005.03.007

[29] Kovacs JZ, Andersen K, Pauls JR, Garcia CP, Schossig M, Schulte K, Bauhofer W. Analyzing the quality of carbon nanotube dispersions in polymers using scanning electron microscopy. Carbon. 2007;**45**:1279-1288. DOI: 10.1016/j.carbon.2007.01.012

[30] Battistella M, Cascione M, Fiedler B, Wichmann MHG, Quaresimin M, SChulte K, Fracture behaviour of fumed silica/epoxy nanocomposites. Composites Part A —Applied Science and Manufacturing. 2008;**39**:1851-1858. DOI: 10.1016/j.compositesa.2008.09.010

[31] Khaderbad MA, Tjoa W, Oo TZ, Wei J, Sheri M, Mangalampalli R, Rao VR, Mhaisalkar SG, Mathews N. Facile fabrication of graphene devices through metalloporphyrin induced photocatalytic reduction. RSC Advances. 2012;**2**:4120-4124. DOI: 10.1039/C2RA00792D

[32] Bo Z, Shuai X, Mao S, Yang H, Qian J, Chen J, Yan J, Cen K. Green preparation of reduced graphene oxide for sensing and energy storage applications. Scientific Reports. 2014;**4**:4684. DOI: 10.1038/srep04684

[33] Stankovich S, Dikin DA, Piner RD, Kohlhaas KA, Kleinhammes A, Jia Y, Wu Y, Nguyen SBT, Ruoff RS. Synthesis of graphene-based nanosheets via chemical reduction of exfoliated graphite oxide. Carbon. 2007;**45**:1558-1565. DOI: 10.1016/j.carbon.2007.02.034

[34] Schniepp HC, Li JL, McAllister MJ, Sai H, Herrera-Alonso M, Adamson DH, Prud'homme RK, Car R, Saville DA, Aksay IA. Functionalized single graphene sheets derived from splitting graphite oxide. Journal of Physical Chemistry B. 2006;**110**:8535-8539. DOI: 10.1021/jp060936f

[35] Si YC, Samulski ET. Synthesis of water soluble graphene. Nano Letters. 2008;**8**:1679-1682. DOI: 10.1021/nl080604h

[36] Kwon J, Kim H. Comparison of the properties of waterborne polyurethane/multiwalled carbon nanotube and acid-treated multiwalled carbon nanotube composites prepared by in situ polymerization. Journal of Polymer Science: Part A: Polymer Chemistry. 2005;**43**:3973-3985. DOI: 10.1002/pola.20897

[37] Xiong J, Zheng Z, Qin X, Li M, Li H, Wang X. The thermal and mechanical properties of a polyurethane/multi-walled carbon nanotube composite. Carbon. 2006;**44**:2701-2707. DOI: 10.1016/j.carbon.2006.04.005

[38] Li J, Wang X, Yang C, Yang J, Wang Y, Zhang J. Toughening modification of polycarbonate/poly(butylene terephthalate) blends achieved by simultaneous addition of elastomer particles and carbon nanotubes. Composites Part A: Applied Science and Manufacturing. 2016;**90**:200-210. DOI: 10.1016/j.compositesa.2016.07.006

[39] Wang YH, Shi YY, Dai J, Yang JH, Huang T, Zhang N, Peng Y, Wang Y. Morphology and property changes of immiscible polycarbonate/poly(L-lactide) blends induced by carbon nanotubes. Polymer International. 2013;**62**:957-965. DOI: 10.1002/pi.4383

[40] Valentini L, Biagiotti J, Lopez-Manchado MA, Santucci S, Kenny JM. Effects of carbon nanotubes on the crystallization behavior of polypropylene. Polymer Engineering and Science. 2004;**44**:303-311. DOI: 10.1002/pen.20028

[41] Bhattacharyya AR, Sreekumar TV, Liu T, Kumar S, Ericson LM, Hauge RH, Smalley RE. Crystallization and orientation studies in polypropylene/single wall carbon nanotube composite. Polymer. 2003;**44**:2373-2377. DOI: 10.1016/S0032-3861(03)00073-9

[42] Anand KA, Agarwal US, Joseph R. Carbon nanotubes induced crystallization of poly(ethylene terephthalate). Polymer. 2006;**47**:3976-3980. DOI: 10.1016/j.polymer.2006.03.079

[43] Sun G, Chen G, Liu Z, Chen M. Preparation, crystallization, electrical conductivity and thermal stability of syndiotactic polystyrene/carbon nanotube composites. Carbon. 2010;**48**:1434-1440. DOI: 10.1016/j.carbon.2009.12.037

[44] Mitchell CM, Krishnamoorti R. Non-isothermal crystallization of in situ polymerized poly(ε-caprolactone) functionalized-SWNT nanocomposites. Polymer. 2005;**46**:8796-8804. DOI: 10.1016/j.polymer.2005.05.101

[45] Garcia-Gutierrez MC, Hernandez JJ, Nogales A, Panine P, Rueda DR, Ezquerra TA. Influence of shear on the templated crystallization of poly(butylene terephthalate)/single wall carbon nanotube nanocomposites. Macromolecules. 2008;**41**:844-851.DOI: 10.1021/ma0713512

[46] Cai Y, Chen L, Yang H, Gou J, Cheng L, Yin X, Yin H. Mechanical and electrical properties of carbon nanotube buckypaper reinforced silicon carbide nanocomposites. Ceramics International. 2016;**42**:4984-4992. DOI: 10.1016/j.ceramint.2015.12.011

[47] Dresselhaus MS, Dresselhaus G, Saito R. Physics of carbon nanotubes. Carbon. 1995;**33**:883-891. DOI: 10.1016/0008-6223(95)00017-8

[48] Chae HG, Liu J, Kumar S. Carbon nanotubes-enabled materials. In: O'Connell MJ, editor. Carbon Nanotubes Properties and Applications. Boca Raton: CRC Press Taylor and Francis Group; 2006. p. 213-253

[49] Cooper CA, Young RJ, Halsall M. Investigation into the deformation of carbon nanotubes and their composites through the use of Raman spectroscopy. Composites Part A: Applied Science and Manufacturing. 2001;**32A**:401-411. DOI: 10.1016/S1359-835X(00)00107-X

[50] Dumitrica T, Hua M, Yakobson BI. Symmetry-, time-, and -dependent strength of carbon nanotubes. Proceedings of National Academy of Sciences of the United States of America. 2006;**103**:6105-6109. DOI: 10.1073/pnas.0600945103

[51] Monthioux M. Filling single-wall carbon nanotubes. Carbon. 2002;**40**:1809-1823. DOI: 10.1016/S0008-6223(02)00102-1

[52] Slonczewski JC, Weiss PR. Band structure of graphite. Physical Review. 1958;**109**:272. DOI: 10.1103/PhysRev.109.272

[53] Novoselov KS, Geim AK, Morozov SV, Jiang D, Zhang Y, Dubonos SV, Grigorieva IV, Firsov AA. Electric field effect in atomically thin carbon films. Science. 2004;**306**:666-669. DOI: 10.1126/science.1102896

[54] Lee C, Wei X, Kysar JW, Hone J. Measurement of the elastic properties and intrinsic strength of monolayer graphene. Science. 2008;**321**:385-388. DOI: 10.1126/science.1157996

[55] Balandin AA, Ghosh S, Bao W, Calizo I, Teweldebrhan D, Miao F, Lau CN. Superior thermal conductivity of single-layer graphene. Nano Letters. 2008;**8**:902-907. DOI: 10.1021/nl0731872

[56] Hernández JJ, García-Gutiérrez MC, Nogales A, Rueda DR, Kwiatkowska M, Szymczyk A, Roslaniec Z, Concheso A, Guinea I, Ezquerra TA. Influence of preparation procedure on the conductivity and transparency of SWCNT-polymer nanocomposites. Composites Science and Technology. 2009;**69**:1867-1872. DOI: 10.1016/j.compscitech.2009.04.002

[57] Szymczyk A, Roslaniec Z, Zenker M, Garcia-Gutierez MC, Hernandez JJ, Rueda DR, Nogales A, Ezquerra TA. Preparation and characterization of nanocomposites based on COOH functionalized multi-walled carbon nanotubes and on poly(trimetylene terephthalate. eXPRESS Polymer Letters. 2011;**5**:977-995. DOI: 10.3144/expresspolymlett.2011.96

[58] Paszkiewicz S, Szymczyk A, Špitalský Z, Soccio M, Mosnáček J, Ezquerra TA, Rosłaniec Z. Electrical conductivity of PET/expanded graphite nanocomposites prepared by in situ polymerization. Journal of Polymer Science. Part B: Polymer Physics. 2012;**50**:1645-1652. DOI: 10.1002/polb.23176

[59] Paszkiewicz S. Multifunctional polymer nanocomposites based on thermoplastic polyester. In: Farrukh MA, editor. Functionalized Nanomaterials. Croatia: InTech; 2016. DOI: 10.5772/63186

[60] Watanabe M. Nagaoka K, Kanba M, Shinohara I. Ionic conductivity of polymeric solid electrolytes based on (polypropoylene oxide) and poly(tetramathylene oxide). Polymer Journal. 1982;**14**:877-886.DOI: 10.1295/polymj.14.877

[61] Kuo SW, Chang FC. POSS related polymer nanocomposites. Progress in Polymer Science. 2011;**36**:1649-1696. DOI: 10.1016/j.progpolymsci.2011.05.002

[62] Ayandele E, Sarkar B, Alexandridis P. Polyhedral oligomeric silsesquioxane (POSS)-containing polymer nanocomposites. Nanomaterials. 2012;**2**:445-475. DOI: 10.3390/nano2040445

[63] Nezakati T, Tan A, Seifalian AA. Enhancing the electrical conductivity of a hybrid POSS-PCL/graphene nanocomposite polymer. Journal of Colloid Interface Science. 2014;**435**:145-155. DOI: 10.1016/j.jcis.2014.08.020

[64] Zhang SM, Lin L, Deng H, Gao X, Bilotti E, Peijs T, Zhang Q, Fu Q. Synergistic effect in conductive networks constructed with carbon nanofillers in different dimensions. eXPRESS Polymer Letters. 2012;**6**:159-168. DOI: 10.3144/expresspolymlett.2012.17

[65] Sandler J, Shaffer MSP, Prasse T, Bauhofer W, Schulte K, Windle AH. Development of a dispersion process for carbon nanotubes in an epoxy matrix and the resulting electrical properties. Polymer. 1999;**40**:5967-5971.DOI: 10.1016/S0032-3861(99)00166-4

[66] Fu F, Scogna RC, Zhou W, Brand S, Fischer JE, Winey KI. Nanotube networks in polymer nanocomposites: rheology and electrical conductivity. Macromolecules. 2004;**37**:9048-9055. DOI: 10.1021/ma049164g

[67] Stauffer D. Scaling theory of percolation clusters. Physics Reports. 1979;**54**:1-74. DOI: 10.1016/0370-1573(79)90060-7

[68] Stauffer D, Aharony A. Introduction to percolation theory. 1st ed. London: Taylor and Francis; 1992

[69] Paszkiewicz S. Polymer hybrid nanocomposites containing carbon nanoparticles. In situ synthesis and physical properties [PhD dissertation]. Szczecin: West Pomeranian University of Technology; 2014. 190 p

[70] www.sigma-aldrich.com [Internet]. [Accessed: 2017-02-5]

[71] www.nanocs.com [Internet]. [Accessed: 2017-02-7]

[72] Zeynalov EB., Friedrich JF. Antioxidative activity of carbon nanotube and nanofiber. The Open Materials Science Journal. 2008;**2**:28-34. DOI: 10.2174/1874088X00802010028

[73] Szymczyk A, Rosłaniec Z. Degradacja i stabilizacja termoplastycznych elastomerów. Polimery. 2006;**51**:627-642

[74] Fakirov S, Roslaniec Z. Handbook of Condensation Thermoplastic Elastomers. Chapter 3 Polyester Thermoplastic Elastomers: Synthesis, Properties, and Some Applications. Weinheim: Wiley-VCH Verlag GmbH & Co. KGaA; 2005

Chapter 5

Thermoplastic Elastomers Based on Block, Graft, and Star Copolymers

Weiyu Wang, Wei Lu, Nam-Goo Kang,
Jimmy Mays and Kunlun Hong

Additional information is available at the end of the chapter

http://dx.doi.org/10.5772/intechopen.68586

Abstract

In this book chapter, we focus on recent advances in thermoplastic elastomers based on synthetic polymers from the aspects of polymer architectures such as linear block, graft, and star copolymers. The first section is an introduction that covers a brief history and classification of thermoplastic elastomers (TPEs). The second section summarizes ABA triblock copolymers synthesized by various methods for TPE applications. The third section reviews TPEs based on graft copolymers, and the fourth section reviews TPEs based on star copolymers. The differences between TPE research in academia and industry are addressed in the last section as a perspective, with a view toward the generation of new, advanced, commercially viable TPEs.

Keywords: thermoplastic elastomers, living/controlled polymerization, polymer architecture, functional polymers, mechanical properties

1. Introduction

Thermoplastic elastomers (TPEs) are biphasic synthetic polymer materials consisting of a continuous soft rubbery matrix physically cross-linked by glassy plastic domains [1, 2] (**Figure 1**). Such materials have the elasticity of a conventional rubber but are suitable for high-throughput plastic-processing techniques such as injection molding and melt extrusion without requiring a curing process [3, 4]. This feature allows TPEs to be manufactured on a large scale using short production time, which makes TPEs one of the most commonly used polymeric materials in many fields [5].

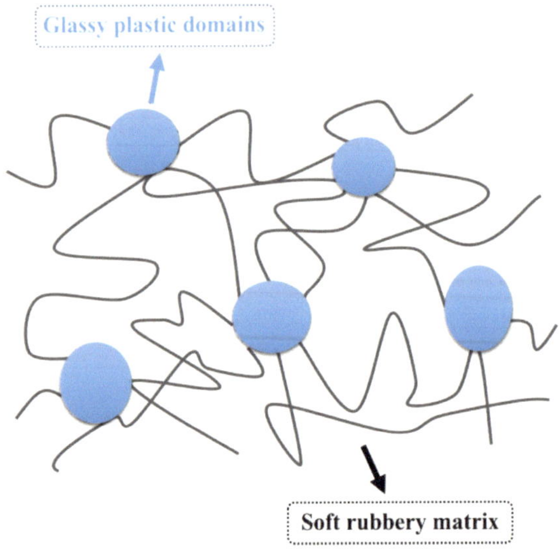

Figure 1. Structure illustration of thermoplastic elastomers.

Commercially available TPEs, based on chemical composition and morphology, can be categorized into eight different groups: (1) styrenic block copolymers (SBCs), (2) polymer blends by dynamic vulcanization (TPVs), (3) polyolefin-based thermoplastic elastomers (TPOs), (4) halogen-containing polyolefins, (5) thermoplastic polyurethane elastomers (TPUs), (6) polyamide-based thermoplastic elastomers (COPA), (7) polyether ester elastomers (COPE), and (8) ionomeric thermoplastic elastomers. These have been extensively reviewed in many handbooks [5–8].

Starting from the 1990s, many fascinating polymers with various functionalities, well-defined structures, and advanced macromolecular architectures were prepared thanks to developments in living/controlled polymerization techniques such as living anionic [9–11]/cationic polymerization [12], atomic transfer radical polymerization (ATRP) [13], ring-opening metathesis polymerization (ROMP) [14], reversible addition-fragmentation chain-transfer polymerization (RAFT) [15], nitroxide-mediated radical polymerization (NMRP) [16], and so on. Many of these new polymers have great potential to be used as thermoplastic elastomers.

Along with innovations in synthetic polymer chemistry, this chapter summarizes recent advances in thermoplastic elastomers based on synthetic polymers from the aspect of polymer architectures including (1) ABA-type triblock polymers, (2) graft polymers, and (3) star-branched polymers.

2. ABA triblock copolymer-type TPEs

2.1. Polymers synthesized by anionic polymerization

The most common ABA triblock copolymer-type TPEs are polystyrene-b-polyisoprene-b-polystyrene (SIS) and polystyrene-b-polybutadiene-b-polystyrene (SBS) triblock copolymers, designed and synthesized by Milkovich and Holden from Shell Development Company in 1965 [17]. With proper composition, PI forms a continuous rubber matrix, which is physically cross-linked by rigid component PS due to the thermodynamic incompatibility between these two components. In a dynamic mechanical analysis of SIS with temperature ramp/frequency sweep, SIS behaves like a glassy plastic with a high storage modulus (*G*') when the temperature is below the glass transition temperature of PI (T_g ~ −56°C). As the temperature increases but remains lower than the T_g of PS (95°C), the polyisoprene chains start to move and *G*' reaches the rubbery plateau value. This temperature range is considered as the service temperature range where such polymers act as elastomer with typical stress-stain behavior. When the temperature is above 95°C, the polymer enters the melt-flow zone and behaves as a viscous liquid.

As many applications benefit from low-cost SBCs or styrenic-based TPEs (S-TPEs), high-temperature applications and other advanced consumptions of S-TPEs, such as in tire rubber, are largely limited by the relatively low glass transition temperature of PS. When the service conditions approach 95°C, softening of PS domains dramatically reduces the tensile stress of S-TPEs. One major research interest in the field of anionic polymerization is to increase the upper service temperature of S-TPEs without changing the polymerization procedure, which has already existed in pilot plants for almost 50 years [18, 19]. These efforts mainly explored anionic polymerization of polymers with higher glass transition temperatures. Such polymers include the following:

2.1.1. *Styrene derivatives*

Styrene derivative polymers include polystyrene with functionalities at α- or para-position: poly(α-methyl styrene) (PMS, T_g ~173°C) [20], poly(α-methyl p-methyl styrene) (PMMS, T_g ~183°C) [21], poly(tert-butyl styrene) (PtBS, T_g ~130°C) [22], and poly(p-adamantyl styrene) (P-AdmS, T_g ~203°C) [23, 24].

For the anionic polymerization of α-methyl styrene and its derivative α-methyl p-methyl styrene, the bulky methyl group at the α-position results in a low monomer ceiling temperature. In order to achieve quantitative yield, polymerization of these monomers requires low polymerization temperature (−78°C) in polar solvent (THF), which is not desirable in large-scale industry application [19]. High T_g polystyrene derivatives with bulky pendent groups such as tert-butyl or adamantyl at the para-position will cause phase blending with polydienes due to the lipophilic nature of the tert-butyl or adamantyl group. In order to increase the strength of phase separation and generate effective physical cross-linking, high overall molecular weight is required for polybutadiene/poly(tert-butyl styrene) (PtBS, T_g ~130°C) systems [22].

2.1.2. *Methacrylate derivatives*

Polymers of methacrylate derivatives include syndiotactic poly(methyl methacrylate) (sPMMA, T_g ~120°C), poly(ethyl methacrylate) (PEMA, T_g ~90°C), poly(tert-butyl methacrylate) (PtBMA, T_g ~116°C), poly(isobornyl methacrylate) (PIBMA, T_g ~202°C) [25], and poly(1-adamantyl acrylate) (P-AdmA, T_g ~133°C) [26].

Since the glass transition temperature of poly(alkyl methacrylate) depends both on tacticity and on the size of alkyl substituents [25–28], incorporating methacrylate derivatives with different tacticities as the hard segment in ABA-type triblock copolymers could tune the service condition over a large temperature range [28]. When using polydienes as the elastic matrix, methacrylate derivatives were initiated in THF at −78°C through a difunctional polydiene anion, which was synthesized in a hydrocarbon solvent since anionic polymerization of butadiene or isoprene in polar solvents forms less cis-1,4 microstructure, and thus dramatically increases the T_g.

In a typical synthesis of all acrylic TPEs such as PMMA-poly(n-butyl acrylate)-PMMA triblock copolymers, PMMA-poly(tert-butyl acrylate)-PMMA precursor was first synthesized by sequential anionic polymerization of MMA, tert-butyl acrylate, and MMA in THF at −78°C. By transalcoholysis with n-butanol of the precursor, PMMA-poly(n-butyl acrylate)-PMMA triblock copolymer was prepared with PMMA as the rigid domain and poly(n-butyl acrylate) (PnBA) as the rubbery matrix [29, 30].

The polymerization of the abovementioned monomers requires low polymerization temperature in a polar solvent. However, anionic polymerization on an industry scale is generally carried out in hydrocarbon solvent at mild temperature [18]. Thus, a high T_g polymer system that can be synthesized in hydrocarbon solvent at mild temperature is ideal for large-scale application. To follow this endeavor, the anionic polymerization of a third group of monomers was explored:

2.1.3. Rigid-conjugated diene monomers

Polymer prepared by rigid-conjugated diene monomers includes poly(1,3-cyclohexadiene) (PCHD) and polybenzofulvene (PBF) (**Table 1**). One feature of anionic polymerization of conjugated dienes is that the microstructure of the resulting polymer varies with different initiation systems. 1,3-Cyclohexadiene demonstrated controlled anionic polymerization behavior with three different initiation systems: n-butyllithium/tetramethyl-ethylenediamine (TMEDA), n-butyllithium/1,2-dimethoxyethane (DME), or sec-butyllithium/1,4-diazabicyclo[2.2.2]-octane (DABCO) [31–33]. Resulting poly(1,3-cyclohexadiene) (PCHD) has 55, 75, and 90% of 1,4-addition, respectively. T_gs of these polymers decreased from 155 to 110°C as the percentage of

Name	Monomer structure	Polymer structure
1,3-Cyclohexadiene	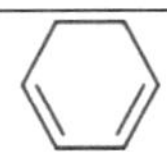	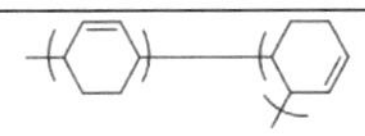 1,4-addition 1,2-addition
benzofulvene	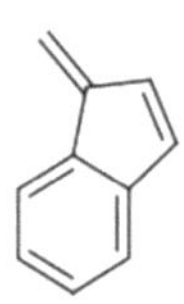	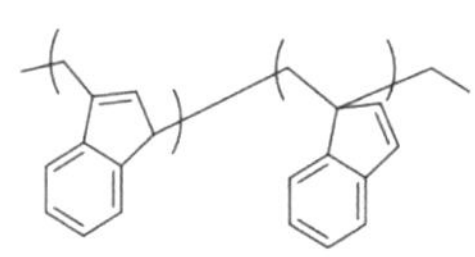 1,4-addition 1,2-addition

Table 1. Monomer and polymer structure of 1,3-cyclohexadiene (CHD) and benzofulvene (BF).

1,2-microstructure decreased. PCHD-PB-PCHD triblock copolymer with 30 wt% of PCHD exhibited 10.2 MPa ultimate stress with a relatively low strain at break of 290% [34]. This might be due to side reactions during anionic polymerization of CHD. By partial hydrogenation of PB without saturated PCHD, ultimate stress increased to 14.0 MPa with better strain at break of 570%, indicating a stronger physical cross-linking. The end block PCHD of this triblock copolymer can completely hydrogenated into polycyclohexylene, a polyolefin with T_g above 231°C [35]. The completely hydrogenated triblock copolymers displayed 10.0-MPa tensile stress at 600% strain without breaking.

Benzofulvene (BF), the polymer from which was first synthesized by Ishizone, is another interesting conjugated diene monomer that undergoes living anionic polymerization in both THF and benzene [36–39]. The resulting PBF has a T_g of 160°C when polymerizing in THF, and 145°C in benzene. The relatively high T_g and the ability to synthesize PBF-PI diblock copolymer in hydrocarbon solvent at room temperature make benzofulvene an ideal candidate to prepare high-temperature thermoplastic elastomer.

By using a difunctional lithium anionic initiator, we synthesized a serious of PBF-PI-PBF triblock copolymer (FIF) via sequential living anionic polymerization with 14, 22, and 31 vol% of PBF [39]. In dynamic mechanical analysis (**Figure 2a**), all samples showed two T_gs, respectively, at −56°C for PI, and 145°C for PBF. For FIF with 14 vol% of PBF, the polymer displayed 1390% strain at break with 14.3 MPa ultimate stress (**Figure 2b**). These mechanical properties are competitive with Kraton D1112P [40], a widely used commercial SIS triblock copolymer-type thermoplastic elastomer.

Another interesting feature of BF is that by using different additive or solvent during the polymerization, the microstructure of the resulting polymer can be tuned from 24% (benzene as the solvent), 41% (THF as the solvent), to 98% (1,2-dimethoxyethane as the additive and benzene as the solvent). The T_g of PBF with these three polymers is increased linearly from 152, 162, to 199°C as the percentage of 1,2-addition increases. Such properties open new opportunities to prepare TPEs with tunable upper service temperature. The chemical structures and T_gs of the abovementioned rigid and soft components have been summarized in **Table 2**.

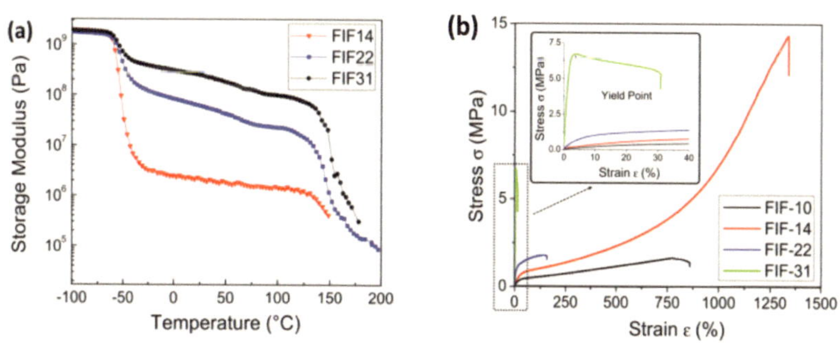

Figure 2. (a) Dynamic mechanical analysis of FIF, (b) tensile test of FIF. (Reprinted with permission from Ref. [40]. Copyright 2016 American Chemical Society).

	Name	Structure	T_g
Styrene derivatives	Polystyrene (PS)	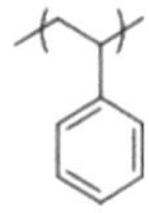	95°C
	Poly(α-methyl styrene) (PMS)	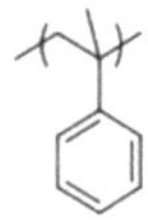	173°C
	Poly(α-methyl p-methyl styrene) (PMMS)	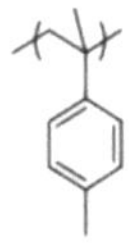	183°C
	Poly(tert-butyl styrene) (PtBS)	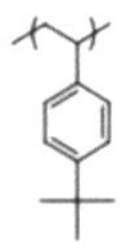	130°C
	Poly(p-adamantyl styrene) (P-AdmS)		203°C
Methacrylate derivatives	Poly(methyl methacrylate) (sPMMA)	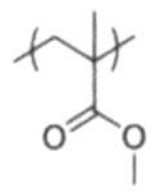	120°C
	Poly(ethyl methacrylate) (PEMA)	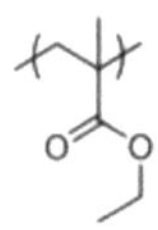	90°C
	Poly(tert-butyl methacrylate) (PtBMA)	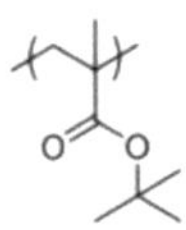	116°C
	Poly(isobornyl methacrylate) (PIBMA)	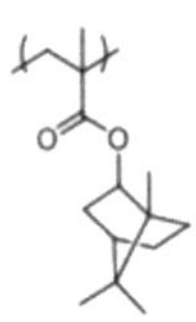	202°C
Rigid conjugated polydienes	Poly(1,3-cyclohexadiene) (PCHD)	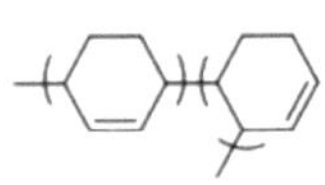	140–150°C
	Polybenzofulvene (PBF)	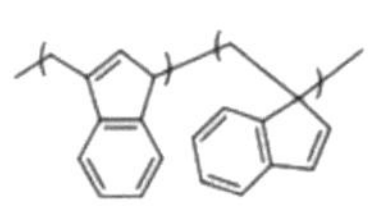	140–195°C

	Name	Structure	T_g
Soft segments	Polyisoprene (PI)	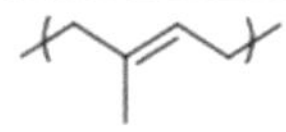	–56°C
	Polybutadiene (PB)		–90°C
	Poly(n-butyl acrylate) (PnBA)	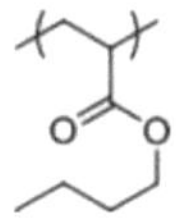	–40°C

Table 2. Hard and soft segments of ABA-type TPEs synthesized by anionic polymerization.

2.2. Block copolymers synthesized by cationic polymerization

Since PI or PB was mainly used as the elastic domains for TPEs synthesized by living anionic polymerization, poor resistance to UV/oxidation can become another issue for PI or PB containing TPEs. A renaissance in living cationic polymerization [12] advanced many research toward TPEs with better UV/oxidation stability and higher UST by employing isobutylene as the elastic block. Many cationically synthesized TPEs used polyisobutylene (PIB) as the elastic middle block due to its softness and chemical resistance. Triblock copolymer PS-PIB-PS prepared by sequential living cationic polymerization through a difunctional initiator displayed an ultimate tensile stress of 26 MPa, which was competitive with commercial Kraton SIS TPEs [41, 42].

Another feature that distinguishes cationic polymerization from anionic polymerization is the ability to control the polymerization of high T_g monomers such as p-chlorostyrene (pCS) [43], indene (ID) [44], and acenaphthylene (ACP) [44, 45]. Triblock copolymers using PpCS (T_g ~129°C), PID (T_g ~225°C), or PACP (T_g ~250°C) as the hard segment and PIB as the soft segment were successfully prepared by cationic polymerization and showed stress-strain behavior similar to typical TPEs. Notice that PpCS is a polar polymer with weather and flame resistance. Indene is potentially a very cost-effective monomer for high-temperature applications.

2.3. Block copolymers synthesized by ring-opening transesterification polymerization

Poly(lactide) (PLA, T_g ~60°C) is an amorphous biodegradable polymer synthesized by ring-opening transesterification polymerization (ROTEP) from racemic D,L-lactide, whereas isotactic poly(L-lactide) (PLLA) and poly(D-lactide) (PDLA) are semicrystalline polymers (T_m ~170°C). Blends of PLLA and PDLA can form stereocomplex crystals, which further improve chemical resistance with higher melting temperature (T_m ~203°C) [46]. Preparing polymers from renewable resource materials instead of from petroleum resources has been a lasting goal of chemists for many decades. Monomers including 3-hydroxybutyrate (HA), menthide (MD), 6-methyl-ε-caprolactone (MCL), ε-caprolactone (CL), β-methyl-δ-valerolactone (MCL), and ε-decalactone (DL) potentially could be produced from sustainable resources [47]. These monomers undergo ring-opening transesterification polymerization (ROTEP), yielding biodegradable elastic polymers [48, 49].

Since ROTEP generated polymers with hydroxyl functionality on both ends, the resulting polymers could be directly used as a macroinitiators to polymerize lactide, producing various types of biodegradable ABA triblock copolymer TPEs. When poly(3-hydroxybutyrate) (PHA) was used as elastic block, TPEs had strain at break lower than 200% [50]. Using polymenthide (PM) as elastic block, the strain at break was largely improved to 960% compared to PHA system. With diethylene glycol as a difunctional initiator and $ZnEt_2$ as the catalyst, α, ω-functionalized polymenthide (HO-PM-OH) was prepared via ring-opening transesterification polymerization (ROTEP). This difunctional PM was used as the initiator for ROTEP of (±)-lactide to yield PLA-PM-PLA triblock copolymers used as TPEs (**Figure 3**). Sample PLLA-PM-PLLA (13-33-13) displayed a strain at break of 765% with ultimate tensile strength of 19.5 MPa [51, 52]. With 30 vol% of poly(6-methyl-ε-caprolactone) (PMCL) as the elastic block, 1880% strain at break was achieved with 10.2 MPa ultimate stress [53].

2.4. Block copolymers prepared by controlled radical polymerization

Starting from the late 1990s, tremendous progress has been achieved in the field of controlled radical polymerization such as atomic transfer radical polymerization (ATRP) [13, 54], reversible addition-fragmentation chain-transfer polymerization (RAFT) [15, 55], and nitroxide-mediated radical polymerization (NMRP) [16]. These techniques open up various opportunities to prepare functionalized polymers with predictable molecular weight, narrow molecular-weight distribution, and complicated macromolecular architectures [56]. Controlled polymerization was achieved for many monomers such as acrylonitrile [57], acrylamide [58], and vinyl amide [59], which cannot be controllably polymerized by anionic or cationic mechanisms.

Many block, star, grafted, and brush polymers with different functionalities have been prepared by ATRP [60]. However, ABA-type block copolymers synthesized by ATRP have received limited success for TPE applications mainly due to two reasons: (1) relatively broad distribution of the hard block reduces the strength of phase separation and (2) unavoidable diblock copolymer mixture in triblock copolymers acts as plasticizer diminishing the phase boundary [25, 30]. Significantly lower tensile stress and strain were observed for PMMA-PnBA-PMMA triblock copolymers prepared by ATRP compared with triblock copolymers prepared by anionic polymerization followed by transalcoholysis [30]. The copolymerization

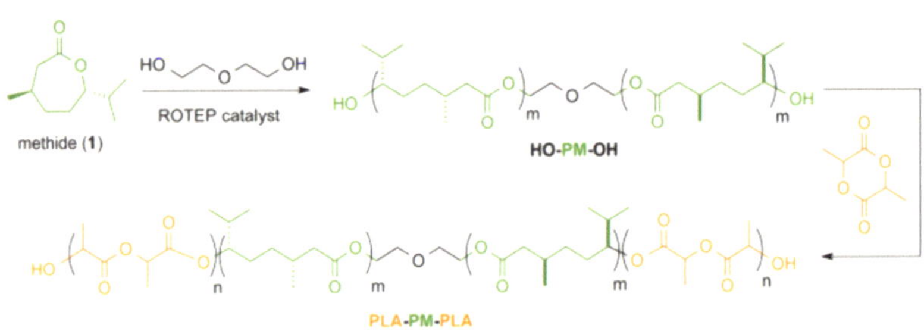

Figure 3. ROTEP to synthesize PLA-PM-PLA. (Reprinted with permission from Ref. [49]. Copyright 2014 American Chemical Society).

of methyl methacrylate with α-methylene-γ-butyrolactone as glassy block was necessary to improve the tensile properties of triblock copolymers with poly(n-butyl acrylate) as elastic block [58]. However, the ultimate stress was still lower than 3.2 MPa with strain at break of 650%.

Poly[2, 5-bis[(n-hexogycarbonyl)]styrene] (PMPCS) is a mesogen-jacketed liquid crystalline polymer with a T_g of about 120°C. As a new type of rod-coil-rod TPE based on PMPCS and PnBA, tensile tests showed 1050% strain at break with 3.2-MPa ultimate stress [61]. Poly2,5-bis[(n-hexogycarbonyl)]styrene (PHCS) is an amorphous polymer with a T_g of about −10°C due to long-chain alkyl substitution at the 2- and 5- positions of styrene (**Figure 4**). Poly(4-vinylpyridine) (P4VP) is a high T_g polymer that can complex with Zn^{2+}. Tuning stress-strain properties, glass transition temperature and morphology of TPEs based on P4VP-PHCS-P4VP was achieved by adding different amounts of $Zn(ClO_4)_2$ [62].

In order to minimize undesired chain transfer and termination reactions, controlled radical polymerization needs to maintain a very low radical concentration. This increases the reaction time compared to conventional free radical and ionic polymerization [63]. Radical segregation effect introduced by (mini)emulsion polymerization in heterogeneous system, on the other hand, reduced the reaction time and suppressed radical termination [64, 65]. Combining emulsion polymerization with RAFT, PS-PnBA-PS triblock copolymers with different molecular weight and composition were prepared in shorter reaction time [66]. By varying weight percentage of PS from 20.2 to 71.5%, the ultimate tensile strength was in the range from 3.0 to 12.5 MPa and strain at break was in the range from 90 to 1300%. It was also found that by using a poly[styrene-alt-(maleic anhydride)] (PSM) as a macro-chain-transfer agent in emulsion polymerization for PS-PnBA-PS [67], ultimate stress increased whereas strain at break decreased as the percentage of PSM increased. Another TPE based on PS and poly(lauryl acrylate) was prepared by a solution RAFT polymerization process [68]. Ultimate stress was lower than 1 MPa and strain at break was lower than 280%. An interesting ABA triblock copolymer was prepared by RAFT polymerization based on P4VP as a hard segment and random copolymer of PnBA and poly(acrylamide) (PAM) as the elastic block. The PAM moiety in the middle block cross-linked the elastic domain through hydrogen-bonding association [69].

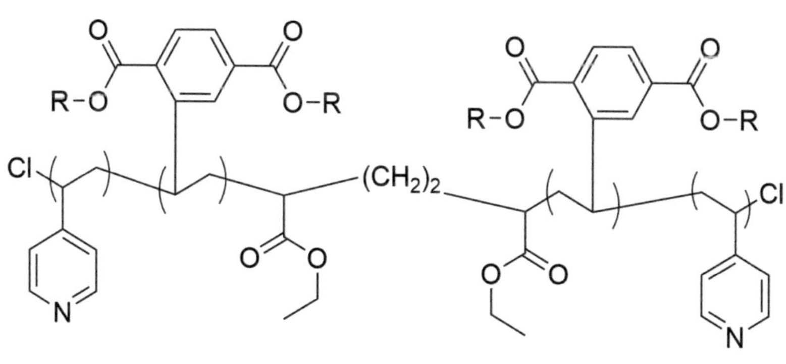

Figure 4. Chemical structure of P4VP-PHCS-P4VP.

3. Graft copolymer-type TPEs

As an important class of commercial polymeric materials, graft copolymers are composed of a polymer backbone with polymer side chains attached to it. Graft polymers can be prepared by three strategies: (1) "Grafting onto," where both polymer backbone and side chain are presynthesized and then through the end functionalities on side chain and in-chain functionality on backbone, side chains are grafted onto the polymer backbones. (2) "Grafting from," where multifunctional polymer backbones serve as the macroinitiator and initiated the polymerization of side-chain monomers to graft from the backbone. (3) "Grafted through" or "macromonomer approaches," where polymer side chains having a polymerizable end group are synthesized, and those macromonomers are subsequently polymerized to form the backbone creating graft polymer [70–73].

By using anionic polymerization followed by polycondensation, Mays and coworkers prepared a series of graft copolymers with regular spaced trifunctional, tetrafunctional, and hexafunctional junction points where PI was the backbone and PS was the side chain [74, 75]. Structure-property relationship of these graft copolymers was elucidated by characterizing morphology [76, 77] and mechanical properties [78–80] of grafted polymers with different compositions (14–23 vol% of PS) and architectures (trifunctional, tetrafunctional, and hexafunctional junction points). From their research, multigraft polymers with tetrafunctional junction points showed 1550% strain at break which is 500% higher than that for the commercial product Kraton 1102. This superelasticity is a consequence of having the PI backbone anchored by multiple PS physical cross-links (**Figure 5**). Both tetra- and hexafunctional multigraft polymers displayed higher elasticity than commercial TPEs like Kraton or Styroflex. Polymers with more functionalities at one junction point had higher tensile stress and modulus.

Inspired by this work, the same group prepared graft all-acrylic TPEs based on PMMA side chain and PnBA backbone [81]. The PMMA macromonomers were synthesized by living anionic polymerization and copolymerized with nBA by RAFT polymerization. Similar to other linear and star all-acrylic TPEs, low modulus and stress were found in PnBA-g-PMMA graft polymers due to high entanglement molecular weight of PnBA and phase blending

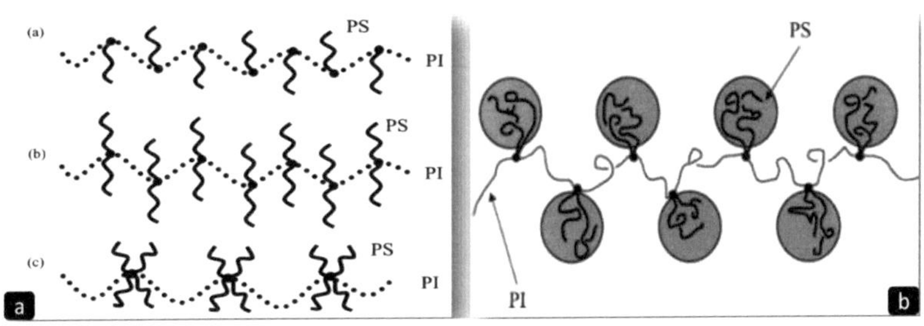

Figure 5. (a) Multigrafted copolymers based on PI backbone and PS branches. (Reprinted with permission from Ref. [75]. Copyright 2002 American Chemical Society.) (b) Chain conformation of multigrafted copolymers in microphase-separated state. (Reprinted with permission from Ref. [78]. Copyright 2001 American Chemical Society).

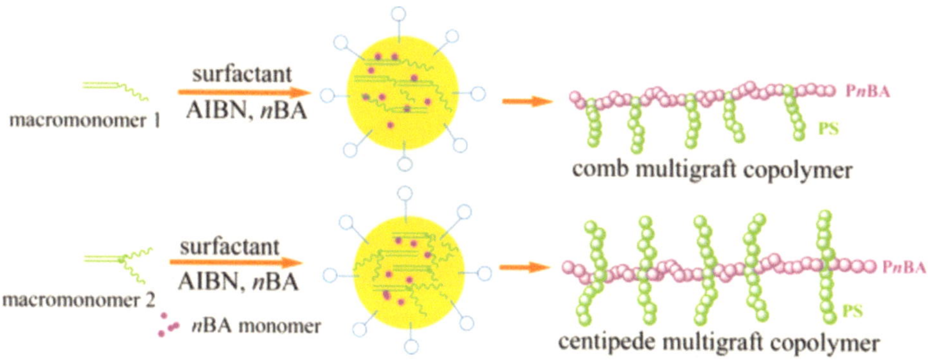

Figure 6. Scheme for emulsion polymerization route to superelastomers. (Reprinted with permission from Ref. [82]. Copyright 2014 American Chemical Society).

between PMMA and PnBA. Zhang and Mays further extended the versatilities of graft polymer architecture by a cost-efficient process combining (mini)emulsion polymerization with anionic polymerization or ATRP to prepare trifunctional- and tetrafunctional-grafted copolymers with PS or PMMA as side chain, and PI or PnBA as the backbone [82–85]. In a typical procedure (**Figure 6**), a hydroxyl end-functionalized PS (PS-OH) was first prepared by living anionic polymerization. Through esterification reaction, the end group of PS-OH was converted into a polymerizable styrene group as the PS macromonomer for emulsion polymerization.

4. Star-branched copolymer-type TPEs

Star-branched polymers are polymers with more than two arms radiating from the same core. If these arms have different chemical compositions or molecular weights, the star polymer is named miktoarm (mixed-arm) star polymer. Generally, star polymers are prepared by two methods: (1) "Arm-first," where polymer arms are synthesized first and coupled onto a core decorated with appropriate reaction sites. (2) "Core first," where polymer arms are grown from a multifunctional initiator [86, 87].

When more than two PS-b-PI diblock copolymers are connected at the same core through the end of PI end blocks, such $(PS\text{-}b\text{-}PI)_x$ star-branched polymers displayed mechanical properties similar to SIS linear triblock TPEs. By using an arm-first divinylbenzene-linking strategy, Bi and Fetters [88] prepared polystyrene-polydiene star block copolymers with number of arms up to 29. They found that these star copolymers had superior tensile properties compared to linear triblock copolymers of similar composition. The enhancement of tensile strength saturated when the number of arms larger than six. Morphological analysis indicated multi-arm star polymers had smaller PS domain size as compared with linear polymers with the same molecular weight [89]. Thus, star polymers had more condensed physical cross-links per unit volume, which were attributed to their higher tensile strength. Another reason for better tensile strength was that the core in star polymers acted as permanent cross-links due to covalent chemical linkage. Besides better tensile stress of star polymers, the intrinsic viscosity of star polymers was lower than their linear analogs.

Confirmed by both experiments [90] and theory [91], the morphological dependence of block copolymers could be decoupled from chemical composition by varying chain architecture. Progress in self-consistent field theory (SCFT) [92] facilitated the ability to design TPEs based on nonlinear architectures such as miktoarm star polymer with superior mechanical properties [93]. For SIS triblock copolymer, over 36 vol% of PS component leads to lamellar morphology which is unfavorable for TPE applications [94]. For $A(BA')_4$ miktoarm star polymer with one A block and four BA' blocks emanating from the same core, Fredrickson [93, 94] predicted a stable morphology, of cylindrical A phase hexagonally dispersed in B matrix with a volume fraction of A polymer up to 70%. As shown in **Figure 7a**, asymmetric miktoarm star polymer $S(IS')_3$ contains one long PS chain and three PS'-PI chains connecting at the same core. For $S(IS')_3$ with 50 vol% of PS, a stable cylindrical morphology was observed (**Figure 7b**) where lamellar morphology was typically observed for an SIS triblock copolymer with the same composition [94]. The high volume fraction of PS enabled these new types of TPE with a higher modulus, strength toughness, and recoverable elasticity, while SIS' with 50 vol% of PS yield at low elongation indicated its thermoplastic nature (**Figure 7c**). By blending with PS homopolymers, a new stiff TPE (modulus was 99.2 MPa) with aperiodic "bricks and mortar" mesophase morphology was achieved with up to 82 wt% of PS [95]. Using similar miktoarm star polymer by blending with PS, a lamellar morphology with up to 97 wt% of PS was observed by Shi [96].

For the "core-first" strategy: developing multifunctional anionic initiators received limited success mainly because of the poor solubility of such initiators in hydrocarbon solvents [97]. However, multifunctional initiators for cationic polymerization are possible. $(PpCS\text{-}PIB)_8$ Eight arms star polymers were prepared through a calixarene core with eight initiation sites [98]. $(PMMA\text{-}PIB)_3$ Three arms star polymers were prepared by a trifunctional cationic initiator followed by ATRP of MMA [99]. For the "arm-first" strategy: at the end of living cationic polymerization, vinyl functionality was introduced by reacting the living cation of with allyltrimethylsilane. The vinyl end functionality further reacted with Si-H on cyclosiloxane by

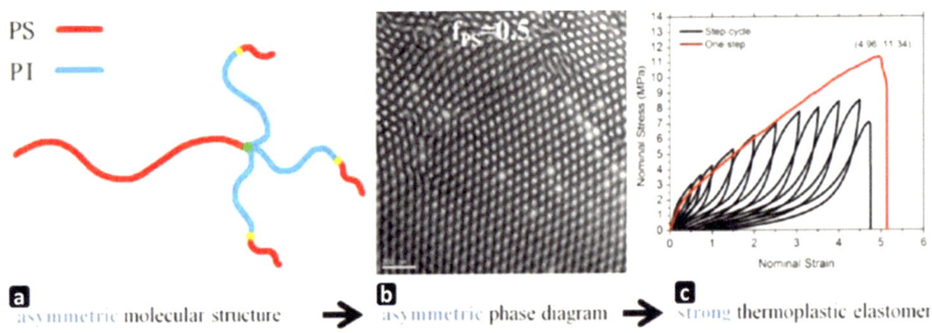

Figure 7. (a) Structure of $S(IS')_3$ miktoarm star copolymer-type TPEs. (b) TEM of $S(IS')_3$ miktoarm star copolymer with 50 vol% PS. (c) Stress-strain curve of $S(IS')_3$. (Reprinted with permission from Ref. [94]. Copyright 2014 American Chemical Society).

Pt-catalyzed hydrosilylation and produced star polymers with different number of arms based on different numbers of Si-H on cyclosiloxane [100–102]. Similar to arm-first divinylbenzene-linking strategy for anionic polymerization, 1,4-cyclohexane dimethanol divinyl ether was applied as the linking agent for arm-first cationic polymerization to prepare star polymers with poly(2-admantyl vinyl ether) as hard segment and poly(n-butyl vinyl ether) as elastic segment [103].

By using trifunctional ATRP initiator for "core-first" strategy, three arms star polymers with PMMA [104], polyacrylonitrile (PAN) [105], and PS [106] as glassy segment, PnBA as elastic segment were prepared for TPE properties evaluation. As an all-acrylic TPE, three arms star $(PMMA\text{-}PnBA)_3$ with 36% of PMMA showed 11-MPa ultimate stress with 545% strain at break. $(PAN\text{-}PnBA)_3$ Star polymers displayed ultimate tensile stress from 6.3 to 12.7 MPa as the strain at break in the range from 382 to 700%. Phase separation between PAN and PnBA was retained when the temperature belows 250°C . As the temperature further raised up to 280°C, the PAN domain started to cross-link chemically, and the storage modulus of these materials dropped when the temperature was close to 300°C. With multifunctional ATRP initiator of 10 and 20 initiation sites, 10 arms and 20 arms PMBL/PnBA star polymers were prepared for high-temperature TPE applications [107]. The highest ultimate tensile stress achieved was 7.8 MPa. Strain at break was lower than 140%.

5. Perspective

The past 60 years has witnessed rapid development of thermoplastic elastomers from discoveries in the laboratory to widely applied commodities involved in everyone's daily life. Starting from the twenty-first century, progress made in different polymerization techniques has advanced to new types of TPEs with various chemical compositions and macromolecular architectures. However, each polymerization technique has both merits and weaknesses.

Kraton styrenic thermoplastic elastomers are the most commercially successful polymeric materials synthesized by living anionic polymerization. The disadvantage of S-TPEs is obvious: low service temperature and poor UV/oxidation resistance. All-acrylic TPEs show better chemical resistance; however, the mechanical properties of these materials are much lower than those of S-TPEs.

Cationic polymerization was used to prepare PIB-based TPEs showing higher service temperature with better chemical resistance. The problem for cationic polymerization is the low polymerization temperature, which is not favorable for industrial applications. Low polymerization temperature also limits large-scale production of (methyl) acrylate-based TPEs by anionic polymerization.

Ring-opening transesterification polymerization produced biodegradable polymers from sustainable resources. However, most metal-catalyzed ROTEPs need toxic tin as the catalyst. Atomic transfer radical polymerization needs to reduce the radical concentration in order to control the polymerization. Polymers prepared by ATRP generally contain residual metal catalyst. Terminating the reaction at low conversion is necessary for block polymers preparation by ATRP.

Well-defined PI-g-$(PS)_n$ (n = 1–3) showed great mechanical properties competitive with Kraton products. However, these anionically prepared polymers required laborious synthetic procedures. As one of the most favorable polymerization techniques in industry, emulsion polymerization offers many benefits: polymers with high weight average molecular could be prepared quickly in water as the reaction medium. Particles of polymers could be directly applied for coating and painting without purification. Recent research using macromonomer approaches to synthesize PI-g-PS by a combination of anionic polymerization and emulsion polymerization opens up opportunities to prepare thermoplastic elastomers with highly tunable mechanical properties by a cost-efficient strategy. However, the PS macromonomer was prepared by anionic polymerization. Living anionic polymerization required oxygen- and moisture-free environment in order to retain the reactivity of chain-end anion. Thrilling opportunities are waiting if PS macromonomer could be prepared by all emulsion process with more than one branch point in the same macromonomer.

Acknowledgements

This work was supported by the U.S. Department of Energy, Office of Science, Basic Energy Science, Materials Sciences and Engineering Division. Part of the synthesis and characterization was conducted at the Center for Nanophase Materials Sciences, which is a DOE Office of Science User Facility.

Author details

Weiyu Wang[1]*, Wei Lu[2], Nam-Goo Kang[2], Jimmy Mays[2,3] and Kunlun Hong[1]

*Address all correspondence to: wangw1@ornl.gov

1 Center for Nanophase Materials Sciences, Oak Ridge National Laboratory, Oak Ridge, Tennessee, USA

2 Department of Chemistry, University of Tennessee, Knoxville, Tennessee, USA

3 Chemical Sciences Division, Oak Ridge National Laboratory, Oak Ridge, Tennessee, USA

References

[1] Bonart R. Thermoplastic elastomers. Polymer. 1979;**20**(11):1389-1403. DOI: 10.1016/0032-3861(79)90280-5

[2] Spontak RJ, Patel NP. Thermoplastic elastomers: Fundamentals and applications. Current Opinion in Colloid & Interface Science. 2000;**5**(5):333-340. DOI: 10.1016/S1359-0294(00)00070-4

[3] Shanks RA. General purpose elastomers: Structure, chemistry, physics and performance. In: Visakh PM, Thomas S, Chandra AK, Mathew AP, editors. Advances in Elastomers I. Berlin Heidelberg: Springer Science & Business Media; 2013. pp. 11-45. DOI: 10.1007/978-3-642-20925-3_2

[4] Walker BM, Rader CP, editors. Handbook of Thermoplastic Elastomers. 2nd ed. New York, NY: Van Nostrand Reinhold; 1988. p. 430. DOI: 10.1002/pol.1989.140270914

[5] Drobny JG, editor. Handbook of Thermoplastic Elastomers. 2nd ed. Elsevier; May 30, 2014. William Andrew; 2014. p. 464. DOI: 10.1016/B978-081551549-4.50002-5

[6] Bhowmick AK, Stephens H, editors. Handbook of Elastomers. 2nd ed. New York, NY: CRC Press; 2000.

[7] Fakirov S, editor. Handbook of Condensation Thermoplastic Elastomers. Weinheim, Germany: WILEY-VCH Verlag GmbH & Co. KGaA; 2005.

[8] Legge NR, Holden G, Schroeder HE, editors. Thermoplastic Elastomers: A Comprehensive Review. New York, NY: Hanser University Press; 1988. p. 574. DOI: 10.1002/pol.1989.140270710

[9] Hirao A, Goseki R, Ishizone T. Advances in living anionic polymerization: From functional monomers, polymerization systems, to macromolecular architectures. Macromolecules. 2014;**47**(6):1883-1905. DOI: 10.1021/ma401175m

[10] Hadjichristidis N, Hirao A, editors. Anionic Polymerization: Principles, Practice, Strength, Consequences and Applications. 1st ed. Japan: Springer; 2015. p. 1082. DOI: 10.1007/978-4-431-54186-8

[11] Goodwin A, Goodwin KM, Wang W, Yu YG, Lee JS, Mahurin SM, Dai S, Mays JW, Kang NG. Anionic polymerization of oxadiazole-containing 2-vinylpyridine by precisely tuning nucleophilicity and the polyelectrolyte characteristics of the resulting polymers. Macromolecules. 2016;**49**(17):6213-6225. DOI: 10.1021/acs.macromol.6b00875

[12] Aoshima S, Kanaoka S. A renaissance in living cationic polymerization. Chemical Reviews. 2009;**109**(11):5245-5287. DOI: 10.1021/cr900225g

[13] Matyjaszewski K, Tsarevsky NV. Macromolecular engineering by atom transfer radical polymerization. Journal of the American Chemical Society. 2014;**136**(18):6513-6533. DOI: 10.1021/ja408069v

[14] Khosravi E, Szymanska-Buzar T, editors. Ring Opening Metathesis Polymerisation and Related Chemistry: State of the Art and Visions for the New Century. 1st ed. The Netherlands: Springer; 2012. p. 488. DOI: 10.1007/978-94-010-0373-5

[15] Moad G, Rizzardo E, Thang SH. Radical addition–fragmentation chemistry in polymer synthesis. Polymer. 2008;**49**(5):1079-1131. DOI: 10.1016/j.polymer.2007.11.020

[16] Hawker CJ, Bosman AW, Harth E. New polymer synthesis by nitroxide mediated living radical polymerizations. Chemical Reviews. 2001;**101**(12):3661-3688. DOI: 10.1021/cr990119u

[17] Holden G, Milkovich R. Block Polymers of Monovinyl Aromatic Hydrocarbons and Conjugated Dienes [Internet]. August 9, 1966. Available from: https://www.google.com/patents/US3265765 [Accessed February 12, 2017]

[18] Handlin DL, Trenor S, Wright K. Applications of thermoplastic elastomers based on styrenic block copolymers. In: Matyjaszewski K, Gnanou Y, Leibler L, editors. Macromolecular Engineering: Precise Synthesis, Materials Properties, Applications. Weinheim, Germany: Wiley-VCH; 2007. pp. 2001-2031. DOI: 10.1002/9783527631421.ch1

[19] Hsieh H, Quirk RP, editors. Anionic Polymerization: Principles and Practical Applications. New York, NY: CRC Press; March 15, 1996. p. 744

[20] Fetters LJ, Morton M. Synthesis and properties of block polymers. I. Poly-α-methylstyrene-polyisoprene-poly-α-methylstyrene. Macromolecules. 1969;**2**(5):453-458. DOI: 10.1021/ma60011a002

[21] Bolton JM, Hillmyer MA, Hoye TR. Sustainable thermoplastic elastomers from terpene-derived monomers. ACS Macro Letters. 2014;**3**(8):717-720. DOI: 10.1021/mz500339h

[22] Fetters LJ, Firer EM, Dafauti M. Synthesis and properties of block copolymers. 4. Poly (p-tert-butylstyrene-diene-p-tert-butylstyrene) and poly (p-tert-butylstyrene-isoprene-styrene). Macromolecules. 1977;**10**(6):1200-1207. DOI: 10.1021/ma60060a008

[23] Kobayashi S, Matsuzawa T, Matsuoka SI, Tajima H, Ishizone T. Living anionic polymerizations of 4-(1-adamantyl) styrene and 3-(4-vinylphenyl)-1,1'-biadamantane. Macromolecules. 2006;**39**(18):5979-5986. DOI: 10.1021/ma060977+

[24] Kobayashi S, Kataoka H, Ishizone T, Kato T, Ono T, Kobukata S, Ogi H. Synthesis and properties of new thermoplastic elastomers containing poly [4-(1-adamantyl) styrene] hard segments. Macromolecules. 2008;**41**(14):5502-5508. DOI: 10.1021/ma7028743

[25] Yu JM, Dubois P, Jérôme R. Poly [alkyl methacrylate-b-butadiene-b-alkyl methacrylate] triblock copolymers: Synthesis, morphology, and mechanical properties at high temperatures. Macromolecules. 1996;**29**(26):8362-8370. DOI: 10.1021/ma960886k

[26] Lu W, Huang C, Hong K, Kang NG, Mays JW. Poly (1-adamantyl acrylate): Living anionic polymerization, block copolymerization, and thermal. Macromolecules. 2016;**49**(24):9406-9414. DOI: 10.1021/acs.macromol.6b01732

[27] Yu JM, Dubois P, Teyssié P, Jérôme R. Syndiotactic poly (methyl methacrylate)(sPMMA)–polybutadiene (PBD)–sPMMA triblock copolymers: Synthesis, morphology, and mechanical properties. Macromolecules. 1996;**29**(19):6090-6099. DOI: 10.1021/ma9603950

[28] Yu JM, Dubois P, Jérôme R. Synthesis and properties of poly [isobornyl methacrylate (IBMA)-b-butadiene (BD)-b-IBMA] copolymers: New thermoplastic elastomers of a large service temperature range. Macromolecules. 1996, Nov 4;**29**(23):7316-7322. DOI: 10.1021/ma960710i

[29] Varshney SK, Kesani P, Agarwal N, Zhang JX, Rafailovich M. Synthesis of ABA type thermoplastic elastomers based on polyacrylates. Macromolecule. 1999;**32**(1):235-237. DOI: 10.1021/ma971428u

[30] Tong JD, Jérôme R. Synthesis of poly (methyl methacrylate)-b-poly (n-butyl acrylate)-b-poly (methyl methacrylate) triblocks and their potential as thermoplastic elastomers. Polymer. 2000;**41**(7):2499-2510. DOI: 10.1016/S0032-3861(99)00412-7

[31] Natori I. Synthesis of polymers with an alicyclic structure in the main chain. Living anionic polymerization of 1,3-cyclohexadiene with the n-butyllithium/N,N,N',N'-tetramethylethylenediamine system. Macromolecules. 1997;**30**(12):3696-3697. DOI: 10.1021/ma9712110

[32] Hong K, Mays JW. 1,3-Cyclohexadiene polymers. 1. Anionic polymerization. Macromolecules. 2001;**34**(4):782-786. DOI: 10.1021/ma0015626

[33] Bornani K, Wang X, Davis JL, Wang X, Wang W, Hinestrosa JP, Mays JW, Kilbey II SM. Impact of chain microstructure on solution and thin film self-assembly of PCHD-based semi-flexible/flexible diblock copolymers. Soft Matter. 2015;**11**(32):6509-6519. DOI: 10.1039/C5SM01245G

[34] Imaizumi K, Ono T, Natori I, Sakurai S, Takeda K. Microphase-separated structure of 1,3-cyclohexadiene/butadiene triblock copolymers and its effect on mechanical and thermal properties. Journal of Polymer Science Part B: Polymer Physics. 2001;**39**(1):13-22. DOI: 10.1002/1099-0488(20010101)39:1<13::AID-POLB20>3.0.CO;2-K

[35] Natori I, Imaizumi K, Yamagishi H, Kazunori M. Hydrocarbon polymers containing six-membered rings in the main chain. Microstructure and properties of poly (1,3-cyclohexadiene). Journal of Polymer Science Part B: Polymer Physics. 1998;**36**(10):1657-1668. DOI: 10.1002/(SICI)1099-0488(19980730)36:10<1657::AID-POLB7>3.0.CO;2-M

[36] Kosaka Y, Kitazawa K, Inomata S, Ishizone T. Living anionic polymerization of benzofulvene: Highly reactive fixed transoid 1,3-diene. ACS Macro Letters. 2013;**2**(2):164-167. DOI: 10.1021/mz4000078

[37] Kosaka Y, Goseki R, Kawauchi S, Ishizone T. Living anionic polymerization of benzofulvene in hydrocarbon solvent. Macromolecular Symposia. 2015;**350**(1):55-66. DOI: 10.1002/masy.201400024

[38] Kosaka Y, Kawauchi S, Goseki R, Ishizone T. High anionic polymerizability of benzofulvene: New exo-methylene hydrocarbon Monomer. Macromolecules. 2015;**48**(13):4421-4430. DOI: 10.1021/acs.macromol.5b00944

[39] Wang W, Schlegel R, White BT, Williams K, Voyloy D, Steren CA, Goodwin A, Coughlin EB, Gido S, Beiner M, Hong K. High temperature thermoplastic elastomers synthesized by living anionic polymerization in hydrocarbon solvent at room temperature. Macromolecules. 2016;**49**(7):2646-2655. DOI: 10.1021/acs.macromol.5b02642

[40] MatWeb Material Property Data. Kraton® D1112P (SIS) Linear Block Copolymer [Internet]. Available from: http://www.matweb.com/search/datasheet.aspx?matguid=ec777ddb6a3540ae977fb8f9a53c4d64 [Accessed February 12, 2017], 1996

[41] Kaszas G, Puskas JE, Kennedy JP, Hager WG. Polyisobutylene-containing block polymers by sequential monomer addition. II. Polystyrene–polyisobutylene–polystyrene triblock polymers: Synthesis, characterization, and physical properties. Journal of Polymer Science Part A: Polymer Chemistry. 1991;**29**(3):427-435. DOI: 10.1002/pola.1991.080290316

[42] Cao X, Faust R. Polyisobutylene-based thermoplastic elastomers. 5. Poly (styrene-b-isobutylene-b-styrene) triblock copolymers by coupling of living poly (styrene-b-isobutylene) diblock copolymers. Macromolecules. 1999;**32**(17):5487-5494. DOI: 10.1021/ma990370b

[43] Kennedy JP, Kurian J. Living carbocationic polymerization of p-halostyrenes. III. Syntheses and characterization of novel thermoplastic elastomers of isobutylene and p-chlorostyrene. Journal of Polymer Science Part A: Polymer Chemistry. 1990;**28**(13):3725-3738. DOI: 10.1002/pola.1990.080281316

[44] Kennedy JP, Midha S, Tsunogae Y. Polyisobutylene-containing block polymers by sequential monomer addition. VIII: Synthesis, characterization, and physical properties of poly (indene-b-isobutylene-b-indene) thermoplastic elastomers. Macromolecules. 1993;**26**(3):429-435. DOI: 10.1021/ma00055a004

[45] Fodor Z, Kennedy JP. Polyisobutylene-containing block polymers by sequential monomer addition. Polymer Bulletin. 1992;**29**(6):697-704. DOI: 10.1007/BF01041157

[46] Pan P, Inoue Y. Polymorphism and isomorphism in biodegradable polyesters. Progress in Polymer Science. 2009;**34**(7):605-640. DOI: 10.1016/j.progpolymsci.2009.01.003

[47] Belgacem MN, Gandini A, editors. Monomers, Polymers and Composites from Renewable Resources. Elsevier; 2011. p. 560

[48] Ajellal N, Carpentier JF, Guillaume C, Guillaume SM, Helou M, Poirier V, Sarazin Y, Trifonov A. Metal-catalyzed immortal ring-opening polymerization of lactones, lactides and cyclic carbonates. Dalton Transactions. 2010;**39**(69):8363-8376. DOI: 10.1039/C001226B

[49] Hillmyer MA, Tolman WB, Aliphatic polyester block polymers: Renewable, degradable, and sustainable. Accounts of Chemical Research. 2014;**47**(8):2390-2396. DOI: 10.1021/ar500121d

[50] Hiki S, Miyamoto M, Kimura Y, Synthesis and characterization of hydroxy-terminated [RS]-poly (3-hydroxybutyrate) and its utilization to block copolymerization with l-lactide to obtain a biodegradable thermoplastic elastomer. Polymer. 2000;**41**(20):7369-7379. DOI: 10.1016/S0032-3861(00)00086-0

[51] Wanamaker CL, O'Leary LE, Lynd NA, Hillmyer MA, Tolman WB. Renewable-resource thermoplastic elastomers based on polylactide and polymenthide. Biomacromolecules. 2007;**8**(11):3634-3640. DOI: 10.1021/bm700699g

[52] Wanamaker CL, Bluemle MJ, Pitet LM, O'Leary LE, Tolman WB, Hillmyer MA. Consequences of polylactide stereochemistry on the properties of polylactide-polymenthide-polylactide thermoplastic elastomers. Biomacromolecules. 2009;**10**(10):2904-2911. DOI: 10.1021/bm900721p

[53] Martello MT, Hillmyer MA. Polylactide–poly (6-methyl-ε-caprolactone)–polylactide thermoplastic elastomers. Macromolecules. 2011;**44**(21):8537-8545. DOI: 10.1021/ma201063t

[54] Matyjaszewski K, Davis TP, editors. Handbook of Radical Polymerization. Weinheim, Germany: WILEY-VCH Verlag GmbH & Co. KGaA;2003. p. 920. DOI: 10.1002/0471220450

[55] Fan F, Wang W, Holt AP, Feng H, Uhrig D, Lu X, Hong T, Wang Y, Kang NG, Mays J, Sokolov AP. Effect of molecular weight on the ion transport mechanism in polymerized ionic liquids. Macromolecules. 2016;**49**(12):4557-4570. DOI: 10.1021/acs.macromol.6b00714

[56] Matyjaszewski K, Tsarevsky NV. Nanostructured functional materials prepared by atom transfer radical polymerization. Nature Chemistry. 2009;**1**(4):276-288. DOI: 10.1038/nchem.257

[57] Dong H, Tang W, Matyjaszewski K. Well-defined high-molecular-weight polyacrylonitrile via activators regenerated by electron transfer ATRP. Macromolecules. 2007;**40**(9):2974-2977. DOI: 10.1021/ma070424e

[58] Thomas DB, Sumerlin BS, Lowe AB, McCormick CL. Conditions for facile, controlled RAFT polymerization of acrylamide in water. Macromolecules. 2003;**36**(5):1436-1439. DOI: 10.1021/ma025960f

[59] Keddie DJ. Chemical Society Reviews. A guide to the synthesis of block copolymers using reversible-addition fragmentation chain transfer (RAFT) polymerization. Chemical Society Reviews. 2014;**43**(2):496-505. DOI: 10.1039/C3CS60290G

[60] Matyjaszewski K. Architecturally complex polymers with controlled heterogeneity. Science. 2011;**333**(6046):1104-1105. DOI: 10.1126/science.1209660

[61] Yi Y, Fan X, Wan X, Li L, Zhao N, Chen X, Xu J, Zhou QF. ABA type triblock copolymer based on mesogen-jacketed liquid crystalline polymer: Design, synthesis, and potential as thermoplastic elastomer. Macromolecules. 2004;**37**(20):7610-7618. DOI: 10.1021/ma0400463

[62] Liu X, Zhao RY, Zhao TP, Liu CY, Yang S, Chen EQ. An ABA triblock containing a central soft block of poly [2, 5-di (n-hexogycarbonyl) styrene] and outer hard block of poly (4-vinylpyridine): Synthesis, phase behavior and mechanical enhancement. RSC Advances. 2014;**4**(35):18431-18441. DOI: 10.1039/C4RA01652A

[63] Braunecker WA, Matyjaszewski K. Controlled/living radical polymerization: Features, developments, and perspectives. Progress in Polymer Science. 2007;**32**(1):93-146. DOI: 10.1016/j.progpolymsci.2006.11.002

[64] Butté A, Storti G, Morbidelli M. Miniemulsion living free radical polymerization by RAFT. Macromolecules. 2001;**34**(17):5885-5896. DOI: 10.1021/ma002130y

[65] Delaittre G, Charleux B. Kinetics of in-situ formation of poly (acrylic acid)-b-polystyrene amphiphilic block copolymers via nitroxide-mediated controlled free-radical emulsion polymerization. Discussion on the effect of compartmentalization on the polymerization rate. Macromolecules. 2008;**41**(7):2361-2367. DOI: 10.1021/ma702498u

[66] Luo Y, Wang X, Zhu Y, Li BG, Zhu S. Polystyrene-block-poly (n-butyl acrylate)-block-polystyrene triblock copolymer thermoplastic elastomer synthesized via RAFT emulsion polymerization. Macromolecules. 2010;**43**(18):7472-7481. DOI: 10.1021/ma101348k

[67] Zhan X, He R, Zhang Q, Chen F. Microstructure and mechanical properties of amphiphilic tetrablock copolymer elastomers via RAFT miniemulsion polymerization: Influence of poly [styrene-alt-(maleic anhydride)] segments. RSC Advances. 2014;**4**(93):51201-51207. DOI: 10.1039/C4RA06185C

[68] Wang S, Vajjala Kesava S, Gomez ED, Robertson ML. Sustainable thermoplastic elastomers derived from fatty acids. Macromolecules. 2013;**46**(18):7202-7212. DOI: 10.1021/ma4011846

[69] Hayashi M, Matsushima S, Noro A, Matsushita Y. Mechanical property enhancement of ABA block copolymer-based elastomers by incorporating transient cross-links into soft middle block. Macromolecules. 2015;**48**(2):421-431. DOI: 10.1021/ma502239w

[70] Feng C, Li Y, Yang D, Hu J, Zhang X, Huang X. Well-defined graft copolymers: from controlled synthesis to multipurpose applications. Chemical Society Reviews. 2011;**40**(3):1282-1295. DOI: 10.1039/B921358A

[71] Uhrig D, Mays J. Synthesis of well-defined multigraft copolymers. Polymer Chemistry. 2011;**2**(1):69-76. DOI: 10.1039/C0PY00185F

[72] Ito S, Goseki R, Ishizone T, Hirao A. Synthesis of well-controlled graft polymers by living anionic polymerization towards exact graft polymers. Polymer Chemistry. 2014;**5**(19):5523-5534. DOI: 10.1039/C4PY00584H

[73] Uhrig D, Schlegel R, Weidisch R, Mays J. Multigraft copolymer superelastomers: Synthesis morphology, and properties. European Polymer Journal. 2011;**47**(4):560-568. DOI: 10.1016/j.eurpolymj.2010.10.030

[74] Iatrou H, Mays JW, Hadjichristidis N. Regular comb polystyrenes and graft polyisoprene/polystyrene copolymers with double branches ("centipedes"). Quality of (1,3-phenylene) bis (3-methyl-1-phenylpentylidene) dilithium initiator in the presence of polar additives. Macromolecules. 1998;**31**(19):6697-6701. DOI: 10.1021/ma980738p

[75] Uhrig D, Mays JW. Synthesis of combs, centipedes, and barbwires: Poly (isoprene-graft-styrene) regular multigraft copolymers with trifunctional, tetrafunctional, and hexafunctional branch points. Macromolecules. 2002;**35**(19):7182-7190. DOI: 10.1021/ma020427l

[76] Beyer FL, Gido SP, Büschl C, Iatrou H, Uhrig D, Mays JW, Chang MY, Garetz BA, Balsara NP, Tan NB, Hadjichristidis N. Graft copolymers with regularly spaced, tetrafunctional branch points: Morphology and grain structure. Macromolecules. 2000;**33**(6):2039-2048. DOI: 10.1021/ma991141s

[77] Duan Y, Thunga M, Schlegel R, Schneider K, Rettler E, Weidisch R, Siesler HW, Stamm M, Mays JW, Hadjichristidis N. Morphology and deformation mechanisms and tensile properties of tetrafunctional multigraft copolymers. Macromolecules. 2009;**42**(12):4155-4164. DOI: 10.1021/ma900414h

[78] Weidisch R, Gido SP, Uhrig D, Iatrou H, Mays J, Hadjichristidis N. Tetrafunctional multigraft copolymers as novel thermoplastic elastomers. Macromolecules. 2001;**34**(18):6333-6337. DOI: 10.1021/ma001966y

[79] Staudinger U, Weidisch R, Zhu Y, Gido S, Uhrig D, Mays J, Iatrou H, Hadjichristidis N, Wiley Online Library. Mechanical properties and hysteresis behaviour of multigraft copolymers. Macromolecular Symposia. 2006;**233**(1):42-50. DOI: 10.1002/masy.200690027

[80] Schlegel R, Wilkin D, Duan Y, Weidisch R, Heinrich G, Uhrig D, Mays JW, Iatrou H, Hadjichristidis NS. Stress softening of multigraft copolymers. Polymer. 2009;**50**(26):6297-6304. DOI: 10.1016/j.polymer.2009.10.026

[81] Goodwin A, Wang W, Kang NG, Wang Y, Hong K, Mays J. All-acrylic multigraft copolymers: Effect of side chain molecular weight and volume fraction on mechanical behavior. Industrial & Engineering Chemistry Research. 2015;**54**(39):9566-9576. DOI: 10.1021/acs.iecr.5b02560

[82] Wang W, Wang W, Lu X, Bobade S, Chen J, Kang NG, Zhang Q, Mays J. Synthesis and characterization of comb and centipede multigraft copolymers PnBA-g-PS with high molecular weight using miniemulsion polymerization. Macromolecules. 2014;**47**(21):7284-7295. DOI: 10.1021/ma501866t

[83] Wang W, Wang W, Li H, Lu X, Chen J, Kang NG, Zhang Q, Mays J. Synthesis and characterization of graft copolymers poly (isoprene-g-styrene) of high molecular weight by a combination of anionic polymerization and emulsion polymerization. Industrial & Engineering Chemistry Research. 2015;**54**(4):1292-1300. DOI: 10.1021/ie504457e

[84] Li H, Wang W, Li C, Tan J, Yin D, Zhang H, Zhang B, Yin C, Zhang Q. Synthesis and characterization of brush-like multigraft copolymers PnBA-g-PMMA by a combination of emulsion AGET ATRP and emulsion polymerization. Journal of Colloid and Interface Science. 2015;**453**:226-236. DOI: 10.1016/j.jcis.2015.04.051

[85] Li H, Wang W, Tan J, Li C, Zhang Q. Synthesis and characterization of graft copolymers PnBA-g-PS by miniemulsion polymerization. RSC Advances. 2015;**5**(56):45459-45466. DOI: 10.1039/C5RA06502J

[86] Hadjichristidis N. Synthesis of miktoarm star (μ-star) polymers. Journal of Polymer Science Part A: Polymer Chemistry. 1999;**37**(7):857-871. DOI: 10.1002/(SICI)1099-0518(19990401)37:7<857::AID-POLA1>3.0.CO;2-P

[87] Khanna K, Varshney S, Kakkar A. Miktoarm star polymers: Advances in synthesis, self-assembly, and applications. Polymer Chemistry. 2010;**1**(8):1171-1185. DOI: 10.1039/C0PY00082E

[88] Bi LK, Fetters LJ. Synthesis and properties of block copolymers. 3. Polystyrene-polydiene star block copolymers. Macromolecules. 1976;**9**(5):732-742. DOI: 10.1021/ma60053a010

[89] Bi LK, Fetters LJ. Domain morphology of star block copolymers of polystyrene and polyisoprene. Macromolecules. 1975;**8**(1):90-92. DOI: 10.1021/ma60043a026

[90] Lai C, Russel WB, Register RA, Marchand GR, Adamson DH. Phase behavior of styrene–isoprene diblock derivatives with varying conformational asymmetry. Macromolecules. 2000;**33**(9):3461-3466. DOI: 10.1021/ma991156q

[91] Milner ST. Chain architecture and asymmetry in copolymer microphases. Macromolecules. 1994;**27**(8):2333-2335. DOI: 10.1021/ma00086a057

[92] Vavasour JD, Whitmore MD. Self-consistent field theory of block copolymers with conformational asymmetry. Macromolecules. 1993;**26**(25):7070-7075. DOI: 10.1021/ma00077a054

[93] Lynd NA, Oyerokun FT, O'Donoghue DL, Handlin Jr DL, Fredrickson GH. Design of soft and strong thermoplastic elastomers based on nonlinear block copolymer architectures using self-consistent-field theory. Macromolecules. 2010;**43**(7):3479-3486. DOI: 10.1021/ma902517v

[94] Shi W, Lynd NA, Montarnal D, Luo Y, Fredrickson GH, Kramer EJ, Ntaras C, Avgeropoulos A, Hexemer A. Toward strong thermoplastic elastomers with asymmetric miktoarm block copolymer architectures. Macromolecules. 2014;**47**(6):2037-2043. DOI: 10.1021/ma402566g

[95] Shi W, Hamilton AL, Delaney KT, Fredrickson GH, Kramer EJ, Ntaras C, Avgeropoulos A, Lynd NA, Demassieux Q, Creton C. Aperiodic "bricks and mortar" mesophase: A new equilibrium state of soft matter and application as a stiff thermoplastic elastomer. Macromolecules. 2015;**48**(15):5378-5384. DOI: 10.1021/acs.macromol.5b01210

[96] Shi W, Hamilton AL, Delaney KT, Fredrickson GH, Kramer EJ, Ntaras C, Avgeropoulos A, Lynd NA. Creating extremely asymmetric lamellar structures via fluctuation-assisted unbinding of miktoarm star block copolymer alloys. Journal of the American Chemical Society. 2015;**137**(19):6160-6163. DOI: 10.1021/jacs.5b02881

[97] Matmour R, Gnanou Y. Synthesis of complex polymeric architectures using multilithiated carbanionic initiators—Comparison with other approaches. Progress in Polymer Science. 2013;**38**(1):30-62. DOI: 10.1016/j.progpolymsci.2012.08.003

[98] Jacob S, Kennedy JP. Synthesis and characterization of novel octa-arm star-block thermoplastic elastomers consisting of poly (p-chlorostyrene-b-isobutylene) arms radiating from a calix [8] arene core. Polymer Bulletin. 1998;**41**(2):167-174. DOI: 10.1007/s002890050348

[99] Keszler B, Fenyvesi GY, Kennedy JP. Novel star-block polymers: Three polyisobutylene-b-poly (methyl methacrylate) arms radiating from an aromatic core. Journal of Polymer Science Part A: Polymer Chemistry. 2000;**38**(4):706-714. DOI: 10.1002/(SICI)1099-0518(20000215)38:4<706::AID-POLA5>3.0.CO;2-D

[100] Shim JS, Asthana S, Omura N, Kennedy JP. Novel thermoplastic elastomers. I. Synthesis and characterization of star-block copolymers of PSt-b-PIB arms emanating from cyclosiloxane cores. Journal of Polymer Science Part A: Polymer Chemistry. 1998;**36**(17):2997-3012. DOI: 10.1002/(SICI)1099-0518(199812)36:17<2997::AID-POLA1>3.0.CO;2-1

[101] Shim JS, Kennedy JP. Novel thermoplastic elastomers. II. Properties of star-block copolymers of PSt-b-PIB arms emanating from cyclosiloxane cores. Journal of Polymer Science Part A: Polymer Chemistry. 1999;**37**(6):815-824. DOI: 10.1002/(SICI)1099-0518(19990315)37:6<815::AID-POLA17>3.0.CO;2-5

[102] Shim JS, Kennedy JP. Novel thermoplastic elastomers. III. Synthesis, characterization, and properties of star-block copolymers of poly (indene-b-isobutylene) arms emanating from cyclosiloxane cores. Journal of Polymer Science Part A: Polymer Chemistry. 2000;**38**(2):279-290. DOI: 10.1002/(SICI)1099-0518(20000115)38:2<279::AID-POLA2>3.0.CO;2-8

[103] Imaeda T, Hashimoto T, Irie S, Urushisaki M, Sakaguchi T. Synthesis of ABA-triblock and star-diblock copolymers with poly (2-adamantyl vinyl ether) and poly (n-butyl vinyl ether) segments: New thermoplastic elastomers composed solely of poly (vinyl ether) backbones. Journal of Polymer Science Part A: Polymer Chemistry. 2013;**51**(8):1796-1807. DOI: 10.1002/pola.26561

[104] Dufour B, Koynov K, Pakula T, Matyjaszewski K. PBA–PMMA 3-arm star block copolymer thermoplastic elastomers. Macromolecular Chemistry and Physics. 2008; **209**(16):1686-1693. DOI: 10.1002/macp.200800151

[105] Dufour B, Tang C, Koynov K, Zhang Y, Pakula T, Matyjaszewski K. Polar three-arm star block copolymer thermoplastic elastomers based on polyacrylonitrile. Macromolecules. 2008;**41**(7):2451-2458. DOI: 10.1021/ma702561b

[106] Pakula T, Koynov K, Boerner H, Huang J, Lee HI, Pietrasik J, Sumerlin B, Matyjaszewski K. Effect of chain topology on the self-organization and the mechanical properties of poly (n-butyl acrylate)-b-polystyrene block copolymers. Polymer. 2011;**52**(12):2576-2583. DOI: 10.1016/j.polymer.2011.04.021

[107] Juhari A, Mosnáček J, Yoon JA, Nese A, Koynov K, Kowalewski T, Matyjaszewski K. Star-like poly (n-butyl acrylate)-b-poly (α-methylene-γ-butyrolactone) block copolymers for high temperature thermoplastic elastomers applications. Polymer. 2010;**51**(21):4806-4813. DOI: 10.1016/j.polymer.2010.08.017

Chapter 6

Fabrication and Properties of Rubber Nanofiber from Electrospinning

Wirach Taweepreda

Additional information is available at the end of the chapter

http://dx.doi.org/10.5772/intechopen.69373

Abstract

Electrospun nanofibers of rubber blending with poly(vinyl chloride) (PVC) were prepared by electrospinning. Various concentrations of rubber solution were added to the PVC solution for several purposes of the electrospun nanofibers. The pristine and rubber plasticized nanofibers were characterized by field emission scanning electron microscopy (FE-SEM), differential thermal analysis (DTA), and thermogravimetry analysis (TGA), respectively. Thermal properties from DTA and TGA results indicated that the addition of rubber decreased glass transition temperature (T_g) of PVC and increased heat resistance of PVC nanofiber. The dielectric constant of plasticized nanofibers measured using precision LCR meter indicated that the membranes were dense.

Keywords: PVC, rubber, electrospinning, nanofiber

1. Introduction

1.1. Elastomeric materials

Elastomers are important polymers having various applications in daily life. Elastomers can be classified by application purposes such as general and special. Elastomers are often mixed together to achieve superior properties which arise from the individual systems. Blending is macroscopically homogeneous mixture of two or more different species of polymers which is carried out for several purposes [1], such as improving the physical and mechanical properties of the first elastomer and obtaining good processing characteristics of rubber compound. The most important characteristic of a blending of two (or more) polymers is the phase behavior. Polymer blends (like low-molecular-weight solvents) can exhibit miscibility or phase separation and various levels of mixing in between the extremes and polymer blends as a mixture of at least two macromolecular substances, polymer, or copolymer, in which the ingredient content is above 2 wt% [2].

Epoxidized natural rubber (ENR) is a modified molecule of natural rubber with the chemical name cis-1,4polyisoprene. There structuring using chemicals such as peroxy acid will react with double bonds of the natural rubber molecules to form ring oxygen that is replaced; the epoxidation reaction of natural rubber is shown in **Figure 1**. In the beginning, formic acid reacts with hydrogen peroxide to form peroxy formic acid [3, 4]. Then, the peroxy formic acid reacts with double bond of natural rubber is arranged as a ring epoxide and formic acid is a by-product as shown in **Figure 1**.

However, only three types of ENR are considered as commercial standard. These are ENR-10, ENR-25, and ENR-50 where the integers designate 10, 25, and 50 mole% of epoxide incorporated into the natural rubber chain, respectively [4, 5].

Besides the proposed commercial uses shown, ENR has the potential to be further exploited for its usage as advanced materials such as in blends, additives, and fuel cells applications [6]. ENR had been blended with various polymers such as natural rubber [7, 8], styrene-butadiene rubber (SBR) [9], and poly(vinyl chloride) (PVC) [10–12].

Polyurethane (PU) is formed by reacting an isocyanate with a polyol. Both the isocyanates and polyols used to make polyurethanes contain on average two or more functional groups per molecule. Most polyurethanes are thermosetting polymers that do not melt when heated.

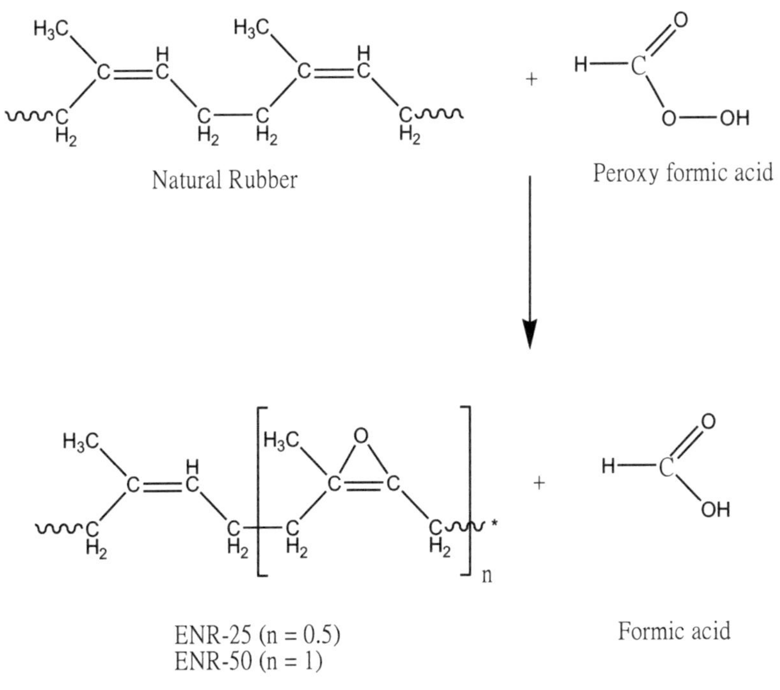

Figure 1. The reaction epoxidation of natural rubber.

Polyurethane has the flexibility and strength of rubber and the hardness of plastic. Many commercially available PU can be used to make good electrospinning solutions.

1.2. Electrospinning

Electrostatic spinning orelectrospinning is a unique process to prepare electrospun fibers with diameters in the range of micrometers to nanometers that depends on the type of polymer and processing conditions [13, 14].

Electrospinning is the process that occurred over 60 years. In the year 1934, Formhals was a patented process and used as electrical instruments in synthetic fibers [15]. In the year 1969, Taylor published work describing the phenomena occurring at the time. Polymer solution is ejected from a metal needle at high-voltage power supply. As the electrostatic charge on the droplet overcomes the surface tension, the low-molecular-weight fluid breaks up into small droplets, named a Taylor cone [16].

The basic principles of electrospinning are very simple. There are mainly three components to fulfill the process as follows: a high-voltage power supply, a syringe with a metal needle, and a collector. The polymer solution with surface tensionis encapsulated within a syringe, forming a hemispherical drop at the end of the capillary tip which is connected to an anode that connects to a high-voltage power supply, usually between 5 and 30 kV, which supports the ejection of a liquid jet followed by solvent evaporation as the jet travels through the air leaving behind ultrafine polymer fibers collected on a grounded cathode-connected metallic collector. The interaction between the liquid jet surface and external electric field is the cause of bending or a spiral track of the liquid jet which not only resulted in the electrospinning jet being elongated up to ultrafine fibers but also randomly deposited nonwoven electrospun mats [17–19]. Surface area and pore size of electrospun mats are larger and smaller, respectively, than commercial nonwoven fabrics. So, they are used in a wide variety of applications of electrospun mats, for example, tissue engineering, drug delivery system [20], wound-dressing [21] membrane [22], and conductive fiber [23].

It is well known that the morphology and size of electrospun fibers depends on various processing parameters such as the solution properties, controlled variables, ambient conditions, viscosity, and surface tension. With the control of these parameters, optimal nanofibers can be prepared. Electrospinning parameters and their effects on fiber morphology are as follows:

1.3. Solvents used for electrospinning

The polymer solution, used for electrospinning, usually was prepared by dissolving in different solvents. The solvent type is important on its spinnability both of first and foremost step of electrospinning process. The solvents should have some proper properties which are suitable with polymer types such assurface tension, vapor pressure, and relative density. Thus, the selection of an appropriate solvent system is indispensable which cause successful electrospinning process [24]. It is well known that different solvents may contribute to different surface tensions. The solution viscosity is determined by the concentration of the polymer, but the value of surface tension depends on both the polymer and the solvent [25, 26]. **Table 1** shows physical properties of different solvents.

The ability of the volatility in the solvent such as reported that study PVC/PU electrospinning by solutions for electrospinning were prepared in a mixed solvent of tetrahydrofuran (THF) and *N,N*-dimethylformamide (DMF). It was found that THF has the ability ofa good volatility and DMF has the ability of a poor volatility. It is found that when THF is used, 100% of the fibershavehigh porosity, and when DMF is used 100% of the fibers are smooth and microtexture fiber [27].

Solvents	CAS number	Surface tension at 20°C (mN/m)	Vapor pressure (hPa at 20°C)	Relative density at 20°C (g/dm³)
Acetone	67-64-1	25.2	233	0.79
Acetic acid	64-19-7	27.4	15.4	1.05
Chloroform	67-66-3	27.5	211	1.48
Dichloromethane	75-09-2	26.5	475	1.33
N,N-Dimethylformamide	68-12-2	37.1	3.77	0.94
Ethanol	64-17-5	22.1	59	0.80
Formic acid	64-18-6	37.6	42	1.22
Hexafluoroisopropanol	920-66-1	16.1	269	1.62
Methanol	67-56-1	22.7	128	0.79
Tetrahydrofuran (THF)	109-99-9	26.4	173	0.89
Trifluoroacetic acid	76-05-1	21.9	141	1.48
Water	7732-18-5	72.8	23.39	0.998

Table 1. Physical properties of some common liquids.

2. Fabrication of polymer nanofibers

There are a wide range of polymers that are used in electrospinning and are able to form fine nanofibers within the submicron range and used for varied applications. Electrospun nano fibers have been reported as being from various synthetic polymers, and natural polymers were electrospun from polymer solutions, such as poly(vinyl chloride) [28], poly(ethylene-co-vinyl alcohol) [29], polycarbonates [30], nylon-6 [31], polyurethane [32], polyacrylonitrile [33] polylactide [34], and poly(ε-caprolactone) [35].

Poly(vinyl chloride)is a hard, stiff material with low thermal stability, can be extensively modified with plasticizer and thermal stabilizer to enhance flexibility and thermal stability, respectively. Almost all plasticizers are commonly phthalate ester. The plasticizers interpose between every individual strand of polymer causing breakdown of polymer-polymer interaction. The polymer structure is modified into more porous, flexible, and with less cohesive structurere-sulting in lower glass transition temperature (T_g). The plasticized PVC behaves like a rubber at room temperature. However, plasticizer can slowly leach out with time anddeform PVC at a lower tensile force. Moreover, the phthalate esters are endocrine disrupters or mutagens.

In this chapter, the effects of electrospinning condition including electric voltage, tip-to-collector distance, and PVC solution concentration were studied on the morphology of formation of PVC nanofibers. The physical properties of PVC nanofibers were studied by varying the blending ratio of PVC with polyurethane and PVC with epoxidized natural rubber.

2.1. Electrospinning of poly(vinyl chloride) blending with polyurethane

Poly(vinyl chloride) with a *K*-value of 580 was kindly supplied by Thai Plastic and Chemicals Co., Ltd. The solvents *N,N*-dimethylformamide and tetrahydrofuran were used as received. DMF was mixed with THF at the ratio 50:50 and was used to prepare the PVC solutionwith varying PVC concentration for 10 and 15 wt% at 50°C. Polyurethane (Dow Chemical Co., Ltd.) was dissolved in the same solvent mixtures as PVC solution at a concentration of 10 wt%. The PU solution was directly mixed with PVC solution at room temperature.

The picture of PVC nanofibers obtained from SEM shows the morphological variation that depends on PVC concentration and electrospinning conditions as illustrated in **Figure 2** [36]. PVC nanofibers were prepared from PVC solution at 10 and 15 wt% with an applied voltage of 12 kV and various distances between the syringe tip and the collector (12, 15, and 18 cm). The surface instabilities were induced by electrospun jet motion and charge interaction. Low concentration of PVC solution (10 wt%) led to the formation of beads on the fiber. On increasing the concentration of PVC solution (15 wt%), the beads formation decreasedwith a gradual increase in the average diameter which can be attributed to higher viscosity.

The morphology of PVC nanofiber is modified by blendingwith polyurethane solution at 10 wt% as shown in **Figure 3** [39]. PU blending with PVC smoothens the fiber withoutbeads which was found in PVC nanofiber. PU component can be attributed to higher viscosity and surface tension of polymer blend solution. Viscosity and surface tension play an important role in determining the electrospun nanofiber formability and diameter.

PVC nanofiber prepared from 10 wt% is rigid than that from 15 wt% which corresponds with thermal properties determined using differential thermal analysis (DTA). The nanofiber prepared from 10 wt% PVC has a high exothermal peak at 30 and 275°C, while the nanofiber from 15 wt% PVC shows endothermal peak. The thermal properties of nanofibers were improved after mixing PVC with PU solution. The mixture of PVC with high PU content gave more flexible nanofiber with a lower glass transition temperature (T_g) and heat resistance as illustrated in **Figure 4** [39]. For higher concentration of PVC (15 wt%) with high viscous, it is not easily blended with PU solution which led to heterogeneous solution.

The thermal stability of the nanofibers was determined using thermogravimetric analysis. The thermogram of the PVC and PVC/PU blends as shown in **Figure 5** [39] has two obvious decompositions. The thermogram of the PVC/PU blend nanofiber shows the decomposition temperature (T_1) at 250°C, weight-loss rate is the largest at approximately 300°C. There are two decompositions of PVC/PU nanofiber. The first decomposition occurs at 275°C (T_2) with a weight loss of about 10%, which may be due to the decomposition of PVC in the nanofiber. The second weight loss of about 62% occurs at around 433°C (T_3), indicating that the PVC/PU blend nanofiber has a good thermal stability.

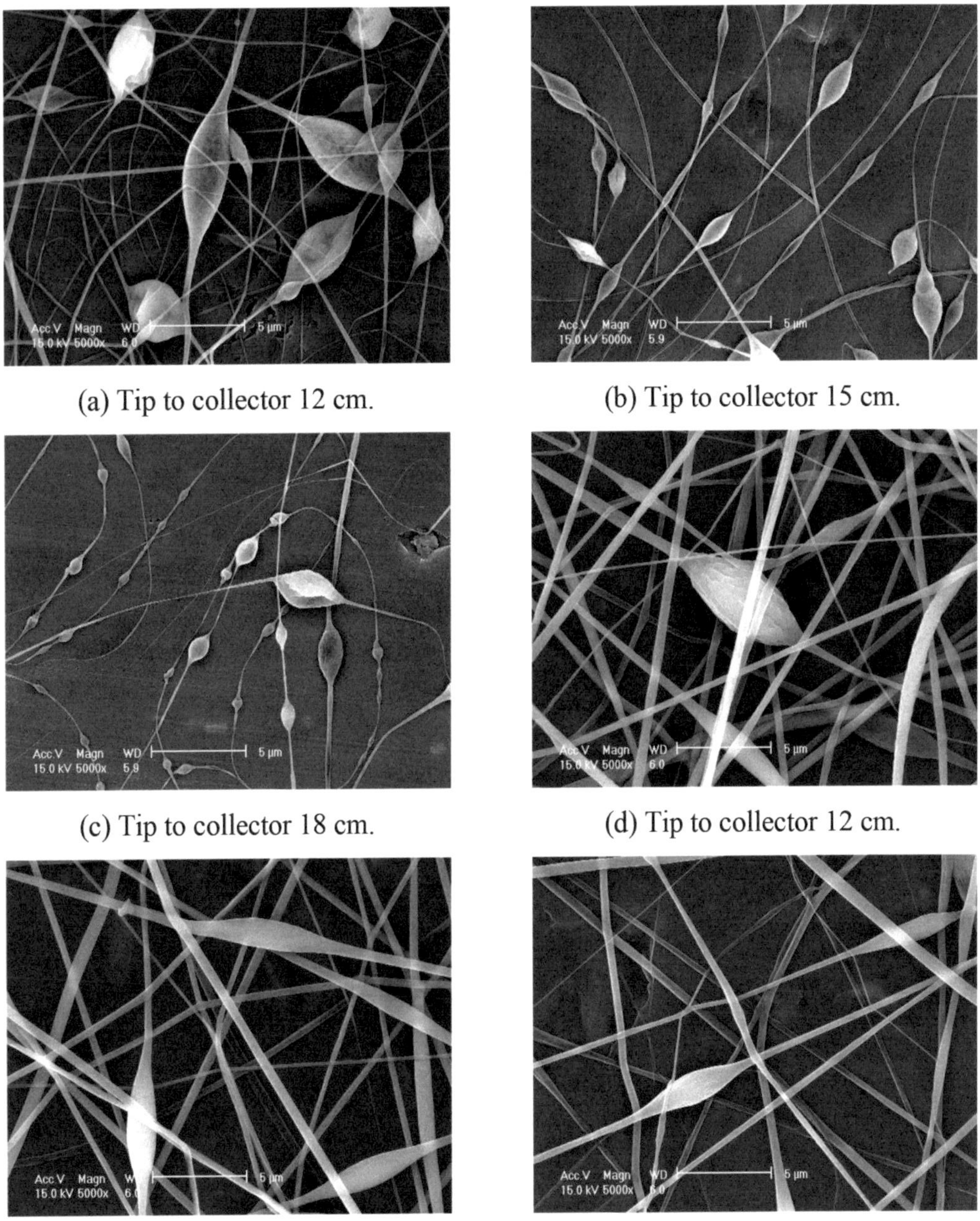

(a) Tip to collector 12 cm.

(b) Tip to collector 15 cm.

(c) Tip to collector 18 cm.

(d) Tip to collector 12 cm.

(e) Tip to collector 15 cm.

(f) Tip to collector 18 cm.

Figure 2. SEM images of electrospun nanofiber from PVC solution concentration of 10 wt% (a–c) and 15 wt% (d–f).

http://dx.doi.org/10.5772/intechopen.69373

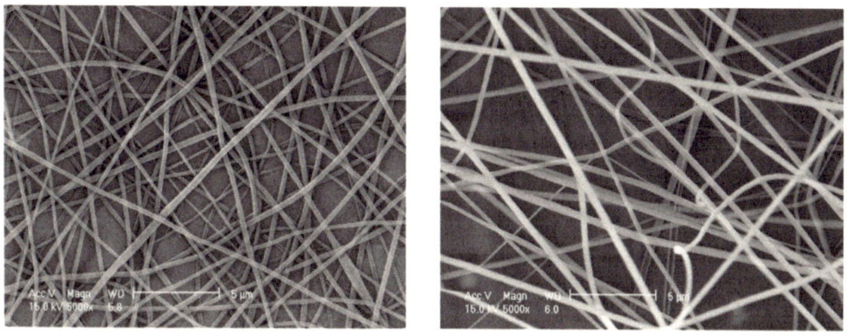

(a) (b)

Figure 3. SEM images of electrospun nanofiber from PVC/PU blending of (a) PVC 10% with PU 10 wt% and (b) PVC 15% with PU 10 wt%.

The permittivity of PVC and PVC/PU blend nanofibers, which were measured over the frequency range from 75 kHz up to 30 MHz at room temperature, is illustrated in **Figure 6**. The dielectric constant of the nanofiber, which was prepared from 10 wt% of PVC mixed with high PU content, reduced the dielectric constant due to denseness of PU phase. On increasing the concentration of PVC solution to 15 wt% blending with PU solution, it was found that the dielectric properties were not changed with PU contents due to the prepared electrospun nanofiber membranes that were dense.

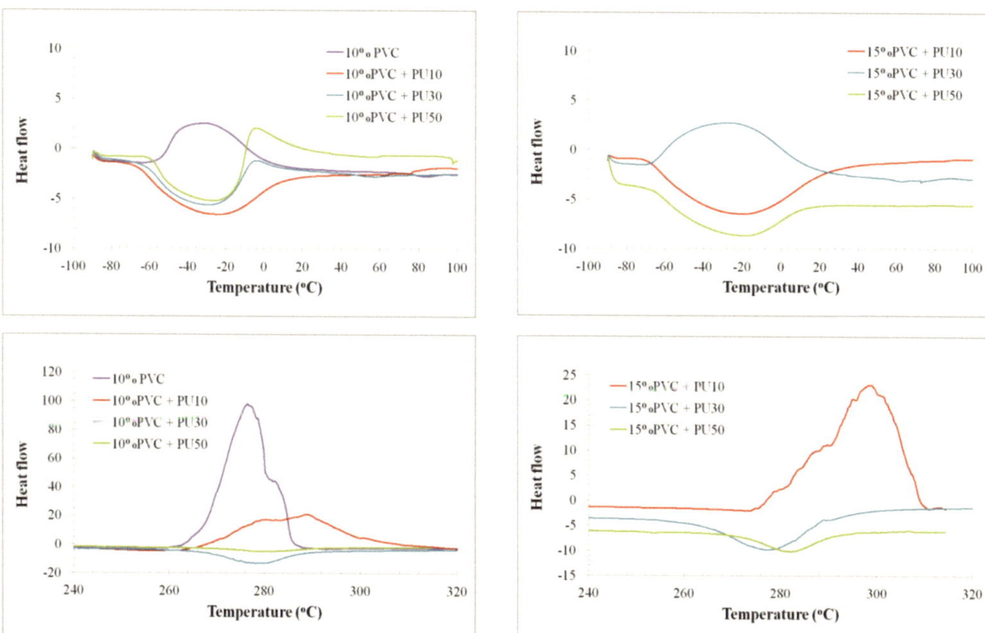

Figure 4. DTA thermogram of nanofibers from PVC and PVC blending with PU by varying PVC concentration and blending ratio.

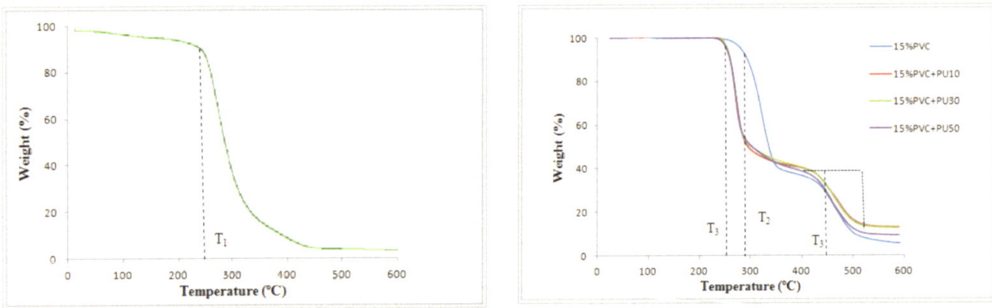

(a) (b)

Figure 5. TGA of nanofibers from (a) PVC and (b) PVC blending with PU at difference ratio.

It was reported that [37] electrospun poly(vinyl chloride)/polyurethane-blended solutions (10/0, 9/1, 8/2, and 7/3 wt%) were prepared by adding PVC and PU in the mixture solvent of THF/DMF (1/9, w/w). It was found that membranes exhibited relatively high tensile strength (9.9 MPa), good air permeability (154.1 mm/s), and excellent abrasion resistance (134 cycles).

As shown in **Figure 7**, when an external load is applied, the hydrogen bonding between C=O groups of PU and the α-hydrogen of PVC, and the dipole-dipole-C=OλCl—C— interactions tend to bear the stress on intermolecular levels, resulting in the elongation of less than 6% in the first linear elastic behavior.

2.2. Electrospinning of poly(vinyl chloride) blending with epoxidized natural rubber [38]

Poly(vinyl chloride) with a *K*-value of 580 was kindly supplied by Thai Plastic and Chemicals Co., Ltd. *N,N*-dimethylformamide was mixed with tetrahydrofuran at the ratio of 50:50 and was used to prepare the solution of PVC and epoxidized natural rubber at 50°C. The various concentrations of ENR solution were added to PVC solution at room temperature. The electrospinning process was operated at a flow rate of 0.5 mL/h with an applied voltage between 12 and 15 kV using e-spinning machine (eS-robot®) capable of generating voltages up to 30 kV at room temperature. The polymer solvent was injected from a syringe tip and

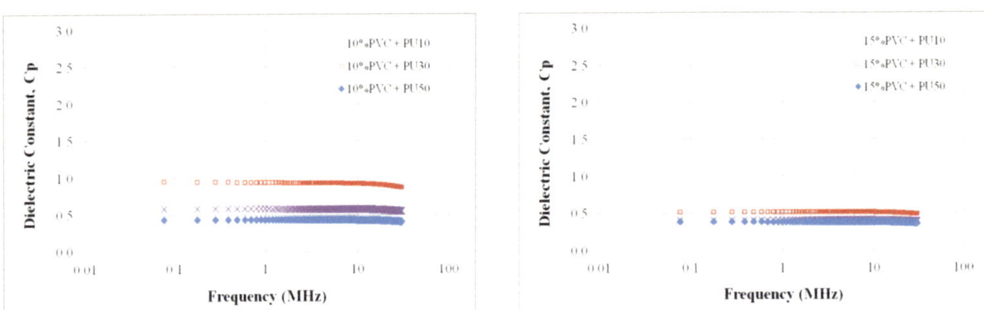

Figure 6. Dielectric properties of nanofibers prepared from PVC blending with PU at difference ratio [38].

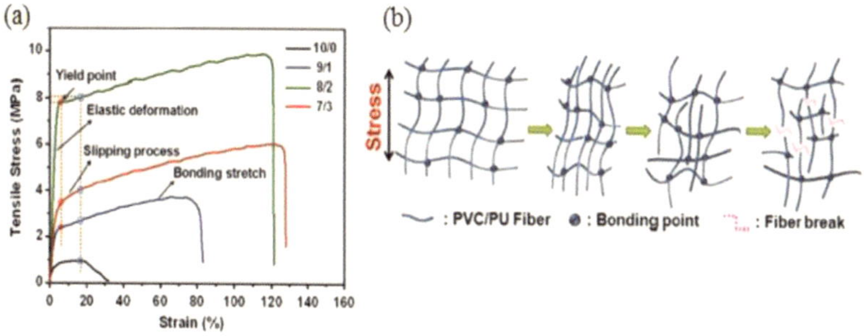

Figure 7. (a) Stress-strain curves of PVC/PU fibrous membranes fabricated from varied weight ratios and (b) a proposed three-step break mechanism of tensile fracture process upon external stress [37].

was deposited on an aluminum collector plate to form the electrospun mats with a thickness of around 0.1 mm.

The morphological structure of PVC/ENR nanofibrous membranes was changed by changing the ENR composition as shown in **Figure 8** [38]. PVC blending with ENR clearly shows that the beads on the fiber disappear as well as the blending of PVC with PU, where ENR composition can be attributed to higher viscosity and surface tension with increasing ENR composition. Viscosity and surface tension play an important role in determining the fiber formability and diameter.

The thermal properties of the PVC nanofibrous membrane and PVC blending with ENR nanofibrous membranes were determined using differential thermal analysis; the results are

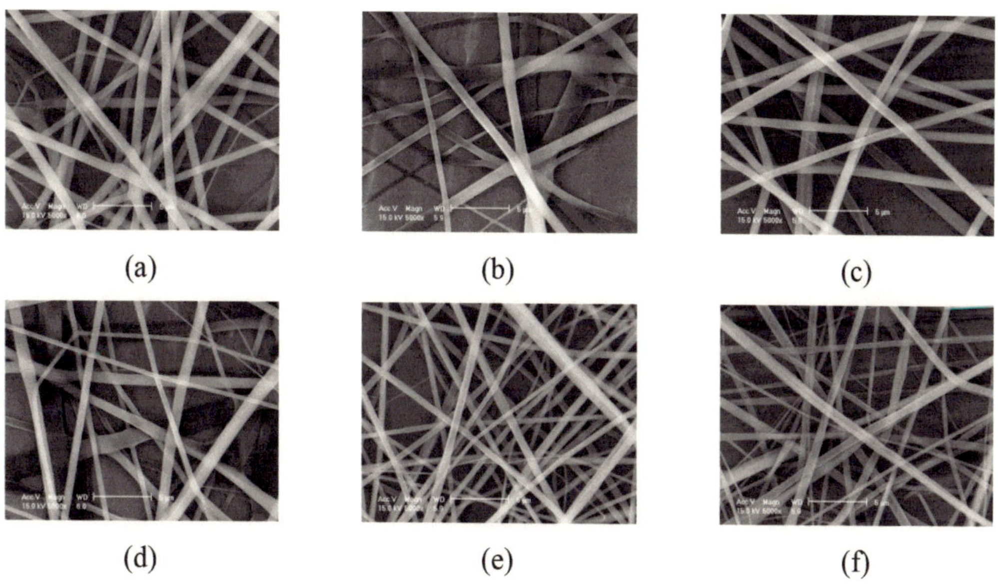

Figure 8. SEM images of electrospun nanofibrous membrane from PVC blending with ENR (a) 0.7, (b) 1.0, (c) 2.2, (d) 3.0, (e) 4.0, and (f) 5.0 by wt.

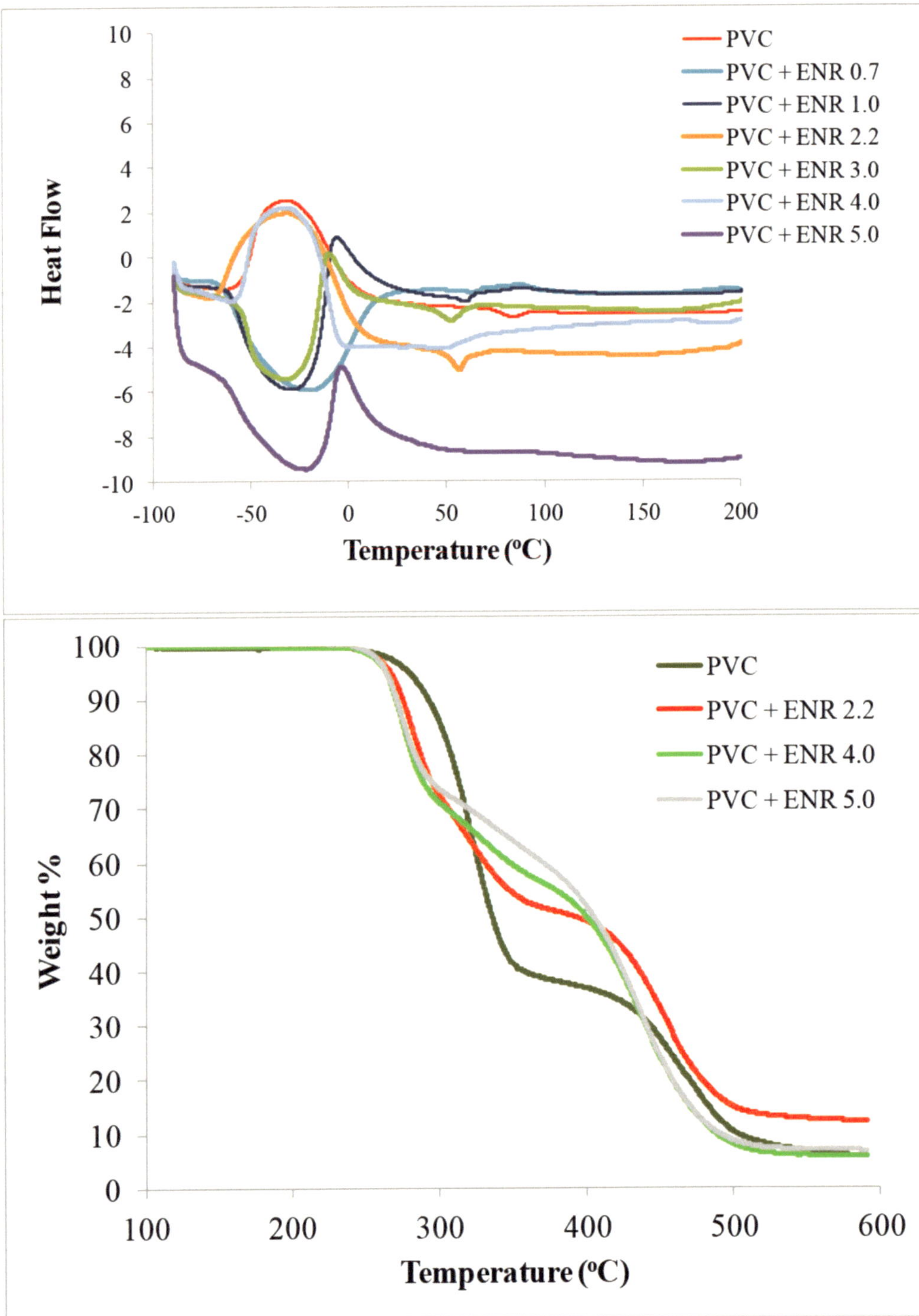

Figure 9. DTA and TGA thermogram of PVC and PVC/ENR blend nanofibrous membrane.

illustrated in **Figure 9** [38]. The PVC nanofibrous membrane, which was prepared from 10 wt% of PVC, is rigid than that from 15 wt% of PVC. The low concentration of PVC solution (10 wt%) shows a high exothermal peak at 30 and 275°C, while a higher concentration of PVC solution (15 wt%) shows an endothermal peak. The thermal properties of nanofibrous membranes were improved after mixing PVC with ENR. The mixture of PVC with high ENR content gave more flexible nanofibrous membranes with a lower glass transition temperature (T_g). For higher concentration of PVC solution (15 wt%) with high viscous not easily to mixed with ENR solution as found in PVC/PU nanofiber. The nanofiber spun from higher concentration of PVC solution blended with rubber solution, which was a heterogeneous solution, needs to be optimized based on the electrospinning condition.

The thermal stability of the nanofibrous membrane was determined using thermogravimetric analysis. The thermogram of the PVC and PVC/ENR blends as shown in **Figure 9** has two obvious decompositions, which is basically according to the reported TGA curve. The TGA curve of the PVC/ENR nanofibrous membrane clearly indicated that the blending of ENR with PVC enhances good thermal stability of the nanofibrous membrane.

3. Conclusions

The nanofibers of PVC and PVC blending with rubber can be prepared from electrospinning method. The electrospinning condition with an applied voltage of 12kV and different distance between the syringe tip and the aluminum collector plate were optimized. The blending of PVC with rubber using either PU or ENR smoothens nanofiber and enhances desired properties of nanofibrous membrane.

Acknowledgements

A research grant from Higher Education Research Promotion and National Research University Project of Thailand, Office of the Higher Education Commission, are gratefully acknowledged. The author would like to thank the School of Nano and Advanced Materials Engineering, Gyeongsang National University, for supportingelectrospinning machine.

Author details

Wirach Taweepreda

Address all correspondence to: wirach.t@psu.ac.th

Department of Materials Science and Technology, Faculty of Science, Prince of Songkla University, Hat-Yai, Songkhla, Thailand

References

[1] Work WJ, Horie K, Hess M, Stepto RFT. Definitions of terms related to polymer blends composites and multiphase polymeric materials. Pure and Applied Chemistry. 2004;**76**:1985-2007. DOI:10.1351/pac200476111985

[2] Lyngaae-Jorgensen J, Utracki LA. Structuring polymer blends with bicontinuous phase morphology. Part II. Tailoring blends with ultralow critical volume fraction. Polymer. 2003;**44**:1661-1669. DOI: 10.1016/S0032-3861(02)00867-4

[3] Saowaroj C, Thongchai N, Onusa S, Anyaporn B. Preparation and characterization of epoxidized natural rubber and epoxidized natural rubber/carboxylated styrene butadiene rubber blends. Journal of Metals Materials and Minerals. 2015;**25**(1):27-36. DOI: 10.14456/jmmm.2015.4

[4] Sadequl AM, Ishiaku US, Poh BT. The effect of accelerator/sulphur ratio on the scorch time of epoxidized natural rubber. European Polymer Journal. 1998;**34**(1):51-57. DOI: 10.1016/S0014-3057(97)00067-0

[5] Poh BT, Ismail H, Quah EH. Fatigue, resilience and hardness properties of unfilled SMR L/ENR 25 and SMR L/SBR blends. Polymer Testing. 2001;**20**(4):389-394. DOI: 10.1016/S0142-9418(00)00048-9

[6] Victor JC, Peter XM. Nano-fibrous poly(l-lactic acid) scaffolds with interconnected spherical macropores. Biomaterials. 2004;**25**(11):2065-2073. DOI: 10.1016/j.biomaterials.2003.08.058

[7] Okwu UN, Okieimen FE. Preparation and properties of thioglycolic acid modified epoxidized natural rubber and its blends with natural rubber. European Polymer Journal. 2001;**37**(11):2253-2258. DOI: 10.1016/S0014-3057(01)00112-4

[8] Ismail H, Shaari SM, Othman N. The effect of chitosan loading on the curing characteristics mechanical and morphological properties of chitosan-filled natural rubber (NR), epoxidised natural rubber (ENR) and styrene-butadiene rubber (SBR) compounds. Polymer Testing. 2011;**30**(7):784-790. DOI: 10.1016/j.polymertesting.2011.07.003

[9] Ismail H, Suzaimah S. Styrene butadiene rubber/epoxidized natural rubber blends: Dynamic properties, curing characteristics and swelling studies. Polymer Testing. 2000;**19**(8):879-888. DOI: 10.1016/S0142-9418(99)00058-6

[10] Chantara TR. Enhancement of PVC/ENR blend properties by electron beam irradiation: Effect of stabilizer content and mixing time. Polymer Testing. 2002;**21**(1):93-100. DOI: 10.1016/S0142-9418(01)00054-X

[11] Chantara TR, Khairul Z. Stabilization of poly(vinyl chloride)/epoxidized natural rubber (PVC/ENR) blends. Polymer Degradation and Stability. 1999;**65**(1):99-105. DOI: 10.1016/S0141-3910(98)00223-7

[12] Chantara TR, Khairul Z. Enhancement of polyvinyl chloride (PVC)/epoxidised natural rubber (ENR) blend properties by electron beam irradiation: Effect of antioxidants. Polymer Degradation and Stability. 1999;**65**(3):481-490. DOI: 10.1016/S0141-3910(99)00039-7

[13] Hong-Ryun J, Wan-Jin L. Electrochemical characteristics of electrospun poly(methyl methacrylate)/polyvinyl chloride as gel polymer electrolytes for lithium ion battery. Electrochimica Acta. 2011;**58**:674-680. DOI: 10.1016/j.electacta.2011.10.015

[14] Keunhyung L, Bongseok L, Chihun K, Hakyong K, Kwanwoo K, Changwoon N. Stress-strain behavior of the electrospun thermoplastic polyurethane elastomer fiber mats. Macromolecular Research. 2005;**13**(5):441-445. DOI: 10.1007/BF03218478

[15] Travis JS, Horst A von R. Electrospinning: Applications in drug delivery and tissue engineering. Biomaterials. 2008;**29**(13):1989-2006. DOI: 10.1016/j.biomaterials.2008.01.011

[16] Royal K, John F, Gary T. The use of AC potentials in electrospraying and electrospinning process. Polymer. 2004;**45**(9):2981-2984. DOI: 10.1016/j.polymer.2004.02.056

[17] Lisa W, Wei S, Yen BT. Investigation of electrospun and film-cast PVC membranes incorporated with aliquat 336 for efficient Cd extraction: A comparative study. Journal of Applied Polymer Science. 2011;**121**(1):327-335. DOI: 10.1002/app.33586

[18] Chunxue Z, Xiaoyan Y, Lili W, Yue H, Jing S. Study on morphology of electrospun poly(vinyl alcohol) mats. European Polymer Journal. 2005;**41**(3):423-432. DOI: 10.1016/j.eurpolymj.2004.10.027

[19] Xinsheng Z, Xiaoshan J, Si C, Kai W, Shuailong M, Li-Juan F. Preparation of high strength ultrafine polyvinyl chloride fibrous membrane and its adsorption of cationic dye. Journal of Polymer Research. 2010;**17**(6):769-777. DOI: 10.1007/s10965-009-9368-6

[20] ChewSY, WenY, Dzenis Y, Leong KW. The role of electrospinning in the emerging field of nanomedicine. Current Pharmaceutical Design. 2006;**12**(36):4751-4770. DOI: 10.2174/138161206779026326

[21] Venugopal J, Ramakrishna S. Applications of polymer nanofibers in biomedicine and biotechnology. Applied Biochemistry and Biotechnology. 2005;**125**(3):147-157. DOI: 10.1385/ABAB:125:3:147

[22] Haiqing L, You-Lo H. Ultrafine fibrous cellulose membranes from electrospinning of cellulose acetate. Journal of Polymer Science Part B: Polymer Physics. 2002;**40**(18):2119-2129. DOI: 10.1002/polb.10261

[23] MacDiarmid AG, JonesWE Jr, Norris ID, Gao J, JohnsonAT Jr, Pinto NJ, Hone J, Han B, Ko FK, Okuzaki H, Llaguno M. Electrostatically-generated nanofibers of electronic polymers. Synthetic Metals. 2001;**119**(1-3):27-30. DOI: 10.1016/S0379-6779(00)00597-X

[24] Nandana B, Subhas CK. Electrospinning: A fascinating fiber fabrication technique. Biotechnology Advances. 2010;**28**(3):325-347. DOI: 10.1016/j.biotechadv.2010.01.004

[25] Darrell HR, Iksoo C. Nanometre diameter fibres of polymer, produced by electrospinning. Nanotechnology. 1996;**7**(3):216-223. DOI: 10.1088/0957-4484/7/3/009

[26] Qingbiao Y, Zhenyu L, Youliang H, Yiyang Z, Shilun Q, Ce W, Yen W. Influence of solvents on the formation of ultrathin uniform poly(vinyl pyrrolidone) nanofibers with electrospinning. Journal of Polymer Science Part B: Polymer Physics. 2004;**42**(20):3721-3726. DOI: 10.1002/polb.20222

[27] Keun HL, Hak YK, Young ML, Douk RL, Nak HS. Influence of a mixing solvent with tetrahydrofuran and N, N-dimethylformamide on electrospun poly(vinyl chloride) non-woven mats. Journal of Polymer Science Part B: Polymer Physics. 2002;**40**(19):2259-2268. DOI: 10.1002/polb.10293

[28] Keun HL, Hak YK, Young JR, Kwan WK, Sun WC. Mechanical behavior of electrospun fiber mats of poly(vinyl chloride)/polyurethane polyblends. Journal of Polymer Science Part B: Polymer Physics. 2003;**41**(11):1256-1262. DOI: 10.1002/polb.10482

[29] El-Refaie K, John ML, Jessica RW, Gary LB, Jamil AM, David GS, Gary EW. Electrospinning of poly(ethylene-co-vinyl alcohol) fibers. Biomaterials. 2003;**24**(6):907-913. DOI: 10.1016/S0142-9612(02)00422-2

[30] Krishnappa RVN, Desai K, Sung C. Morphological study of electrospun polycarbonates as a function of the solvent and processing voltage. Journal of Materials Science. 2003;**38**(11):2357-2365. DOI: 10.1023/A:1023984514389

[31] Hao F, Weidong L, Chyi-Shan W, Richard AV. Generation of electrospun fibers of nylon 6 and nylon 6-montmorillonite nanocomposite. Polymer. 2002;**43**(3):775-780. DOI: 10.1016/S0032-3861(01)00665-6

[32] Demir MM, YilgorI, Yilgor E, Erman B. Electrospinning of polyurethane fibers. Polymer. 2002;**43**(11):3303-3309. DOI:10.1016/S0032-3861(02)00136-2

[33] Yang QB, Li DM, HongYL, Li ZY, Wang C, Qiu SL, Wei Y. Preparation and characterization of a PAN nanofibre containing Ag nanoparticles via electrospinning. Synthetic Metals. 2003;**137**(1-3):973-974. DOI: 10.1016/S0379-6779(02)00963-3

[34] Luu YK, Kim K, Hsiao BS, Chu B, Hadjiargyrou M. Development of a nanostructured DNA delivery scaffold via electrospinning of PLGA and PLA-PEG block copolymers. Journal of Controlled Release. 2003;**89**(2):341-353. DOI: 10.1016/S0168-3659(03)00097-X

[35] Yoshimoto H, Shin YM, Terai H, Vacanti JP. A biodegradable nanofiber scaffold by electrospinning and its potential for bone tissue engineering. Biomaterials. 2003;**24**(12):2077-2082. DOI: 10.1016/S0142-9612(02)00635-X

[36] Taweepreda W, Phatcharasit K, Boonkerd K, Kim J-K. Preparation and properties of electrospun PVC nanofiber. Advanced Materials Research. 2013;**770**:193-196. DOI: 10.4028/www.scientific.net/AMR.770.193

[37] Wang N, Raza A, Si Y, Yu J, Sun G, Ding B. Tortuously structured polyvinyl chloride/polyurethane fibrous membranes for high-efficiency fine particulate filtration. Journal of Colloid and Interface Science. 2013;**398**:240-246. DOI: 10.1016/j.jcis.2013.02.019

[38] Taweepreda W, Phatcharasit K, Boonkerd K, Kim J-K. Electrospun epoxidized natural rubber with poly(vinyl chloride) (ENR-PVC) nanofibrous for PEMFC applications. Advanced Materials Research. 2014;**844**:507-510. DOI: 10.4028/www.scientific.net/AMR.844.507

[39] Taweepreda W, Phatcharasit K, Boonkerd K, Kim J-K. Electrospun condition of poly(vinyl chloride) nanofibrous membrane plasticized with polyurethane. Jurnal Teknologi. 2014;**68**(5):59-62. DOI: 10.11113/jt.v68.3032

Section 2

Elastomers for Natural and Medical Applications

Chapter 7

Castor Oil Polyurethanes as Biomaterials

Said Arévalo-Alquichire and Manuel Valero

Additional information is available at the end of the chapter

http://dx.doi.org/10.5772/intechopen.68597

Abstract

Medical application of polyurethane (PU) elastomers has been contributed significantly to the quality and effectiveness of health care systems. Applications such as nasogastric catheter, sutures, wound dressing, drug delivery system, and insulation on the leads of electronics pacemakers are already in the market. Properties of polyurethanes such as thermal and chemical stability, mechanical performance, and low degradation rate make an outstanding material for that kind of application. More recently, castor oil as polyol source has been investigated due to the needs of friendly environmental sources. In this chapter, we want to approach the advances in polyurethane area from castor oil for application in medicines. A review of castor oil-based polyurethanes, chemical modification, processing techniques, and applications in tissue engineering and medical devices will be made with the objective to understand the current situation, limitations, challenges, and perspective of this type of material.

Keywords: castor oil, polyurethane, biomaterial, elastomer, medicine

1. Introduction

Polyurethanes are an important class of polymers which are used in a surprising array of commercial applications [1]; this special group of heterochain polymers is characterized by the presence of urethane groups, which are esters of carbamic acid, but they are not containing primarily those groups [2]. The polyurethanes include those polymers containing a plurality urethane groups in the molecule backbone, regardless of chemical composition of the rest of the chain. A type of polyurethane may contain aliphatic and aromatic hydrocarbons, esters, ethers, amides, urea, and isocyanurate groups [1]. In **Figure 1**, we can see the variety of functional groups in a polyurethane structure from castor oil and methylene diphenyl diisocyanate against the structure of polyethylene glycol.

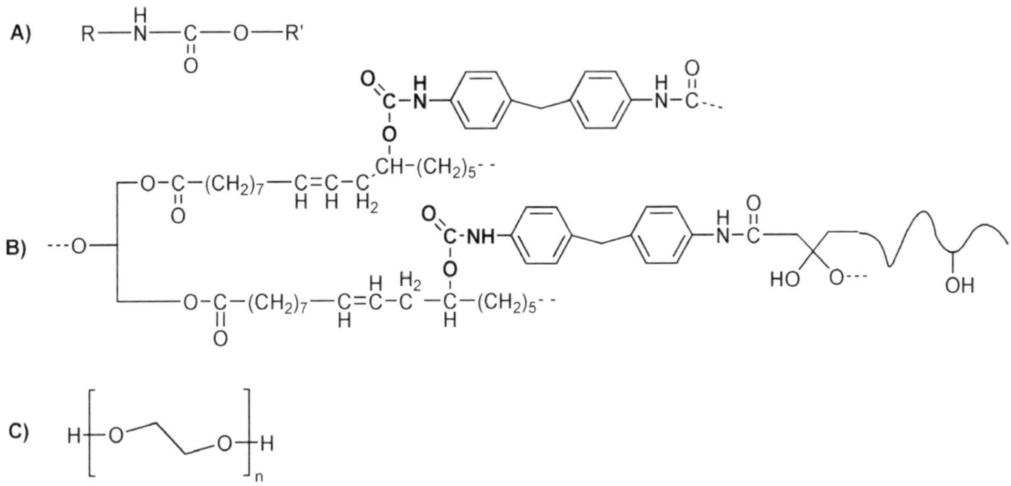

Figure 1. (A) Basic structure of urethane bond. (B) Structure of polyurethane synthetized by the reaction between modified castor oil and diphenylmethane diisocyanate. (C) Structure of polyethylene glycol. Polyurethane structure shows the variety of functional groups in its structure.

The use of polyurethane in the medical field has been documented since 1965 [3]; these products range from nasogastric catheters to insulation on electronic pacemakers. The versatility of being rigid, semirigid or flexible, biocompatible, and hydrolytic stable has abrasion resistance, good physical strength, thermoplasticity or thermosetting and also polyurethanes have been described to resist gamma radiation, oil, bases, and acid. These are some properties which made this type of polymer as an important resource in the medical devices industry [1].

However, current researches have focus on the design of biodegradable formulation for the tissue and organ engineering, drug delivery, and resorbable implants. Biodegradable polyurethanes have a set of design requirements that include the use of biocompatible monomer, tissue-like mechanical response, bioactivity, appropriate degradation rate, and so on [3]. The strong structure-properties relationship of this kind of materials is a good characteristic that allows structure modification to include biodegradable and cell response linkage or functional groups to accommodate the designed criteria.

In this way, the use of castor oil as polyol source has been proved outstanding because it is a biocompatible and hydrophobic monomer which is composed of 90% ricinoleic fatty acid—the unique natural fatty acid with hydroxyl groups in its chain [4]. This characteristic made the polyurethane synthesis easy and also, the reported noncytotoxic activity, elastomeric behavior, and tunable mechanical properties can mimic tissue performance.

According to all information above, in this chapter, we are going to study the recent advances in medical applications of castor oil polyurethanes; first, a brief review of the chemistry of polyurethanes to understand the qualities of these outstanding materials is discussed. Second, an overview of castor oil's applications and advantages is discussed. Next, a review of castor oil chemical modification, filler agents, and scaffold fabrication techniques is discussed. At the end of the chapter, a reflection about perspectives and new challenges of castor oil polyurethanes applications in medicine is described.

2. Chemistry of polyurethanes

The chemistry of urethanes makes use of the reaction of organic isocyanates with compounds containing active hydrogens. For example, hydroxyl groups, amines, and water react to produce urethane linkage, urea groups, and carbon dioxide, respectively.

It is possible to synthesize the urethane groups by various methods, but the most popular one is the reaction between an isocyanate and an alcohol as shown in **Figure 2**. This route was first reported by Wurtz in 1849. But in 1937, Dr. Otto Bayer synthesized the first polyurethane from a diisocyanate and polyester through a polyaddition reaction where the reaction product is exclusively the polymer [2].

As mentioned previously, polyurethanes are an extremely versatile group produced in wide range of densities, cross-link densities, and stiffness. For that reason, the polyurethanes are classified in two main groups: elastic polyurethanes (e.g., flexible foams, elastomers, coating, and adhesive) and rigid polyurethanes (e.g., rigid foams, structural foams, and wood substitutes). This classification is based on the polyol structure. The molecular weight and the number of hydroxyl groups/mol or polyol functionality are two important properties for the arrangement of the structure. Low functional polyol (2–3 hydroxyl groups/mol) and high molecular weight (over 2000 daltons) produce an elastic polyurethane; on the contrary, low molecular weight and high functionality produce a rigid cross-linked polyurethane [1, 2].

Because urethane groups can generate hydrogen bounds, they form a *hard segment* and the rest of the polyol chain is a *soft segment*. The increase in functionality also increases the hydrogen bound interaction and reduces the soft segment mobility, and further reduces the flexibility of the polymer. In **Figure 3**, a representation of the arrangement of each elastic polyurethane and rigid polyurethane has been shown.

Polyurethanes can be prepared by mixing diisocyanates with liquid diol or polyol at fixed NCO/OH ratio being casted in a mold. Isocyanate-terminated prepolymer can be made by mixing with excess of diisocyanate, to further addition of low molecular weight molecules to chain extension. This molecule is usually called *chain extender*. A brief explanation of these components is discussed in the following sections.

2.1. Polyol

Polyols are reactive substances containing at least two reacting groups. These compounds can be hydroxyl or amine-terminated. There are four types of polyols.

2.1.1. Polyether polyols

Polyether polyols are the products of reaction between a simple molecule called initiator and an alkylene oxide. The hydroxyl-terminated polyether polyol functionality depends on the

$$R-N=C=O + HO-R' \longrightarrow R-NHCOO-R'$$

Figure 2. Scheme of reaction between isocyanate and alcohol.

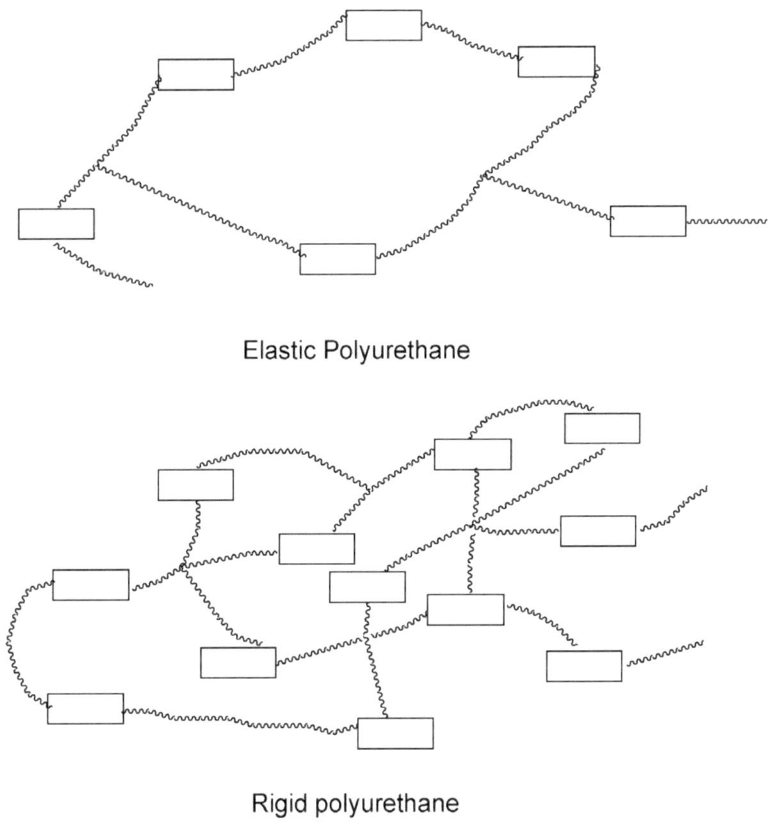

Figure 3. Scheme of elastic and rigid polyurethanes arrangement of hard and soft segments. Adapted from Ref. [2].

functionality of the initiator. If the initiator is a diol, the product will have a functionality of two and, in the same way, if the initiator is a triol, the resultant functionality will be of three [1].

This type also includes active hydrogen-containing polymers such as acrylonitrile or styrene, and also polyureas prepared from the reaction of polyisocyanates and diamines, or polyurethanes prepared by the reaction of polyisocyanates and polyalkanol amines such as triethanol amine. Polyether polyols produce high quality polyurethane foams and elastomers [1]. **Figure 4** shows some structures of reported polyether polyols.

2.1.2. Polyester polyols

Figure 5 shows some structures of polyalkylene glycol esters. They are prepared by condensation polymerization of alkylene glycol and the corresponding diester or diacid.

Vegetable oil-based polyols are included in this category; however, these are produced in most cases by reacting saturated bonds to add hydroxyl groups. Castor oil is a natural vegetable oil-based polyester, and more related information is given in the next section [1].

http://dx.doi.org/10.5772/intechopen.68597

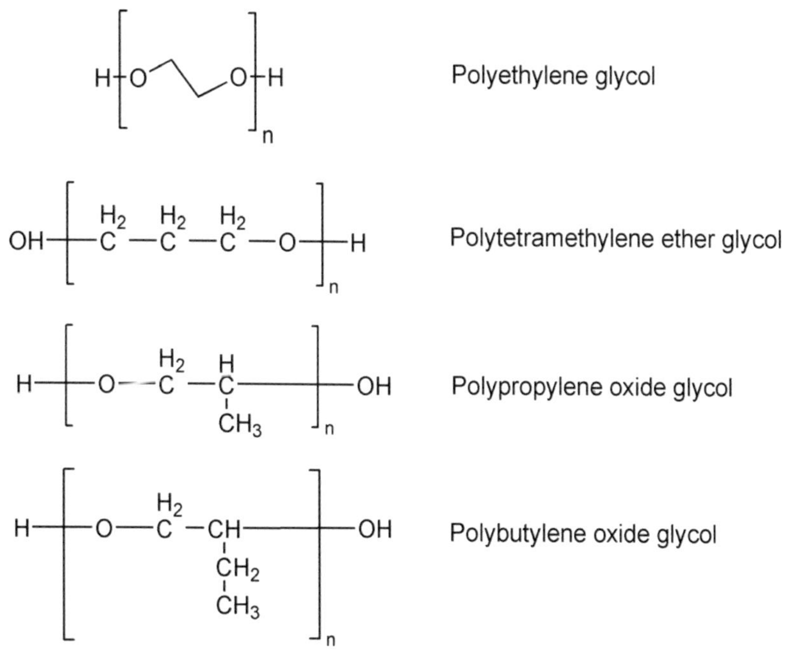

Figure 4. Examples of polyether polyols.

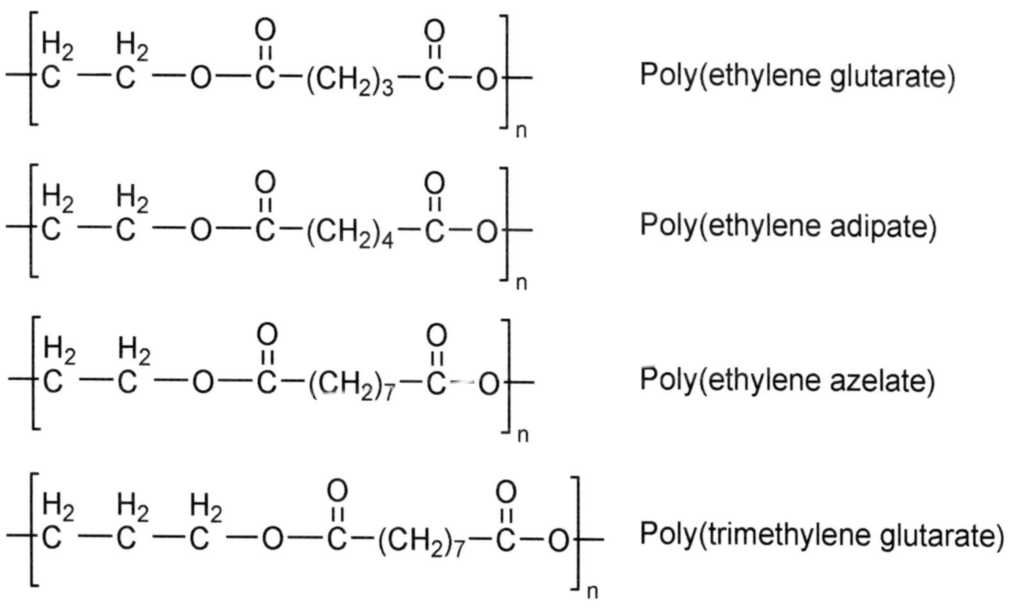

Figure 5. Examples of polyesters polyols.

2.1.3. Polycarbonates

Polycarbonates are prepared by condensation of phosgene or alkylene glycol carbonates. This chemistry is controlled to ensure that resultant contain terminal hydroxyl groups [1]. Polycarbonates-based polyols are characterized by high polarity and strong carbonate bond expecting polyurethanes with good mechanical properties and phase separation [5]. Also, more stability has been observed toward oxidative and hydrolytic biodegradation than polyesters polyols [6]. However, production cost of polycarbonate-based polyols is greater than polyether and polyester. Polycarbonates synthesis is strongly dependent on petroleum. But recently, a new generation of polycarbonates from carbon dioxide has been studied [7].

One important property for polyol analysis is called the hydroxyl value. The quantification of that is described in ASTM D1957 standard.

2.2. Isocyanate

Isocyanate is an important and high reactive organic functional group which is involved in a surprisingly diverse range of chemical reactions. Commercially available organic isocyanates include aliphatic, cycloaliphatic, araliphatic, aromatic, and heterocyclic polyisocyanates [1]. Some examples are presented in **Figure 6**.

The high reactivity of this group can be explained by the electron density among atoms. **Figure 7** shows that how oxygen, which has the highest electron density, has negative charge and carbon with the lowest density has positive charge. Nitrogen is an intermediate negative charge [2].

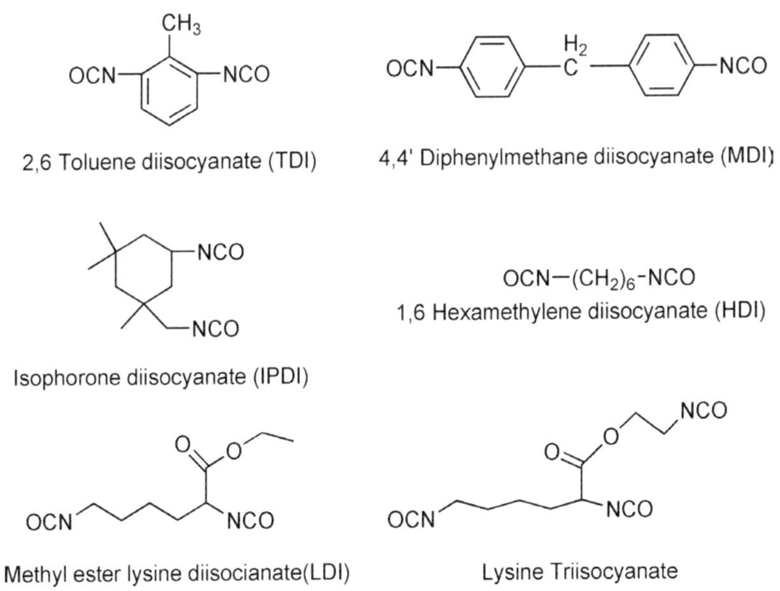

Figure 6. Some commercially available diisocyanates.

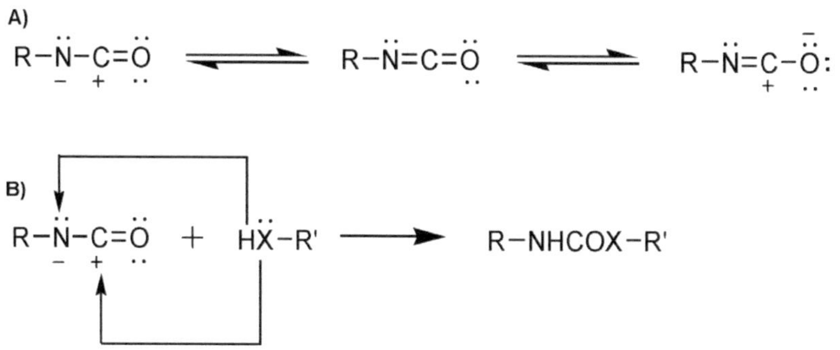

Figure 7. (A) Electron density of isocyanate group, and (B) nucleophilic reaction between isocyanate and active hydrogen compounds. Adapted from Ref. [2].

The reaction between an isocyanate group and active hydrogen is an addition at the carbon-nitrogen double bond. The nucleophilic center of the active hydrogen compounds attacks the electrophilic carbon and the hydrogen is added with the nitrogen atom [2] (**Figure 7**).

As shown in **Figure 6**, there are a lot of commercial isocyanates. However, 4,4-diphenylmethane diisocyanate (MDI) and 2,6-toluene diisocyanate (TDI) are the mostly used compounds in the polyurethane industry [1]. Mechanical performance of MDI- or TDI-based polyurethanes is superior against aliphatic diisocyanates and also, they are more reactive because of their aromatic structure. Despite of this, aromatic structures are not desired in materials for medical applications due to the fact that they could produce toxic degradation products. Recent researches are more focused toward aliphatic diisocyanates for polyurethane synthesis with medical purpose [8].

2.3. Chain extenders

They are low molecular weight reactants that produce familiar elastomeric behavior of the polyurethanes. Chain extenders have molecular weight in the range of 40–300 daltons and can be classified as hydroxyl-terminated or amino-terminated. However, only dysfunctional compounds are considered as chain extender; higher functionalities are considered as crosslinkers [1]. Introduction of those compounds allows to build block copolymer with alternating blocks of hard and soft segments [3].

Introduction of high hard-segment content in the polyurethane backbone determinates the final properties and performance of the polymer. Hard-segment content controls the mechanical properties such as modulus and ultimate strength, in addition to the thermal and hydrolytic stability of the finished product [1].

2.4. Catalyst

Polyurethane synthesis can be catalyzed in the form of any chemical reaction. A catalyst is a material that affects the rate of a reaction but emerges from the reaction unchanged. The catalyst can be considered as the controlling agent of the reaction. It is widely used in urethane foam production; its primary job is to provide the desired reaction profile, which can be measured by cream, rise, gel, and tack-free times to obtain the desired properties of the foam [1]. The science of the catalytic process is out of range in this chapter; please see reference [1] to enhance your knowledge about the effect of catalyst in the process of polyurethane manufacturing.

Catalysts in polyurethanes are tertiary amines and some of them are organotin, lead, and mercury organometallics. Some tertiary amines used as catalyst are: triethylenediamine, dimethylcyclohexylamine, bis(dimethylaminoethyl) ether, and N-methyl-N′-(2-dimethylaminoethyl)-piperazine. And some commonly used as organotin are dibutyltin dilaurate, stannous octoate, and tin(II) 2-ethylhexanoate.

To summarize, chemical composition of polyurethanes is a factor that determinates the final properties response such as mechanical properties, abrasion resistant, and hydrolytic degradation. Also, as we can observe that the variety of sources opens infinite possibilities to design new and alternative polyurethanes which show the important role of this type of materials in the medical field and the potential to be applied in new fields to improve the health system.

3. Castor oil polyol

Castor oil is extracted from seed of *Ricinus communis*. Its chemical composition has been well described by many authors in the past. All of them agree that around 90% is ricinoleic acid triglyceride [9, 10]. In **Figure 8**, we can observe the distribution of compounds in castor

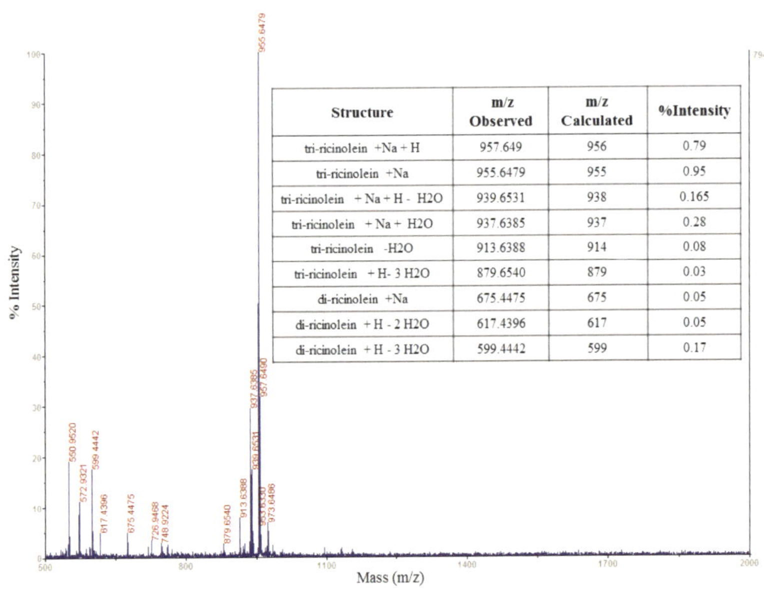

Structure	m/z Observed	m/z Calculated	%Intensity
tri-ricinolein +Na + H	957.649	956	0.79
tri-ricinolein +Na	955.6479	955	0.95
tri-ricinolein + Na + H - H2O	939.6531	938	0.165
tri-ricinolein + Na + H2O	937.6385	937	0.28
tri-ricinolein -H2O	913.6388	914	0.08
tri-ricinolein + H- 3 H2O	879.6540	879	0.03
di-ricinolein +Na	675.4475	675	0.05
di-ricinolein + H - 2 H2O	617.4396	617	0.05
di-ricinolein + H - 3 H2O	599.4442	599	0.17

Figure 8. Mass spectroscopy of castor oil. Adapted from Ref. [9].

Component	Composition
Ricinoleic acid Triglyceride	77%
Ricinoleic acid di and monoglyceride	14%
Minor components	9%

Table 1. Chemical composition of castor oil determinate by mass spectrometry.

oil obtained from mass spectrometry. In **Table 1**, a summary of the results forming mass spectrometry is shown. Only 9% of the sample is having minor components such as stearic and palmitic acid.

Ricinoleic acid is the unique naturally occurring fatty acid with hydroxyl functional group in its structure. The abundance of castor oil makes it as friendly environmental alternative for polyurethane synthesis. The results of polyurethanes are characterized by being flexible and elastomeric due to the long fatty acid chain and these results lead polyurethanes as a thermosetting material because of their trifunctional functionality nature [10].

Recent studies have shown the potential of castor oil in polyurethane industry applications such as flame retardancy foams [11, 12], adhesives [13, 14], surface coating [15, 16], and, of course, biomaterials [8, 17]. For example, castor oil was used for the synthesis of rigid foams with flame retardancy properties; castor oil was alcoholyzed with glycerol and epoxidation of glyceride castor oil was then carried out to react with diethyl phosphate to get the flame retardancy polyols. Improvement in mechanical strength, thermal stability, and limiting oxygen index at low concentration of phosphate compounds was observed [12]. Also, Zhang et al. [11] used castor oil-based glycerides from transesterification reaction with glycerol and pentaerythritol condensed with phthalic anhydride and ammonium polyphosphate as filler flame retardancy agent. Mechanical properties and thermal stability during pyrolysis were enhanced to obtain better results than the commercially available polyol PS-3152-based polyurethane foams.

In the adhesive field, Moghadam et al. [13] developed a group of castor oil polyester polyols by condensation with difference carboxylic acid. These green adhesive showed superior bonding properties like wood to wood specimen and also, some of them showed good chemical resistance. Polyurethane adhesive based on modified castor oil through transesterification reaction with pentaerythritol was reported; this type of adhesive showed better response with regard to properties like physico-mechanical and anticorrosive with the modification, also improvement in properties was observed with the increase of pentaerythritol in the structure [14].

Above examples of castor oil application demonstrated the flexibility of this source; authors used the advantage of hydroxyl groups for urethane formation and also, they made additional modification to enhance polymer performance. In castor oil structure, there are other functional groups susceptible for chemical modification through easy organic and well-reported reactions. In **Figure 9**, we can observe the structure of castor oil highlighting three important parts: ester bond, double bound, and hydroxyl group.

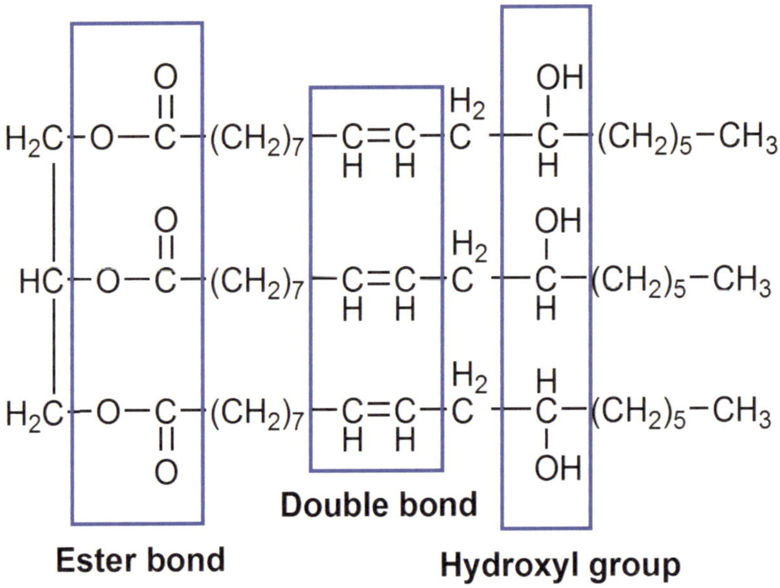

Figure 9. Chemical structure of castor oil. Highlighted are the three main groups related with many chemical reactions. Those groups have been modified by several authors to add desired functional groups and enhance performance of materials.

One of the commons reactions, which has been studied involving the characteristic ester bond of triglyceride, is the transesterification reaction. This type of reaction is widely described in biodiesel manufacturing process where a vegetable oil reacts with a low molecular weight alcohol such as methanol and ethanol. Instead of long linear chains for fuel applications, the aim of castor oil transesterification is to increase the hydroxyl group content, branching of polyols, and finally to increase the hard segments content and network formation of polyurethane structure. For those reasons, polyalcohol compounds such as glycerol, pentaerythritol, and triethanolamine are used in this field.

Application of glycerol in transesterification showed an enhancement on tensile properties, hardness, and chemical attack resistance due to higher increase of cross-linking density. Published works demonstrated an increase of hydroxyl value in glycerides of castor oil [18]. As mentioned above, more hydroxyl groups produce more urethane bonds and these are related with the hydrogen interaction determining cross-linking density. Also, products of transesterification generate more branched polyurethanes which tend to form network arrangements. However, those polymers characterize by low thermal stability [9]. Other works synthesized polyurethanes from modified castor oil obtained by transesterification reaction. Those materials exhibited better mechanical response and thermal stability against glyceride castor oil derivatives. Pentaerythritol by-products have higher hydroxyl value and also higher cross-linking density [18] which is related with an improvement of the performance of the material. In the same way, Dave and Patel [10] use triethanolamine as polyhydroxyl agent in transesterification for interpenetrating networks.

There are few instances that reported the involvement of the saturated bond in castor oil [12]. However, many reactions such as oxidation, polymerization, hydrogenation, epoxidation, halogenation, sulfonation, and other addition reactions could carry out. Additionally, many authors have reported the use of filler agents to enhance some properties [8, 11, 15, 19, 20], these agents are physically bonded with the castor oil and they do not react.

Exploring the chemical structure of the castor oil, we can realize the potential of this source for innovation in the polyurethane field. With the biocompatibility character, it seems obvious why it has excelled in the biomedical field. In the next section, we will study some described approaches of castor oil-based polyurethanes and their derivatives.

3.1. Medical applications

As mention above, medical applications of polyurethanes have contributed significantly to the quality of the care system. More recently, the study of biodegradable formulations for tissue and organ engineering and resorbable implants has been outperformed.

In the context of castor oil, several researches have been carried out studying applications such as wound healing, bioadhesive, cardiac, and bone tissue engineering. To reach adequate degradation rate, mechanical performance, antimicrobial activity, cell-biomaterial interaction, cell viability, and proliferation techniques such as composite reinforced, segmented polyurethane, and scaffold manufacturing are applied. Following a little review of some applied castor oil-based polyurethanes will present highlighting chemical modifications, composites, and scaffolds structures used to achieve desired properties.

Nguyen Dang et al. [21] developed a castor oil-segmented thermoplastic polyurethane with controlled mechanical properties mixing difference soft segments like castor oil, poly(tetrahydrofuran) polyol, and poly(dimethylsiloxane) polyol. They demonstrated that mechanical properties and flexibility of polyurethane can be significantly altered by both incorporated castor oil and poly(dimethylsiloxane)/poly(tetrahydrofuran) polyol ratio during the synthesis. Polyurethanes, also, exhibited a good biocompatibility in 3T3 fibroblast culture. In the same way, Miao et al. [22] synthetized smart polymers with shape memory using castor oil and polycaprolactone triol. Full shape recovery was reached at physiological temperature with various recovery speeds according to the polymer composition. These examples show the use of segmented polyurethanes or block copolymer to control mechanical behavior of materials.

Combining polyester with polyether is a good option for controlled behavior too. Not only mechanical behavior is affected by block copolymerization, adding hydrophilic monomer at reactive mixture could also modify the degradation. Ganji et al. [20] used polyethylene glycol with castor oil to improve degradation rate, because it is a hydrophilic polyol which increases the permeability in bulk as well as surface and makes the material more susceptible to degradation.

Though, there are many different ways to achieve criteria design. As we have already described, alcoholyzed castor oil is an interesting alternative to increase mechanical performance.

Products of transesterification of castor oil with glycerol and pentaerythritol have been used for medical purposes. Du et al. [17] worked with glyceride of castor oil to enhance the mechanical strength of hydroxyapatite/polyurethane scaffolds for bone tissue engineering. Compared to castor oil, the glyceride can provide more hydroxyl groups (from 155 to 288 of the hydroxyl value) resulting in an increase of the scaffold compressive strength (from hundreds of kPa to 4.6 MPa). Likewise, Li et al. [23] reported that the molecular modification of soft segments by glyceride of castor oil can increase the scaffold compressive strength by 48% and the elastic modulus by 96%. Those materials also demonstrated promising prospects for bone repair and regeneration. Also, using pentaerythritol for castor oil modification, Uscategui et al. [24] compared polyurethanes from transesterification reaction with castor oil-based polyurethanes. Though mechanical behavior was increased in alcoholyzed castor oil due to increase of cross-linking density, cell viability of those polymers had lower values than castor oil. But, the observed viability was above 70% which was comparable to commercial polypropylene suture used as positive control producing no toxic effects of materials on L-929 fibroblast.

Introduction of specific bounds or functional groups by chemical modification of castor oil is possible to regulate other properties rather than only mechanical stretch. For example, Sathiskumar et al. [25] had reacted castor oil with sebacic acid, citric acid, and D-mannitol by melt condensation polymerization to introduce ester linkages which are easily cleavable in aqueous medium. The contact angle measurement and hydration test results indicated that the surface of the polymer was hydrophilic and *in vitro* degradation of polymer in PBS solution carried out at physiological conditions indicated that the degradation goes to completion within 21 days. Physical, mechanical, and degradation can be tuned varying the curing conditions of the polyurethanes accordingly to the desired application in human body.

Similarly, Yari et al. [26] synthetized epoxy-terminated polyurethane prepolymer based on castor oil and glycidyltriethylammonium chloride (GTEAC) as a reactive bactericidal agent for wound dressing. Membranes could maintain for a long period the moist environment over the wounds with low exudates. Cytotoxicity analysis of samples against mouse L929 fibroblast and MCA-3D keratinocyte cells showed good cytocompatibility. The membrane containing 50% GTEAC exhibited an effective antibacterial activity while showing acceptable cytocompatibility.

Also, polyols from castor oil transesterification with glycoside starch obtained by glycosylation with ethylene glycol were used to synthetize polyurethanes type bioadhesives with methyl ester lysine diisocyanate (LDI) and isophorone diisocyanate (IPDI). It was determined that polyurethanes obtained from LDI have lower mechanical properties and adhesion, and a higher degree of cross-linking compared to homologs obtained from IPDI. The results of the biodegradable character (contact angle and weight loss) showed that these LDI polyurethanes materials had a greater hydrophobic character due to the presence of the ester bonds in the structure of the hard segment. It was also found that the polyurethanes obtained from LDI have lower cytotoxicity [27].

On the other hand, reinforced composite is another option, wide reported, to achieve design criteria. This process is based on adding some compounds with specific characteristic to get a desired response of materials. These components are not involved in reactions, but are physically bounded with the polyurethane matrix. **Figure 10** shows a method for composite polyurethanes preparation.

In that context, Uscategui et al. [24], additionally, studied the effect of polycaprolactone diol (PCL) incorporation into modified castor oil-based polyols, obtained from transesterification reaction with pentaerythritol. They studied the effect of three concentrations of PCL (5, 10, 20% w/w). Mechanical test shows performance enhancement in some polymers with PCL, however in some cases, largest concentration of PCL (20%) reduces by phase separation of soft and hard segments. At cell viability test, PCL concentration did not have any effect on modified castor oil-based polyols but 20% of PCL reduced the cell viability and lower concentration did not affect significantly in castor oil-based polyurethanes.

Also, Arevalo et al. [8] studied the effect of chitosan on the physico-chemical, mechanical properties, and biological activity on mixtures of PCL and polyurethanes obtained from castor oil. They studied three concentrations of chitosan (0.5, 1, 2% w/w) and two of PCL (5 and 10%). It was found that the incorporation of chitosan enhances the ultimate tensile strength of the polyurethanes and does not affect the strain at fracture in polyurethanes with 5% of PCL. Mechanical stretch suggested applications on the aorta and the skin. Water absorption was increased with the concentration of chitosan in the matrix due to the hydrophilic behavior of chitosan. The results of *in vitro* biocompatibility suggested that polyurethanes do not cause risks to cell viability and can be used in biomedical applications.

Nowadays, with the enhancement of nanotechnology, nanocomposites have become a new field to study in the biomaterials area. In this case nanoparticles can induce morphological and structural changes in the hard and soft segment, also some nanoparticles give additional bioactivities such as antibacterial [19] and cell adhesion or proliferation [28]. Like this, some studies related with bone and cardiac tissue engineering with castor oil-based polyurethanes have been published.

Li et al. [23], also, studied the effect of nanoparticles of hydroxyapatite (HA) into glycerides of castor oil, when nanoHA particles were incorporated into the matrix, the compressive

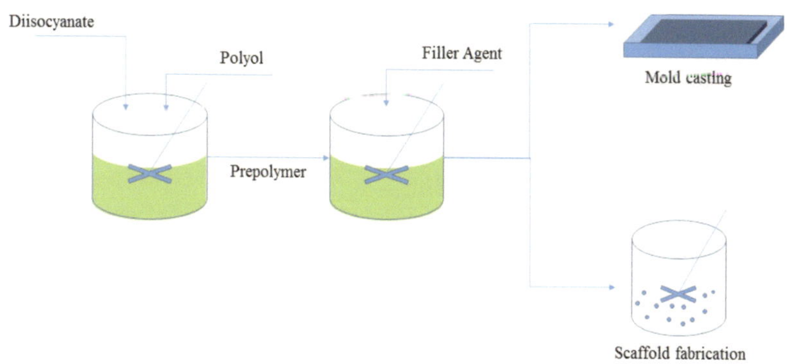

Figure 10. Scheme of composite polyurethane preparation.

strength and elastic modulus further increased by 49–74%, from 2.91 to 4.34 MPa, and from 95 to 165.36 MPa, respectively. Particles improved the interface bounding with the polyurethane and provided effective bioactivity for bonding with bone tissue.

Ganji et al. [20] fabricated nanocomposite scaffolds made of gold nanotubes/nanowires incorporated into biodegradable castor oil-based polyurethanes as alternative for cardiac patches to treat postmyocardial infarction. By incorporating gold nanotubes/nanowires into polyurethane scaffolds, the wired material structure can mimic the electromechanical properties of the myocardium. Cardiomyocyte adhesion and proliferation were strongly increased in response to electrical stimulation on composites. After 4 days of incubation and electrical stimulation on the scaffolds, cardiomyocytes on polyurethane scaffolds samples showed a more native morphology and enhanced proliferation compared to gold-free PU-0. A concentration of 50 ppm of nanoparticles induced optimum cell distribution and spreading, as well as the largest up-regulated expression levels of genes relevant to cardiac differentiation and hypertrophy. In another work, Ganji et al. [29] fabricated porous composite scaffolds for biomedical application based on gold nanotubes/nanowires and mixed with castor oil–polyethylene glycol-based polyurethanes. Addition of 50 or 100 ppm nanostructures had significant effects on thermal, mechanical, and cell attachment of fat-derived mesenchymal stem cells of polyurethanes. Higher cross-link density and better cell attachment and proliferation were observed in polyurethane containing 50 ppm. The results revealed that gold nanotubes/nanowires formed hydrogen bonding with the polyurethane matrix and improved the thermomechanical properties of nanocomposites, as compared with pure PU.

The word "scaffolds" has been mentioned several times in this chapter. Scaffolds are considered as three-dimensional porous structure with the main goal to provide appropriate base for tissue growth and cell proliferation [30]. A lot of techniques are well described in the literature. Some of them are particle leaching, solvent casting/particle leaching, thermally-induced phase separation, melt molding, gas foaming, emulsion freeze-drying, solvent casting/solvent evaporation, electrospinning, 3D printing, and rapid prototyping. However, few of these methods can be applied to castor oil-based polyurethanes because of the cross-linking and thermoset nature of these types of materials, it is not possible to solubilize in any organic solvent so techniques that involve solvents like solvent casting/particle leaching, thermally-induced phase separation are discarded. Additionally, methods such as 3D printing and melt molding require thermoplastic behavior. The pore size of the scaffold should mimic the size of the specific type of cell accordingly with the application; also, pore interconnectivities are desired to allow fluid exchange and cell proliferation.

Methods like particle leaching, gas foaming, and 3D printed sacrificed mold are used to fabricate castor oil-based polyurethane scaffolds.

3.1.1. Particle leaching

Specified diameter particles are added to prepolymer solution; particles remain embedded throughout the polymer matrix after curing period. After immersion in water, particles are leached out leaving a porous structure (**Figure 11**). The shape and size of pores are directly determined by the shape and dimensions of the leachable particles used. Tablet salt, sugar,

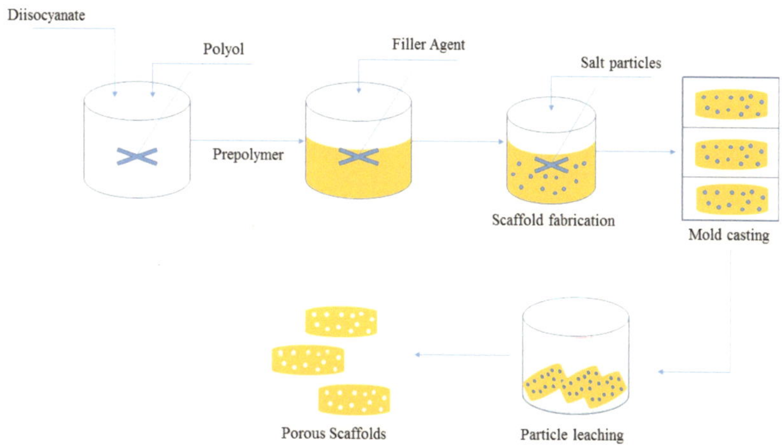

Figure 11. Scheme of particle leaching polyurethane scaffolds preparation.

ammonium chloride, sucrose, starch particles, gelatine and paraffin microsphere have been used. However, it is important to know that particles must be inert during all manufacturing process. Another parameter that influences the structure is the amount of particles added. If the salt content is insufficient, the polymer solution will surround the particles and isolated pores will appear. On the other hand, if the amount of salt added is too high, a deficient structure with voids will be formed due to close geometric packing [30].

This technique was used to obtain porous size between 355 and 600 μm using sieved tablet salt. Scaffolds show cell attachment of H9C2 cardiomyocyte cells [20] and fat-derived mesenchymal stem cells [29]

3.1.2. *Gas foaming*

Gas foaming is a process that is widely used in industry for the preparation of e.g., expanded polystyrene, polyvinyl chloride foams, but it can also be applied for the preparation of scaffolds. Foaming can be carried out by reacting the components or releasing the gas which is a product of the thermal degradation of the gas-foaming agent. In polyurethanes, water can be added to polyol and isocyanate mixture. Water reacts with an excess of isocyanate producing carbon dioxide which is the foaming agent. This technique is rarely used for the fabrication of scaffolds, because it is hard to control pore diameter and the average pore diameter is too large to allow adequate cell proliferation [30]. However, some authors reported the use of nonreactive foaming agents like cyclopentane [11]. It evaporates with the heat of urethane reaction creating the foam. Most of the applications of this method are related with bone tissue engineering. Like this, Du et al. [17] used gas foaming to develop their HA/glyceride of castor oil-based polyurethane scaffolds with an *in situ* reaction with water. Sixty percent of porosity and 500 μm of interconnectivity were reached with that technique. Also, Li et al. [23] used similar process for scaffold fabrication. They reached a porosity between 52 and 57%. They discussed that a porous structure with a pore size of approximately 200–800 μm and a porosity of 57% are appropriate for the growth of cells in bond tissue engineering.

3.1.3. 3D printed sacrificed mold

This technique required a 3D printer. In this technique, basically, a mold is printed with the desired characteristics of pore size and porosity. Then the prepolymers are casted in the mold allowing curing. After that, mold is removed by dissolving the printed mold with a compatible solvent.

One advantage of 3D printing is to directly print porous scaffolds with designed shape and interconnected porosity from a CAD file with a variety of materials such as ceramic, metallic, polymeric, and composites. And the use of sacrificed mold is an opportunity for thermoset polymers. Generally, this technique contain inherent difficulties to form small-sized tubular channels around 250 mm; the pores in the printed scaffold from the guided approach here are actually the shape of the extruded fiber, which can be readily adjusted by shaping the size and geometry of the nozzle [22].

In this context, Miao et al. [22] use 3D printed PLA scaffold serving as a sacrificial mold to get PCL-castor oil-based polyurethanes with shape memory and with a gradient pore structure to biomimic the tissue structure. The pores had a gradient distribution from the top to the bottom as the distance between pores increased from 240 to 560 mm. These authors used the term four-dimensional printing to include the additional shape memory property of their polyurethanes. Four-dimensional (4D) printing is an emerging new concept that refers to the ability of 3D printed objects to change form and/or function after fabrication, thereby offering additional capabilities and/or performance-driven applications. For example, hydrophilic materials have been utilized to 4D fabricate self-evolving structures that perform geometric folding, curling, expansion, and various other programmed shape changes after submersion in water

In this chapter, we have been discussing how the polyol affects the materials response; however it is important to dedicate a place to talk about isocyanates. There are a lot of commercial available isocyanates and the mostly used in polyurethane industry are MDI and TDI because they are high reactive aromatic structure and also, as we described, they produce superior mechanical response. Aromatic structures should be avoided in medical applications because they are related to toxic degradation products. Many studies have demonstrated carcinogenic potential of aromatic diamines under *in vivo* conditions; these diamines are characteristic degradation products of MDI and TDI [8]. In biomedical fields, there are many researches carried out with aliphatic isocyanates as an alternative, for example Valero et al. [31] compared the mechanical and degradation response of castor oil-based polyurethanes from three difference isocyanates: isophorone diisocyanate (IPDI), methyl ester lysine diisocyanate (LDI), and lysine triisocyanate (LTI). Mechanical stretch was similar in IPDI and LDI but higher in LTI-based polyurethanes, due to high functionality isocyanate, and degradation rate was higher in LDI-based polyurethanes because LDI introduces additional ester bound, as we mention, it is high susceptible to hydrolytic degradation.

It can be realized that many authors were introduced in each section of this chapter. It is because they used different methods to study the effect on several properties, so they can identify the right method for the better performance.

To summarize, castor oil-based biomaterials have been tested at *in vivo* models, especially for bone tissue regeneration. This means that castor oil-based polyurethanes achieve many requirements for real medical applications. It is another proof of the potential of this material in the biomaterials and tissue engineering field. Further, some reported examples of animal's model are presented.

Du et al. [17] used female sprague dawley rats with a defect in region of femoral condyle to test the osteogenesis of HA/glyceride of castor oil-based polyurethanes scaffolds. They conclude with the histological studies that the newly formed bone tissue had appeared around the scaffold and in the pores of scaffold. After 8 weeks, there were more new bone formations around and within the porous scaffold, and no adverse inflammatory response could be noticed. In contrast, the amount of new bone was more in cell-seeded scaffold than unseeded scaffold. Although the cell seeded group seemed to perform slightly better than unseeded group, both scaffolds showed good *in vivo* osteogenesis.

Li et al. [23] tested the *in vivo* osteogenesis of HA/glyceride of castor oil-based polyurethanes scaffolds too with New Zealand white rabbits with a defect in femoral condyles. New bone with high density had formed in the surface region, and a bone matrix and trabecula had also grown into the scaffold porous structure. The quantity of new bone at 24 weeks is substantially greater than the quantity of new bone at 12 weeks. New bone preferentially forms on the surface and subsequently grows inward following an osteoconductive pathway.

Nacer et al. [32] used male *rattus norvegicus albinus*, Wistar lineage submitted to bone defect filled with castor oil-based polymer. Three experimental groups were formed with (1) castor oil polymer containing only calcium carbonate; (2) castor oil polymer with calcium carbonate and doped with 5% of silica nanoparticles; and (3) castor polymer with calcium carbonate doped with 10% of silica nanoparticles. The results showed that there was bone growth in all the studied groups, with a greater tendency of growth in the group 1. After 30 days, all the groups presented similar results. After 60 days, a greater amount of fibroblasts, osteoblasts, osteocytes, and osteoclasts in group 3 were observed, with integrated activity of three kinds of cells involved in the bone activation-reabsorption-formation.

4. Conclusion and perspectives

In this chapter, we studied the chemistry of polyurethanes and we understood the versatility of these types of polymers and their importance in industry. Also, we reviewed the properties that made it an outstanding material for biomedical applications. Additionally, we explored the potential of castor oil-based polyols for biomaterials and tissue engineering applications. It is clear that castor oil have many qualities such as biocompatibility, biodegradability, naturally occurring hydroxyl groups, and easy chemical modification which call the attention for this particular field. However, it is clear too that medical devices and tissue engineering require multiples criteria design, as we observed in the review, from several points of view such as mechanical performance, adequate 3D structure, and adequate degradation rate accordingly with the application, nontoxic degradation products, specific hydrophilic and hydrophobic behavior, and so on. These requirements could be satisfied combining as many techniques of chemical modification, composite reinforced, scaffold manufacturing, and polyol-isocyanate composition, as need to achieve all the criteria designs for biomaterials. Also, new technologies advances, like 4D printing, are helping to solve the requirements.

Particularly, castor oil-based polyurethanes have a huge way for innovation in the biomaterials fields. As we saw, many of the applications are related with bone tissue engineering. New fields such as skin, vascular tissues, eyes, and connective tissues could explore and, also, for fabrication of medical devices, sutures, catheters, and wound dressing. Also, in the age of bioactive biomaterials, it is important to include the design criteria with additional properties like antibacterial activity which can increase the performance and the value of the material.

Author details

Said Arévalo-Alquichire and Manuel Valero*

*Address all correspondence to: manuel.valero@unisabana.edu.co

Universidad de La Sabana, Chía, Colombia

References

[1] Szycher M. Szycher's handbook of polyurethanes. 2nd ed. Boca Raton, FL: CRC Press; 2013.

[2] Ionescu M. Chemistry and technology of polyols for polyurethanes. 1st ed. Shrewsbury: Rapra Technology; 2005.

[3] Touchet TJ, Cosgriff-Hernandez EM. 1 – Hierarchal structure–property relationships of segmented polyurethanes. Advances in Polyurethane Biomaterials. 2016;3-22

[4] Medeiros AMMS, Machado F, Rubim JC. Synthesis and characterization of a magnetic bio-nanocomposite based on magnetic nanoparticles modified by acrylated fatty acids derived from castor oil. European Polymer Journal 2015;**71**:152-163. [Internet]. Elsevier Ltd. Available from: http://dx.doi.org/10.1016/j.eurpolymj.2015.07.023

[5] Jofre-Reche JA, García-Pacios V, Costa V, Colera M, Martín-Martínez JM. Role of the interactions between carbonate groups on the phase separation and properties of waterborne polyurethane dispersions prepared with copolymers of polycarbonate diol. Progress in Organic Coatings. Nov 2015;**88**:199-211. [Internet, cited 5-March-2017]. Available from: http://linkinghub.elsevier.com/retrieve/pii/S0300944015300412

[6] Fernández-d'Arlas B, Alonso-Varona A, Palomares T, Corcuera MA, Eceiza A. Studies on the morphology, properties and biocompatibility of aliphatic diisocyanate-polycarbonate polyurethanes. Polymer Degradation and Stability. Dec 2015;**122**:153-160. [Internet, cited 5-March-2017]. Available from: http://linkinghub.elsevier.com/retrieve/pii/S0141391015301221

[7] Orgilés-Calpena E, Arán-Aís F, Torró-Palau AM, Orgilés-Barceló C. Novel polyurethane reactive hot melt adhesives based on polycarbonate polyols derived from CO2 for the footwear industry. International Journal of Adhesion and Adhesives. 2016;**70**:218-224. [Internet, cited 5-March-2017]. Available from: http://linkinghub.elsevier.com/retrieve/pii/S0143749616301440

[8] Arevalo F, Uscategui YL, Diaz L, Cobo M, Valero MF. Effect of the incorporation of chitosan on the physico-chemical, mechanical properties and biological activity on a mixture of polycaprolactone and polyurethanes obtained from castor oil. Journal of Biomaterials Applicatons. 2016;**31**(5):708-720

[9] Valero MF, Pulido JE, Ramírez Á, Cheng Z. Polyurethanes sintetized of polyols obtained from castor oil modified by transesterification with pentaerythritol. Quimica Nova. 2008;**31**(8):2076-2082. [Internet. SBQ; cited 31-March-2014]. Available from: http://www.scielo.br/scielo.php?script=sci_arttext&pid=S0100-40422008000800031&lng=pt&nrm=iso&tlng=pt

[10] Dave VJ, Patel HS. Synthesis and characterization of interpenetrating polymer networks from transesterified castor oil based polyurethane and polystyrene. Journal of Saudi Chemical Society. 2017;**21**(1):18-24. [Internet, King Saud University]. Available from: http://dx.doi.org/10.1016/j.jscs.2013.08.001

[11] Zhang M, Pan H, Zhang L, Hu L, Zhou Y. Study of the mechanical, thermal properties and flame retardancy of rigid polyurethane foams prepared from modified castor-oil-based polyols. Industrial Crops and Products. 2014;**59**:135-143. [Internet, cited 17-Feburary-2016]. Available from: http://www.sciencedirect.com/science/article/pii/S0926669014002817

[12] Zhang L, Zhang M, Hu L, Zhou Y. Synthesis of rigid polyurethane foams with castor oil-based flame retardant polyols. Industrial Crops and Products. 2014;**52**:380-388. [Internet, Elsevier B.V.]. Available from: http://dx.doi.org/10.1016/j.indcrop.2013.10.043

[13] Moghadam PN, Yarmohamadi M, Hasanzadeh R, Nuri S. Preparation of polyurethane wood adhesives by polyols formulated with polyester polyols based on castor oil. International Journal of Adhesion and Adhesives. 2016;**68**:273-282.

[14] Valero MF, Gonzalez A. Polyurethane adhesive system from castor oil modified by a transesterification reaction. Journal of Elastomers and Plastics. 2012;**44**(5):433-442. [Internet, cited 23-March 2014]. Available from: http://apps.webofknowledge.com.ezproxy.unisabana.edu.co/full_record.do?product=UA&search_mode=Refine&qid=8&SID=1DukxKeCqDaN613PBmH&page=1&doc=8&cacheurlFromRightClick=no

[15] Valero MF, Ortegón Y. Polyurethane elastomers-based modified castor oil and poly(e-caprolactone) for surface-coating applications: synthesis, characterization, and *in vitro* degradation. Journal of Elastomers and Plastics. 2015;**47**(4):360-369

[16] Gite VV., Kulkarni RD, Hundiwale DG, Kapadi UR. Synthesis and characterisation of Polyurethane coatings based on trimer of isophorone diisocyanate (IPDI) and monoglycerides of oils. Surface Coatings International Part B Coatings Transactions. 2006;**89**(2):117-122

[17] Du J, Zou Q, Zuo Y, Li Y. Cytocompatibility and osteogenesis evaluation of HA/GCPU composite as scaffolds for bone tissue engineering. International Journal of Surgery. 2014;**12**(5):404-407. [Internet, Elsevier Ltd]. Available from: http://dx.doi.org/10.1016/j.ijsu.2014.03.005

[18] Valero MF, Pulido JE, Ramírez Á, Cheng Z. Determination of crosslink density of polyurethanes obtained from castor oil modified by transesterification. Polímeros. 2009;**19**(1):14-21. [Internet]. Available from: http://www.scielo.br/scielo.php?script=sci_arttext&pid=S0104-14282009000100008&lng=es&nrm=iso&tlng=es

[19] Macocinschi D, Filip D, Paslaru E, Munteanu BS, Dumitriu RP, Pricope GM, et al. Polyurethane-extracellular matrix/silver bionanocomposites for urinary catheters. Journal of Bioactive and Compatible Polymers. 2015;**30**(1):99-113. [Internet]. Available from: http://jbc.sagepub.com/cgi/doi/10.1177/0883911514560661

[20] Ganji Y, Li Q, Susanne E, Böttner M, Selhuber-unkel C, Kasra M. Cardiomyocyte behavior on biodegradable polyurethane / gold nanocomposite scaffolds under electrical stimulation. Materials Science and Engineering C: Biomimetic Materials, Sensors and Systems. 2016;**59**:10-18. [Internet, The Authors]. Available from: http://dx.doi.org/10.1016/j.msec.2015.09.074

[21] Nguyen Dang L, Le Hoang S, Malin M, Weisser J, Walter T, Schnabelrauch M, et al. Synthesis and characterization of castor oil-segmented thermoplastic polyurethane with controlled mechanical properties. European Polymer Journal. 2016;**81**:129-137. [Internet]. Available from: http://dx.doi.org/10.1016/j.eurpolymj.2016.05.024

[22] Miao S, Zhu WJ, Leng J, Castro N, Zhang LG. Four-Dimensional printing hierarchy scaffolds with highly biocompatible smart polymers for tissue engineering applications. Tissue Engineering Part C Methods. 2016;**22**(10):952-963. [Internet]. Available from: http://online.liebertpub.com/doi/10.1089/ten.tec.2015.0542

[23] Li L, Zuo Y, Zou Q, Yang B, Lin L, Li J et al. Hierarchical Structure and Mechanical Improvement of an n-HA/GCO–PU Composite Scaffold for Bone Regeneration. ACS Applied Materials & Interfaces. 2015;**7**(40):22618-22629.

[24] Uscátegui Y, Arévalo F, Díaz L, Cobo M, Valero M. Microbial degradation, cytotoxicity and antibacterial activity of polyurethanes based on modified castor oil and polycaprolactone. Journal of Biomaterials Sciences, Polymer Edition. 2016;**27**(18):1860-1879. [Internet, Taylor & Francis]. Available from: http://dx.doi.org/10.1080/09205063.2016.1239948

[25] Sathiskumar PS, Madras G. Synthesis, characterization, degradation of biodegradable castor oil based polyesters. Polymer Degradation and Stability. 2011;**96**(9):1695-1704. [Internet, Elsevier Ltd]. Available from: http://dx.doi.org/10.1016/j.polymdegradstab.2011.07.002

[26] Yari A, Yeganeh H, Bakhshi H, Gharibi R. Preparation and characterization of novel antibacterial castor oil-based polyurethane membranes for wound dressing application. Journal of Biomedical Materials Research - Part A. 2014;**102**(1):84-96

[27] Valero MF, Díaz LE. Bioadhesive polyurethane obtained from polyols derived of castor oil and starch modified candidates in biomedical applications. Revista Latinoamericana de Metalurgia y Materiales. 2016;**36**(1):9-19. [Internet, Universidad Simón Bolívar; cited 21-Feburary-2017]. Available from: https://www.scopus.com/inward/record.uri?eid=2-s2.0-84964070312&partnerID=40&md5=381ffd43402c9686bf8d4b9fd59bbc68

[28] Depan D, Venkata Surya P, Girase B, Misra R. Organic/inorganic hybrid network structure nanocomposite scaffolds based on grafted chitosan for tissue engineering. Acta Biomaterialia. 2011;**7**(5):2163-2175.

[29] Ganji Y, Kasra M, Salahshour Kordestani S, Bagheri Hariri M. Synthesis and characterization of gold nanotube/nanowire-polyurethane composite based on castor oil and polyethylene glycol. Materials Science and Engineering: C Materials for Biological Application. 2014;**42**:341-349. [Internet, cited 24-May-2016]. Available from: http://www.sciencedirect.com/science/article/pii/S0928493114003063

[30] Janik H, Marzec M. A review: Fabrication of porous polyurethane scaffolds. Materials Science and Engineering C: Biomimetic Materials, Sensors and Systems. 2015;**48**:586-591. [Internet, Elsevier B.V.]. Available from: http://dx.doi.org/10.1016/j.msec.2014.12.037

[31] Valero MF, Díaz LE. Polyurethane networks from pentaerythritol-modified castor oil and lysine polyisocyanates: Synthesis, mechanical, and thermal properties and in vitro degradation. Quimica Nova 2014;**37**(9):1441-1445. [Internet, SBQ; cited 21-Feburary-2017]. Available from: http://www.gnresearch.org/doi/10.5935/0100-4042.20140228

[32] Nacer RS, Poppi RR, Carvalho P de TC de, Silva BAK da, Odashiro AN, Silva IS, et al. Castor oil polyurethane containing silica nanoparticles as filling material of bone defect in rats. Acta Cirurgica Brasileira. 2012;**27**:56-62. [Internet, scielo]. Available from: http://www.scielo.br/scielo.php?script=sci_arttext&pid=S0102-86502012000100010&nrm=iso

Chapter 8

Natural Rubber and its Derivatives

Azanam Shah Hashim and Siew Kooi Ong

Additional information is available at the end of the chapter

http://dx.doi.org/10.5772/intechopen.69661

Abstract

This chapter summarizes various types of chemical modification of natural rubber (NR). These chemically modified NRs are known as NR derivatives. Chemical modification reviewed includes chemical modification without introducing new atom such as cyclized natural rubber and deproteinized natural rubber (DPNR), modification by introducing a chemical group such as hydrogenated natural rubber (HNR), **chlorinated natural rubber (CNR)** and epoxidized natural rubber (ENR) and lastly modification by grafting on NR. Grafting can be carried out using DPNR latex to yield styrene-grafted-NR, methyl methacrylate-grafted-NR and styrene and methyl methacrylate-grafted-NR. The NR derivatives are reviewed in terms of their preparation, mechanism, properties and applications.

Keywords: natural rubber, chemical modification, graft copolymerization

1. Introduction

In this chapter, various types of chemically modified natural rubbers (modified NRs) are reviewed in terms of their preparation, physical, chemical and mechanical properties and their potential applications. These NR derivatives include liquid natural rubber (LNR), cyclized natural rubber, deproteinized natural rubber (DPNR), chlorinated natural rubber (CNR), epoxidized natural rubber (ENR), styrene-grafted NR (SNR) and methyl methacrylate-grafted NR.

Commercial natural rubber (NR) or *cis*-1,4-polyisoprene is obtained from *Hevea brasiliensis* [1]. The NR latex, which is exuded from *H. brasiliensis* bark, is processed in either wet or dry form. Rubber bale is obtained from either field latex or coagulum. The field latex will be coagulated, formed into crepe, dried and packed. Commercial NRs are graded based on dirt content, ash content, nitrogen content, volatile matter content, etc. as specified by International Standards Organisation (ISO). The technically specified rubbers were first introduced into the market by Malaysia in 1965 as the Standard Malaysian Rubbers (SMRs). Currently, commercial NRs are also specified according to Standard Thailand Rubber (STR), Standard Indonesian Rubber (SIR) and Standard Vietnam Rubber (SVR). Due to the high molecular weight of NR as well as its low glass transition temperature (T_g), NR vulcanizates are elastic with high tensile strength and low heat build-up. However, due to the presence of allylic carbons, the vulcanizates have poor abrasion, ageing and weathering resistance. These drawbacks can be overcome by chemical modifications of NR. In chemical modifications, changes in molecular structure occurred. Some examples of chemical modifications are hydrogenation and grafting.

Chemical modifications on NR can be carried out in either solid, solution or latex state. The modifications can be categorized into three categories:

- Modification by bond rearrangement without introducing new atoms. This type of modification will lead to changes in the chemical structure and/or reduction in molecular weight of NR. Cyclization is an example of bond rearrangement.
- Modification by introducing a chemical group such as hydrogen, chloride or epoxy via addition or substitution reactions at the olefinic double bonds. The resultant modified NR will possess the properties of the new chemical groups. For example, epoxidation of NR results in higher T_g and mechanical properties as well as higher hydrophilicity.
- Modification by grafting with one or two molecules, usually vinyl monomers at the allylic carbon. The new grafted polymer can either have low or high molecular weight. The degree of grafting and degree of polymerization measure the efficiency of the modification. Grafting of methyl methacrylate (MMA) onto NR; commercially known as Heveaplus MG results in harder and stiffer copolymer.

2. Modification by bond rearrangement

2.1. Cyclized NR

Cyclization is the simplest and earliest chemical modification of NR [2, 3]. Cyclization can be carried out by treating NR with a proton donor. Examples of the proton donors are sulphuric acid [2, 4–6], sulphonic acid [7], stannic chloride [8] and acid hydrochloric [9]. Modification can be carried out either in solution or in solid or emulsion form [2, 4–9]. For modification in emulsion state, destabilization of the NR latex is crucial and Lewis acid is not suitable. **Figure 1** shows the cyclization mechanism of NR.

Recently, cyclization on deproteinized natural rubber (DPNR) latex was reported by Riyajan et al. [6, 10, 11]. Cyclization on DPNR was reported to be more effective as the presence of the protein can retard and disturb the chemical modification [12]. Cyclized NR in general was reported to be tough, hard and brittle but some elastic behaviour is still maintained. The T_g of cyclized NR increases with higher cyclization where the degree of cyclization is measured from unsaturation of the cyclized NR. For example, the elastomeric behaviour is retained at 30% cyclization but at 60% cyclization, it has a leather-like characteristic. Cyclization up to

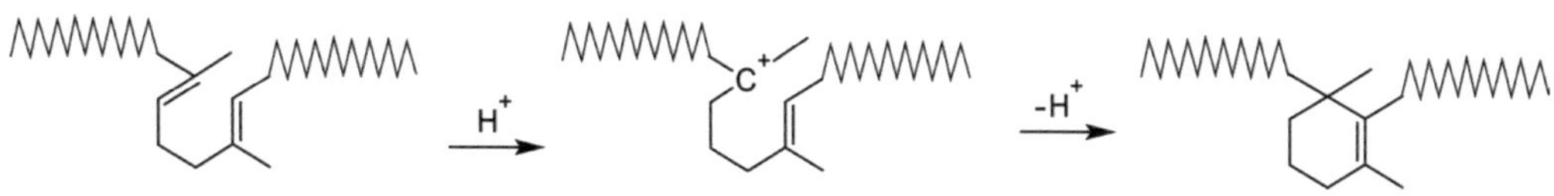

Figure 1. Mechanism of NR cyclization [6].

90% resulted in a powdery form material indicating high brittleness [11]. Cyclized NR is used as adhesives, printing ink as well as paints [10, 13].

2.2. Liquid natural rubber

Liquid natural rubber (LNR) can be derived from depolymerization of NR. The initial objective to prepare LNR was to reduce the internal viscosity of NR. NR with lower viscosity facilitates the addition of additives [14]. This can be achieved via mechanical shearing using either two roll mill or internal mixer. Depending on the milling time and temperature, the average molecular weight in the range of 400,000–1,000,000 Da was achieved [15].

LNR with reactive terminal groups received high attention by researchers as the terminal groups can undergo further chemical reaction or take part during vulcanization. LNR with reactive terminal groups can be prepared via redox couples, photochemical or oxidation. Different preparation methods yield different molecular weights. The oxidation approach yields lowest number average molecular weight, M_n; that is, 2500–3000 Da while the redox couples approach reported M_n in the range of 3000–35,000 Da [16]. The photochemical approach received high attention from researchers. Its range of M_n is of 3000–8400 Da [16].

The only LNR obtained from oxidation is the hydroxyl-terminated LNR. It was prepared by depolymerizing masticated NR in toluene solution under pressure of 200–300 psi and temperature of 150°C with the presence of hydrogen peroxide [17]. LNR prepared via redox couples used both oxidizing and reducing agents either in solution or in emulsion state [16, 18]. LNR with hydroxyl or carbonyl or phenylhydrazone terminal groups was prepared using either hydrogen peroxide or organic peroxide/sulphanilic acid or phenylhydrazine, or phenylhydrazine/ferric chloride/oxygen, or phenylhydrazine/oxygen.

Photochemistry of NR was achieved by ultraviolet (UV) light or sunlight in the presence of compounds such as nitrobenzene and hydrogen peroxide [19, 20]. The former compound will produce carboxyl-terminated LNR, while the latter compound hydroxyl-terminated LNR. **Figure 2** shows an example of light-induced photochemical approach in producing hydroxyl-terminated LNR [20]. Free radicals formed from the bond cleavage under either sunlight or UV light will lead to depolymerization of NR.

Metathesis of NR is a reaction which can induce the cleavage and reforming of the double bonds as well as exchange of substituents in depolymerization of NR. A study by Alimuniar et al. [21] shows that cyclic oligomer of isoprene was produced from intra- and inter-molecular reaction by using tungsten hexachloride and tetramethyl tin in chlorobenzene with purified NR. Depending on the reaction temperature and duration, the molecular weight was in the range of 5800–110,000 Da.

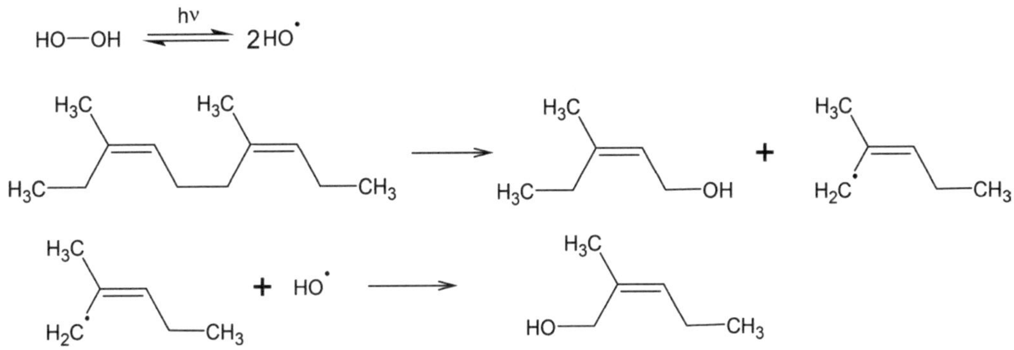

Figure 2. Mechanism of photochemical reaction of NR induced by light [20].

LNR can be used as viscosity modifier, adhesive, tackifier, sealing agent, plasticizer as well as compatibilizer [14, 22]. LNR was reported to be a good compatibilizer in linear low-density polyethylene/natural rubber (LLDPE/NR) blends, polypropylene/natural rubber (PP/NR) blends, high-density polyethylene/natural rubber (HDPE/NR) blends and low-density polyethylene/natural rubber (LDPE/NR) blends. The presence of LNR in HDPE/NR blend changed the morphology from co-continuous to dispersed phase suggesting better degree of compatibility. The tensile strength and elongation at break results were reported to be consistent with the morphological analysis [14].

2.3. Deproteinized natural rubber

Studies on deproteinized natural rubber (DPNR) were initiated due to protein allergy issue on dipped rubber products. DPNR latex is prepared by removing the protein layer covering the rubber particles [23]. The protein layer can be removed by centrifugation, enzymatic degradation or irradiation [24, 25]. Since the protein layer actually acts as stabilizer, surfactant is added during deproteinization. **Figure 3** shows the schematic diagram on removing of protein layer from high ammonia natural rubber (HANR) particle [26].

There are two typical methods used to prepare DPNR latex from field latex or commercial HANR latex. Deproteinization via centrifugation and hydrolysis of protein are the common methods [24]. In centrifugation, the latex is diluted in deionized water containing surfactant. The cream fraction is re-dispersed into deionized water containing sodium dodecyl sulphate (SDS) surfactant. After five times of centrifugation, the total solid content (TSC) of the DPNR obtained is adjusted to 60%. In enzymatic method, the latex is diluted with deionized water and stabilized with surfactant such as SDS. The pH is adjusted to the pH which is suitable for hydrolysis of the enzyme. The range of the pH studied was 5.4–9.2 [27–29]. **Table 1** shows the nitrogen content of the untreated latex and DPNR latex from field latex and HANR latex.

Irradiation of latex was also reported by Makuuchi [25]. In this method, the proteins decomposed during irradiation at 20 kGy. **Table 2** shows the nitrogen content before and after irradiation.

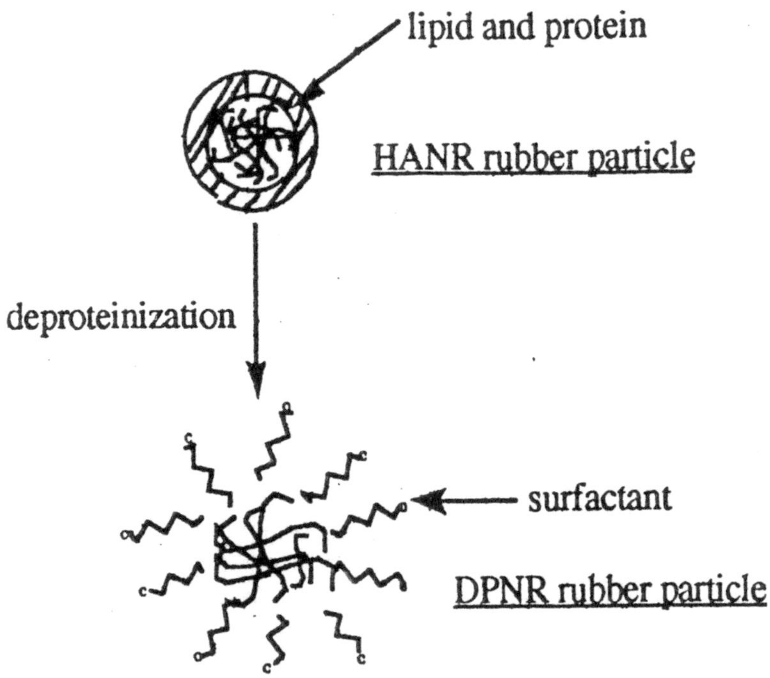

Figure 3. Schematic diagram of deproteinization of rubber particles in HANR latex [26].

Source of latex	Treatment on the latex	Nitrogen content (%)
Field latex	Untreated	0.87
	Five times of centrifugation	0.03
	Enzymatic method	0.008
HANR latex	Untreated	0.56
	Five times of centrifugation	0.03
	Enzymatic method	0.008

Table 1. Nitrogen content of untreated latex and DPNR latex [24].

Treatment type	Nitrogen content (%)
Non-irradiated latex: non-leached	0.26
Non-irradiated latex: leached	0.19
20-kGy-irradiated latex: non-leached	0.53
20-kGy-irradiated latex: leached	0.06

Table 2. Nitrogen content of non-irradiated and irradiated latex [25].

DPNR latex is mainly used to manufacture dipped product such as gloves as lower protein gloves are preferred. A significant amount of protein may cause allergies issue [23, 30]. Chemical modification on DPNR latex with vinyl monomers was also studied as the absence of protein layer was reported to enhance the degree of polymerization, degree of grafting and stability of the grafted NR latex [23, 26].

3. Modification by introducing a chemical group

3.1. Hydrogenated natural rubber

Hydrogenation is a process where hydrogen molecule was added onto NR which results in saturation of NR. **Figure 4** shows the hydrogenation of NR [13].

The purpose of preparing hydrogenated natural rubber (HNR) is to improve NR retention properties while maintaining its T_g. HNR can be prepared in either solution or emulsion forms where the latter has greater interest due to environment and economic reasons [31–33]. There are three methods of hydrogenation: catalytic homogeneous hydrogenation, catalytic heterogeneous hydrogenation and non-catalytic hydrogenation. These methods differ in terms of yield, selectivity, side reactions and/or catalyst poisoning and removal.

Catalytic homogeneous hydrogenation usually uses organotransition metals as the catalyst together with reducing agent such as hydrocarbon activator. Thus, this method of hydrogenation has high selectivity with the absence of microscopic diffusion problems. The catalyst used such as nickel is sensitive to impurities where the catalyst will react with the impurities [31, 34–44]. Catalytic homogeneous hydrogenation results in high yield of hydrogenation but degradation of NR occurs. The degradation can be overcome by using nickel catalyst [13].

Catalytic heterogeneous hydrogenation yields high purity HNR with 100% hydrogenation without any foreign groups or large-scale structural changes [36, 37, 41, 45–47]. Recent study by Kongparakul et al. [48] on metathesis hydrogenation of DPNR using second-generation Grubbs catalyst reported more than 97% hydrogenation.

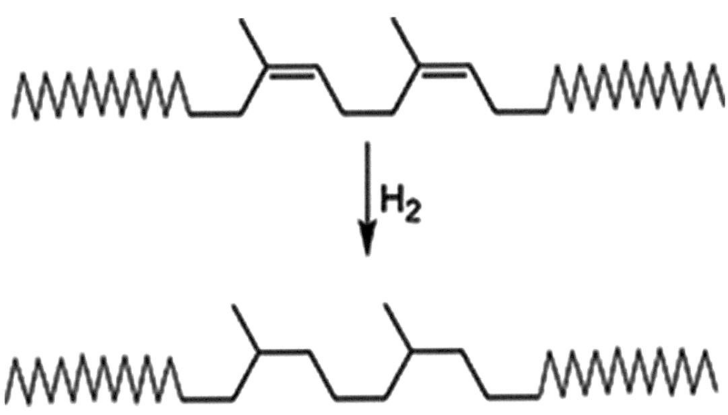

Figure 4. Hydrogenation of NR [13].

Non-catalytic hydrogenation normally is via diimide reduction [37]. Diimide can be obtained from hydrazine [37, 49–56]. Interest in non-catalytic hydrogenation is because of its mild reaction conditions [55, 56]. However, this hydrogenation usually results in chain scission [37]. In recent study by Mahittikul et al. [32], it was observed that hydrogenation of NR latex with diimide generated *in situ* by thermolysis of *p*-toluenesulphonyl hydrazide (TSH) results in >90% hydrogenation without affecting the T_g of the NR.

3.2. Chlorinated natural rubber

Chlorination of NR can be carried out in solution or emulsion forms in the presence of gaseous chlorine [33, 57, 58]. Four reactions of chlorination of NR are addition, substitution, cyclization and crosslinking, and are shown in **Figure 5** [6].

In an early study, chlorinated NR (CNR) was prepared by dissolving NR in a solvent followed by chlorination [59]. Zhong et al. [60] later prepared CNR by passing chlorine gas into pH-adjusted water, followed by the addition of stabilized NR latex. The water pH was adjusted with chloric acid. Chlorination of 60% was reported at where the CNR showed flame resistance, corrosion resistance, better thermal stability, acid-proof, alkaline-proof and wear resistance [60–63]. Due to these properties, CNR found its applications as a raw material for paints (for ships, containers, airport roads, etc.), adhesives, inks and coatings [3, 33]. Surface chlorination, supported by Fourier transform infrared spectroscopy (FTIR) analysis, on vulcanized NR sheets was investigated by Radabutra et al. [64] where the chlorination results in variation of surface roughness and stiffness. The surface roughness enhanced the bonding between CNR sheet and nitrile rubber (NBR) sheet.

3.3. Epoxidized natural rubber

Epoxidation of NR is carried out by converting the carbon-carbon double bond into oxiranes. Epoxidized NR (ENR) was reported since 1920s by Pummere and Burkard [65]; however, its commercial value was only realized in 1980s. Epoxidation of NR latex usually used peracid where the peracid can either be prepared separately or formed *in situ* from a precursor using acid and hydrogen peroxide. **Figures 6** and **7** show the epoxidation of NR with peracid prepared separately and *in situ*, respectively. The latter approach is preferable commercially because of the peracid instability at room temperature and higher temperatures [66].

ENRs with 25 and 50% mol of epoxidation are available commercially under the trade name of Epoxyprene 25 and Epoxyprene 50, respectively. ENR has a higher T_g compared to NR, thus giving higher damping, better air permeability, higher hysteresis, better wet grip as well as higher polarity [3, 66–68]. Therefore, blends of ENR were studied extensively. Blends of ENR with poly(vinyl chloride) (PVC), NR, styrene butadiene rubber (SBR), poly(acrylic acid), poly(ε-caprolactone) (PCL), nitrile rubber (NBR), and polychloroprene (CR) were reported [69–75]. Studies on ENR as pressure-sensitive adhesive were also reported extensively by Hashim and Ong [76]. Epoxidation of NR not only maintains the elasticity of the original NR even at modification higher than 20 mol% but also opens the opportunities to utilize the oxirane groups in various chemical reactions. These reactions included novel crosslinking systems and reinforcement systems.

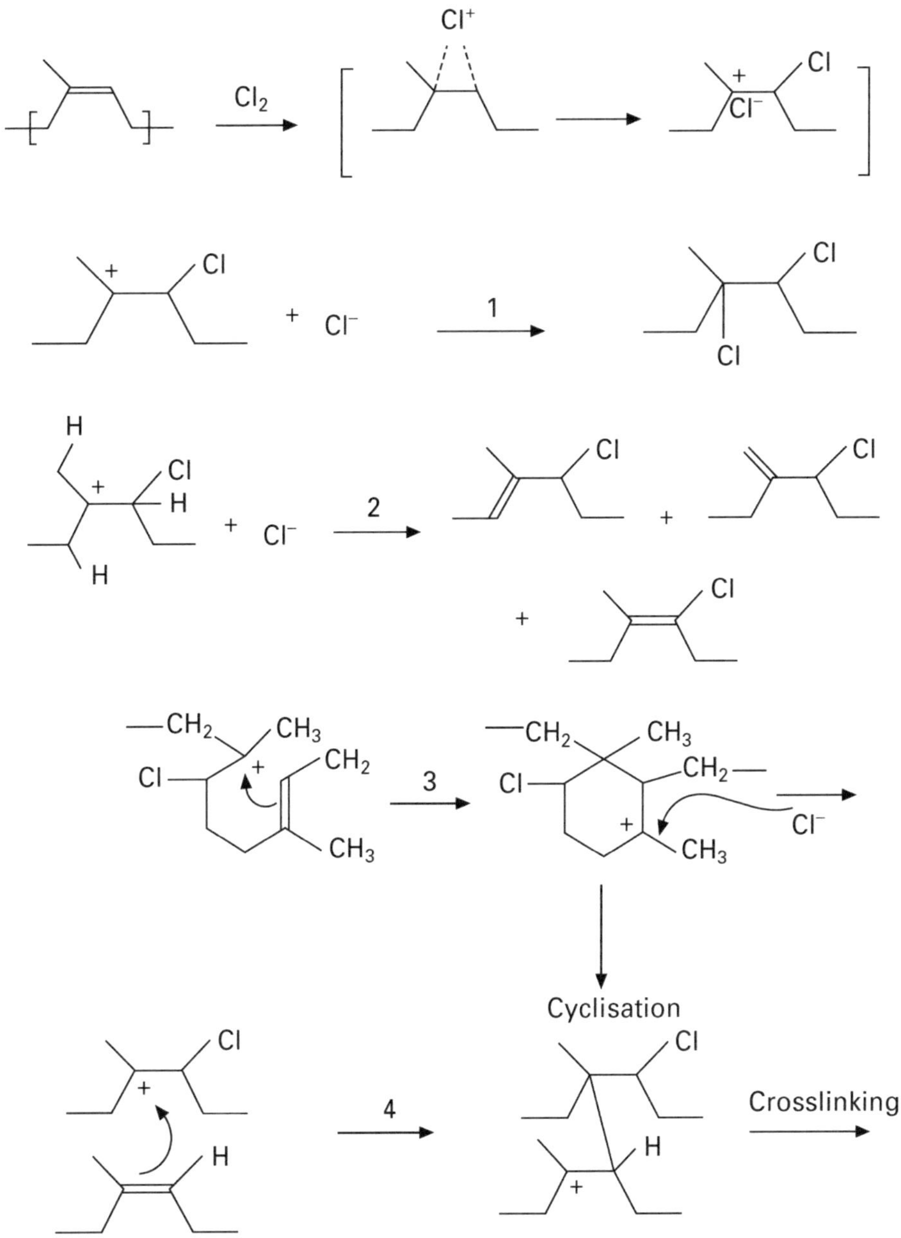

Figure 5. Chlorination of NR: (1) addition of chlorine, (2) substitution chlorination, (3) cyclization, (4) crosslinking [6].

ENR can be vulcanized with typical sulphur-vulcanizing systems, peroxide, amine compounds, aminosilanes and moisture [66, 77–80]. It was reported that in sulphur-vulcanized ENR, the residual acidity of ENR is neutralized with sodium carbonate, magnesium oxide, calcium oxide or calcium stearate [80]. The ENR-amine network formed via ring-opening reaction of *p*-phenylenediamine (catalysed by bisphenol A) was reported to be more rigid, less stretchable and with higher T_g compared to the conventional sulphur-vulcanized ENR

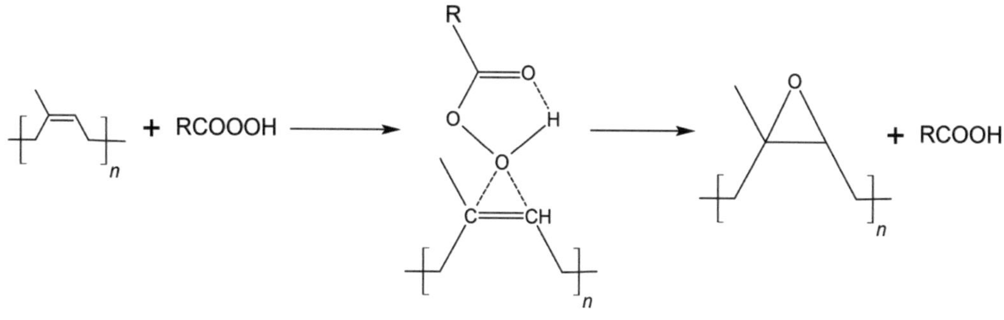

Figure 6. Epoxidation of NR with peracid [66].

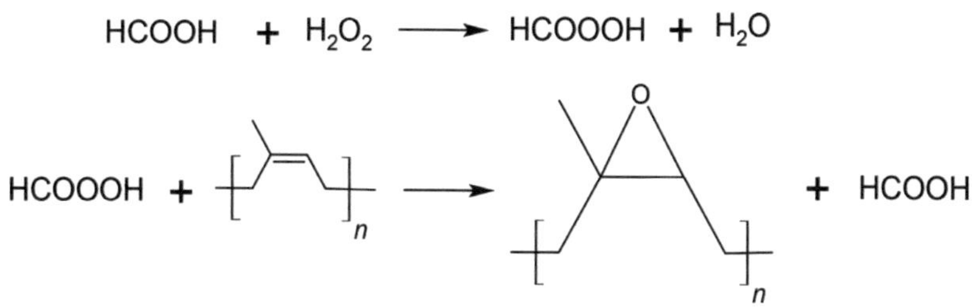

Figure 7. *In situ* epoxidation of NR [66].

at similar physical crosslink density [77, 78]. The curing of ENR with *p*-phenylenediamine is shown in **Figure 8** [77, 78].

Moisture-cured ENR was carried out by first procuring ENR with 3-aminopropyltriethoxysilane (APS) followed by hydrolysis and condensation. The ENR-APS moisture-cured vulcanizate reported lower tensile strength than the sulphur-cured vulcanizate but higher than the peroxide-cured vulcanizate. Partial strain-induced crystallization was demonstrated by moisture-cured vulcanizate [79]. The moisture-cured ENR-APS reaction is shown in **Figure 9** [79].

The precured ENR-APS was also reported to be suitable in preparing sol-gel silica-reinforced ENR vulcanizate via *in situ* sol-gel process using tetraethyl orthosilicate (TEOS) as the silica precursor [79, 81]. The *in situ* sol-gel process to produce sol-gel silica-filled ENR-APS vulcanizate is visualized in **Figure 10** [81].

Tensile properties of sol-gel silica-filled ENR-APS network and silica-filled ENR-sulphur are shown in **Figure 11** [81]. It is obvious that both the sol-gel silica-filled ENR-APS network is stiffer and stronger as compared to the ENR-sulphur system. By assuming that both particle size and dispersion of silica are not the dominant factors, sol-gel silica is a better reinforcing agent [81].

3.4. Epoxidized liquid natural rubber

Epoxidized liquid natural rubber (ELNR) can be produced by the degradation of ENR. This low-molecular-weight ENR can be produced via mechanical breakdown with two roll mill, photo-oxidation with UV radiation, or chemical degradation [82–87].

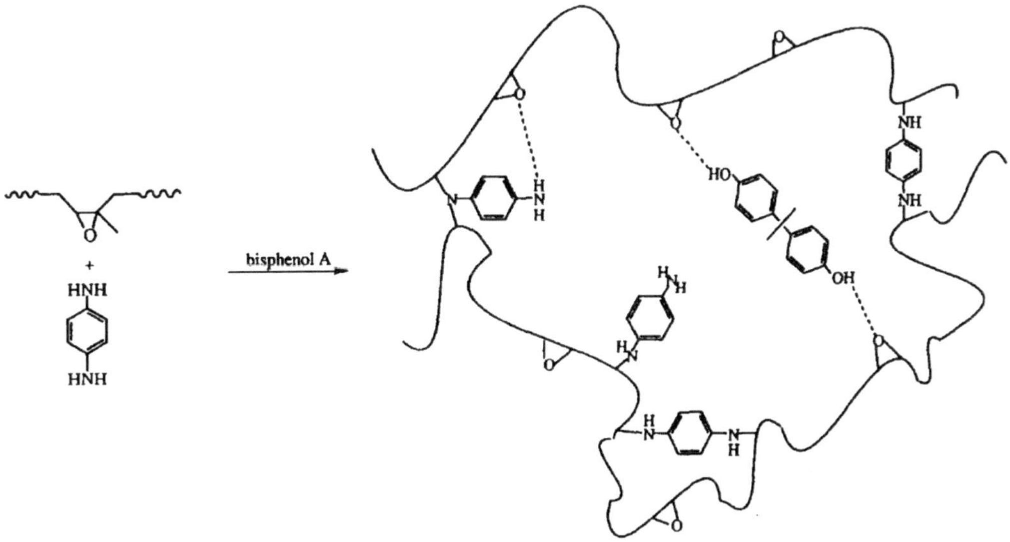

Figure 8. Proposed crosslinking reaction of ENR with *p*-phenylenediamine catalysed with bisphenol A [77, 78].

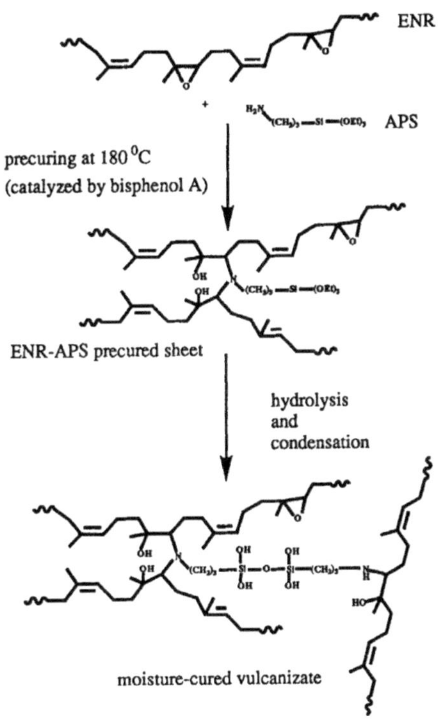

Figure 9. Reaction in obtaining ENR-APS moisture-cure vulcanizate [79].

http://dx.doi.org/10.5772/intechopen.69661

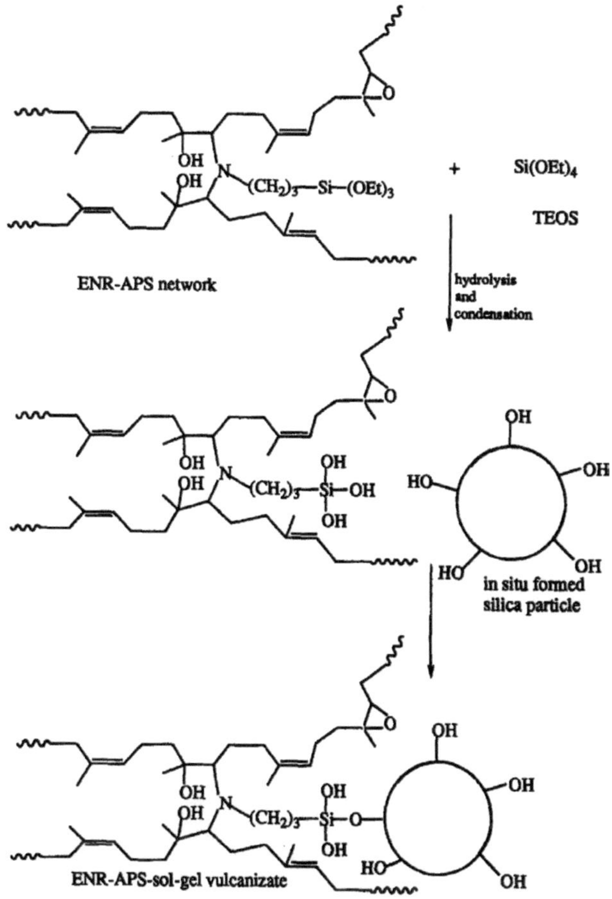

Figure 10. Sol-gel process of ENR-APS network with TEOS to produce silica-filled ENR-APS vulcanizate [81].

A study by Rooshenass et al. [87] observed the degradation of ENR 25 using UV radiation and chemical degradation initiated by potassium peroxodisulphate occurred predominately at the carbon-carbon double bonds. Degradation via mechanical breakdown occurred due to chain scission of the saturated carbon-carbon bonds. The photo-oxidation resulted in more carbonyl and hydroxyl groups on the backbone of the ELNR and fewer double bonds compared with mechanical breakdown and chemical degradation.

Phinyocheep et al. [83] obtained ELNR by reacting ENR in emulsion state using periodic acid. Derouet et al. [82] prepared ELNR by epoxidized LNR where LNR was obtained in the presence of phenyl hydrazine and oxygen and followed by epoxidation via performic acid formed in situ. Apart from the applications mentioned earlier, ELNR is also used as polymeric herbicide [84], compatibilizer in polystyrene/natural rubber blend [86], glass coating [88] and toughening agent [89–91].

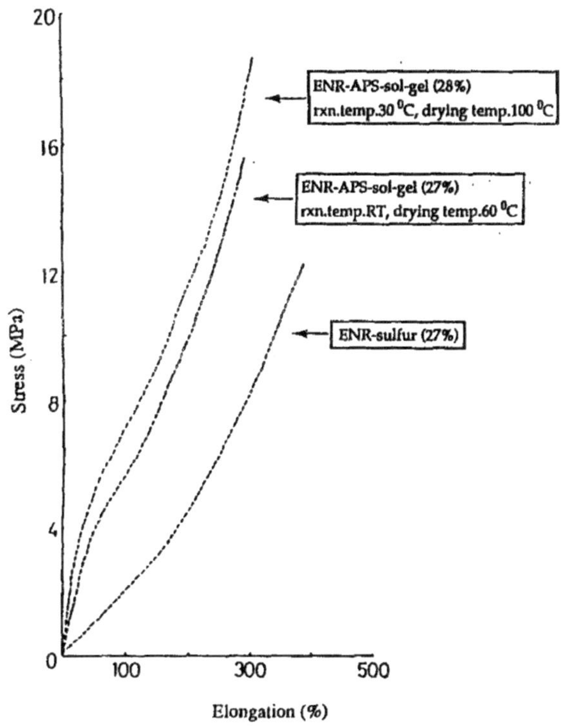

Figure 11. Tensile properties of ENR-APS network and ENR-sulphur network filled with different amounts of silica [81].

ELNR can undergo further reaction to yield hydroxylated LNR (LNR-OH) or hydroxyl-terminated ELNR (ELNR-OH). LNR-OH was obtained when ELNR produced via Na_2WO_4/ CH_3COOH/H_2O_2 catalytic system in toluene underwent higher reaction time, reaction temperature and higher amount of catalyst [86]. **Figure 12** shows the schematic diagram on the reaction of ELNR to LNR-OH. The LNR-OH is used as compatibilizer in polystyrene/natural rubber blend.

ELNR-OH was prepared from oxidative degradation of ELNR with cobalt (II) acetylacetonate in toluene, and the molecular weight depends on the reaction time and cobalt amount [85]. The lowest M_n, approximately 34,000 Da, was reported at 20-h reaction time when the cobalt amount was fixed at 1 wt%. Increasing the cobalt amount leads to greater reduction in M_n where the M_n reported was approximately 20,000 Da at 5-h reaction time and 3 wt% of cobalt. The ELNR-OH produced was used as binder for solid rocket propellant.

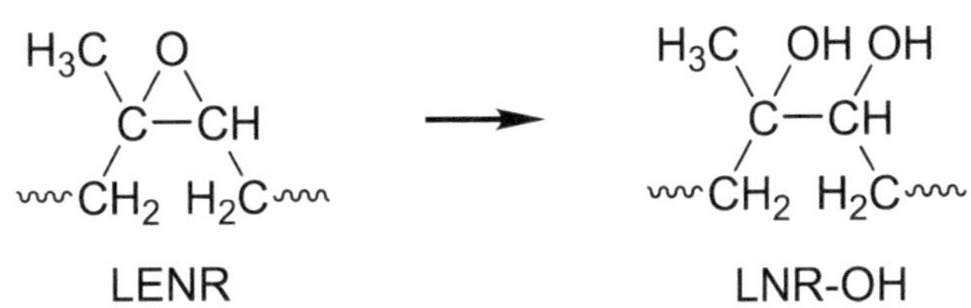

Figure 12. Schematic diagram representing reaction of ELNR to LNR-OH [86].

4. Modification by grafting

4.1. Maleated NR

Maleated NR can be prepared via NR solid phase by adding maleic anhydride using kneader, two roll mill or internal mixer [92–94]. Maleated NR can also be prepared in toluene solution [95]. NR was first masticated to reduce the NR molecular weight prior to dissolving in toluene (100 g NR in 1000 cm^3 toluene). Then, maleic anhydride and benzoyl peroxide were added. The degree of conversion is in between 95 and 97% at 1–10 phr maleic anhydride at 80°C for 2 h and benzoyl peroxide at 3.0 phr. The degree of grafting is reported to be same as the degree of conversion as homopolymerization does not exist.

The proposed mechanism of preparing maleated NR initiated with benzoyl peroxide is shown in **Figure 13** [96]. Maleated NR can used as compatibilizer in polyamide/NR blends [92, 93] and maleated NR/cassava starch blends [94].

4.2. Methyl methacrylate-grafted NR

Methyl methacrylate (MMA) and styrene are the two monomers which are most suitable for grafting with NR [13, 97]. Vinyl monomer-grafted NR can be carried out in solid phase, solution

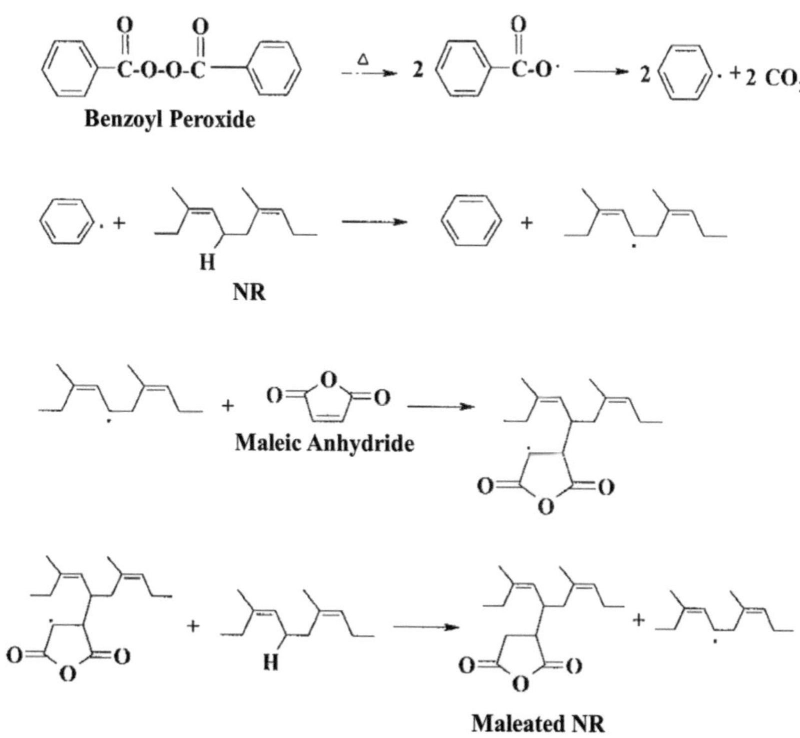

Figure 13. Proposed mechanism of benzoyl peroxide-initiated-maleated NR [96].

phase and emulsion phase. Emulsion phase graft copolymerization is preferable due to economical reason, and it is also more practical than the other phases. The radical-initiated grafting of vinyl monomers onto NR is shown in **Figure 14** [3].

MMA-grafted NR (MMA-NR) is available commercially under the trade name of Heveaplus MG 30 and Heveaplus MG 49 with 30 and 49% of MMA concentration, respectively. Heveaplus MG has the degree of grafting ranging in 60–80%. Heveaplus MG is harder, stiffer, and has higher abrasion resistance and electrical resistance as compared to the unmodified NR. Heveaplus MG is used as an impact modifier as well as a reinforcing agent. The solution or emulsion form of Heveaplus MG is used as an adhesive or a bonding agent between rubber and PVC, leather, textiles and metal [13, 98].

Several literatures have reported on the grafting of MMA onto NR using different initiator systems [99–111]. These studies show that at NR/MMA ratio (w/w), the range is between 50/50 and 80/20 with grafting efficiency between 70 and 95% and degree of conversion of the monomer up to about 96% depending on the polymerization-processing parameters.

A comparative study of *in situ* polymerization of MMA in HANR latex and DPNR latex observed that the latter is more stable and efficient in terms of grafting efficiency. It was

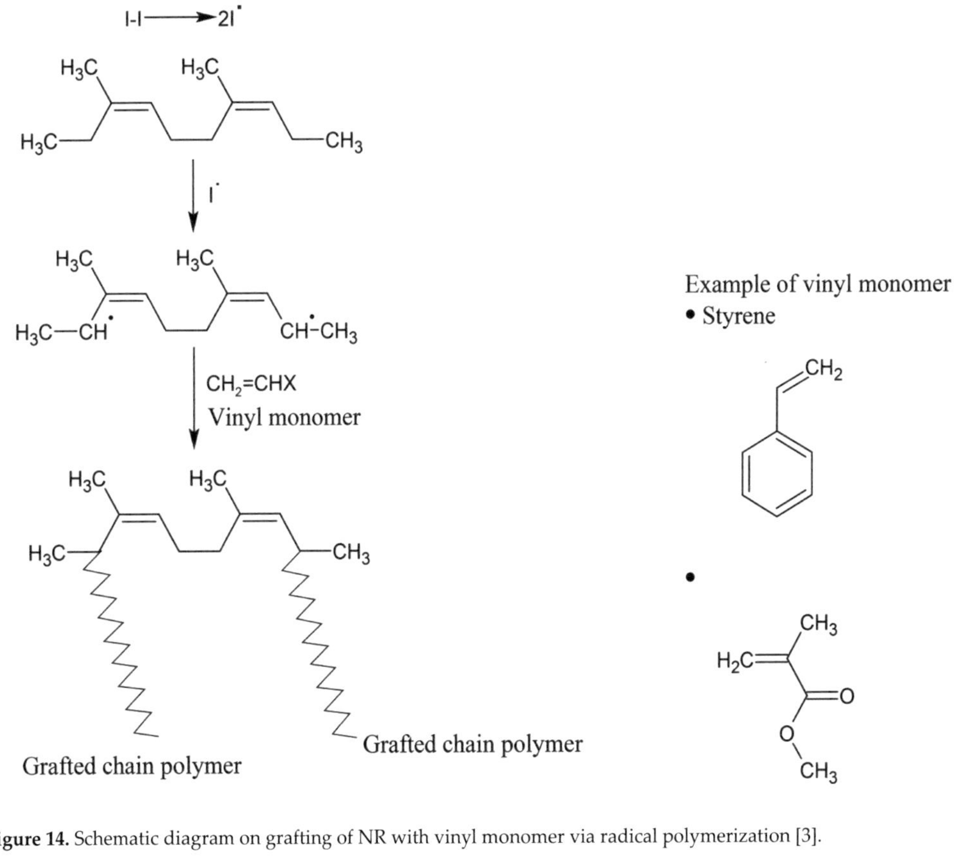

Figure 14. Schematic diagram on grafting of NR with vinyl monomer via radical polymerization [3].

hypothesized that both the radical initiator and MMA diffused better and faster into the rubber particles in using DPNR latex [26]. Further investigation was carried out by Jessy and Hashim [112] on *in situ* polymerization of MMA in DPNR latex via seeded emulsion polymerization. The rubber:water ratio and initiator:monomer ratio were fixed at 0.2109 and 0.05, respectively, with monomer:rubber of 5:95 weight ratio with the reaction temperature of 65°C and the stirring speed of 340 rpm. The grafted DPNR, that is, MNR, had degree of conversion and degree of grafting of 99 and 96%, respectively.

MNR can be used as an impact modifier for poly(methyl methacrylate) (PMMA) [113]. In this study, MNRs (at 2.5, 5.0, 7.5 and 10.0% by weight) were blended with PMMA and compared with PMMA/poly(butyl acrylate) (PMMA/PBA):80/20 blend. The performance of PMMA/PBA:80/20 blend was investigated by Neliappan et al. [114]. **Figure 15** summarizes the fracture toughness, K_{IC} of PMMA, PMMA/MNR blend [113] and PMMA/PBA blend [114]. It is obvious that the presence of both MNR and PBA results in higher K_{IC}. The energy required to fracture the glassy and rubber phases is due to either chain stretching or chain scission or chain pull-out or any of these combinations [115]. It is obvious that that presence of either MNR or PBA results in higher K_{IC} indicating that higher energy is required to fracture both glassy and rubber phases. It is worth noting that with only 10 wt% of MNR content, the K_{IC} is comparable with PMMA toughened with 20 wt% PBA.

4.3. Styrene-grafted NR

Styrene-grafted NR (SNR) has been reported by a comparative study on *in situ* polymerization of styrene in HANR latex and DPNR latex [23]. **Table 3** shows the formulations, reaction

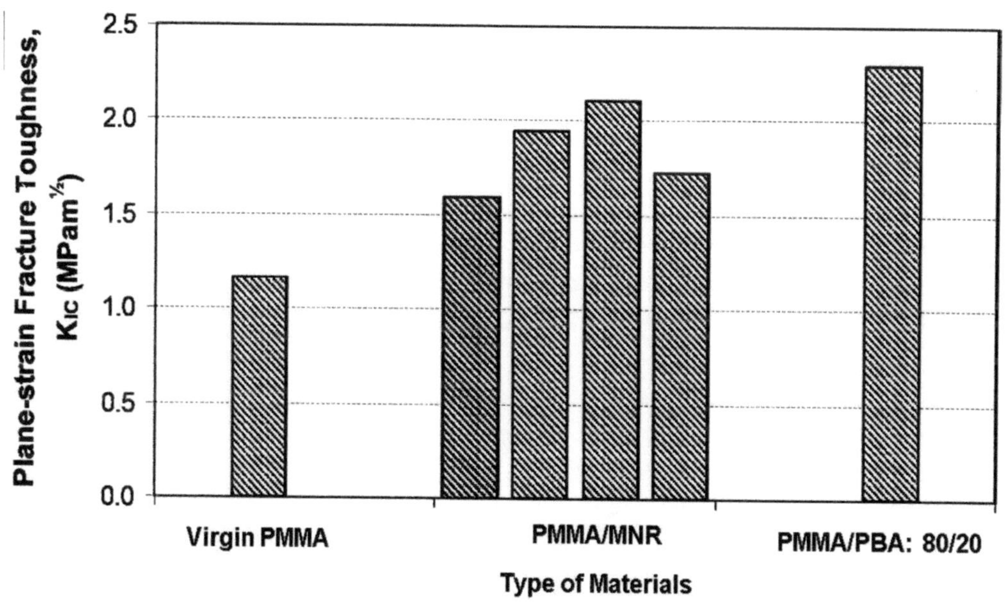

Figure 15. Fracture toughness of virgin PMMA, PMMA/MNR blends and PMMA/PBA: 80/20 blend [113].

	HANR latex			DPNR latex		
Formulation	1	2	3	4	5	6
HANR latex (g)	100.0	100.0	100.0	–	–	–
DPNR latex (g)	–	–	–	100.0	100.0	100.0
Styrene (g)	20.0	20.0	20.0	19.7	19.7	19.7
Initiator (g)	0.40	0.40	0.40	0.39	0.39	0.39
Surfactant (g)	0.00	0.24	0.36	0.00	0.23	0.35
Water (g)	80.0	80.0	80.0	77.0	77.0	77.0
Styrene:rubber ratio	25:75	25:75	25:75	25:75	25:75	25:75
Surfactant concentration (g/water)	0.0	2.0	3.0	0.0	2.0	3.0
Degree of conversion (%)	*	66.7	65.8	97.1	89.8	86.7
Degree of grafting (%)	–	24.4	–	79.1	–	–

*Sample coagulated after 2.5 h.

Table 3. Formulations, reaction conditions and the degree of conversion of polymerization of styrene in HANR latex and DPNR latex [23].

conditions and the respective formulation of the study at 60°C for 10 h. Both the initiator, that is, ammonium peroxy disulphate concentration and TSC were fixed at 2.0% of styrene weight and 40%, respectively. The TSC is the total weight of both styrene and dry rubber.

From the results, it is obvious that the presence of protein/lipid layer in HANR latex rubber particles results in lower conversion. This is because of poorer diffusion of monomer and initiator into the rubber particles. Non-rubber substances such as protein and lipid were found to inhibit free radical polymerization [116]. Consistently, the degree of grafting (via sol-gel method) in DPNR is higher than in HANR as shown by Formulations 4 and 2, respectively.

SNR has a potential application as an impact modifier for PS [117]. As impact modifier for PS, 5–20 wt% of SNR results in 141–1050% increment in impact strength in relation to the pure PS. The presence of 15 wt% of SNR in PS was reported to give a 25% higher impact strength compared to a commercial high-impact polystyrene (HIPS).

SNR can also be used as compatibilizer in polypropylene/natural rubber (PP/NR) blends and PP/PS blend [118–120] as well as a modifier in PP-based binary blend [121]. As a high-molecular-weight compatibilizer in PP/NR blend, 10 wt% of SNR was reported to increase the tensile strength and Young's modulus by 28 and 24%, respectively. Some improvement in elongation at break and flexural properties was also observed.

The effectiveness of SNR as compatibilizer, in comparison with commercial compatibilizers, in PP/PS blend is summarized in **Table 4**. It is obvious that SNR is a better choice compared to the others [120].

To highlight the effect of SNR as a high-molecular-weight modifier, PP/SNR 70/30 blend and PP/NR 70/30 blend were investigated where both blends were dynamically vulcanized [121]. The PP/SNR blend was observed to be stiffer and stronger with higher tensile strength, higher modulus and higher impact strength but with comparable elongation at break as shown in **Table 5** [121]. The tensile properties of the PP/SNR blend are better than those of polyolefin/NR blends [121]. Compared to Santoprene®, a commercial product, the properties are with the range although the modulus is significantly higher indicating a stiffer material.

Compatibilizers	Tensile strength (MPa)		Elongation at break (%)		Impact strength (kJ/m²)	
	0 wt%	7.5 wt%	0 wt%	7.5 wt%	0 wt%	7.5 wt%
SNR	20.3	21.9	7.9	177.5	198.4	437
[a]SEBS		16.0		7.88		304
[b]Surlyn		20.3		4.24		240
[c]EVA		13.17		4.95		250
[d]4 ssa,ssh		17.61		3.06		198

[a]SEBS: styrene-ethylene-butylene-styrene block copolymer.
[b]Surlyn: ionomer from ethylene acid copolymers.
[c]EVA: ethylene-vinyl acetate copolymer.
[d]4 ssa,ssh: 4-Styrenesulfonic acid sodium salt hydrate.

Table 4. Mechanical properties of PP/PS blends using various compatibilizers [120].

Properties	PP/SNR: 70/30 blend	PP/NR: 70/30 blend	*Polyolefin/NR blend	**Santoprene®
Tensile strength (MPa)	25.3	21.7	6–20	4.4–27.6
Elongation at break (%)	341.9	362.4	200	330–600
Modulus at 100% elongation (MPa)	17.0	12.7	–	2.1–10
Impact strength (J/ m²)	150.3	140.8	–	–

*[122].
**[123].

Table 5. Some properties of PP/SNR: 70/30 blend, PP/NR: 70/30 blend, polyolefin/NR blends and Santoprene® [121–123].

4.4. Styrene/methyl methacrylate-grafted NR

Grafting of two monomers could also be carried out in DPNR latex. It should be noted that, from chemical structure point of view, both HANR and DPNR are still NR, except that the latter contains much less protein content and the rubber particles are stabilized by adding surfactant as mentioned earlier. Styrene/methyl methacrylate-grafted NR (SMNR) can be prepared using DPNR latex via seeded emulsion polymerization using ammonium peroxydisulphate ($N_2H_8O_8S_2$) as an initiator resulting in 97.4% degree of conversion [124]. The styrene:MMA:rubber ratio studied was 10:10:80 by weight. The amount of styrene in SMNR was reported to be higher due to higher styrene reactivity [124]. Man et al. [125] later varied the concentration of both styrene and MMA concentration between 10 and 30 wt% fixing the styrene:MMA ratio at 1:1 by weight. It was observed that the highest degree of grafting is at 20 wt% total monomer concentration as shown in **Figure 16**. At 10 wt% total monomer concentration, no grafting was observed indicating that only copolymerization took place. At 20 wt% total monomer concentration, the degree of grafting observed was at approximately 50%. At 30 wt% total monomer concentration, the degree of grafting reduced to 30% due to copolymerization between the two monomers is more pronounced. This was evidenced from the ^{1}H NMR results where similar styrene and MMA content was detected in the grafted rubber.

Based on the thermogravimetric analysis (TGA), the decomposition temperature of SMANR at different monomer concentrations was consistently higher than the DPNR. Both the SMNR gum and N330-filled SMNR vulcanizates are stiffer and less elastic compared to the DPNR gum and filled vulcanizates, respectively [125].

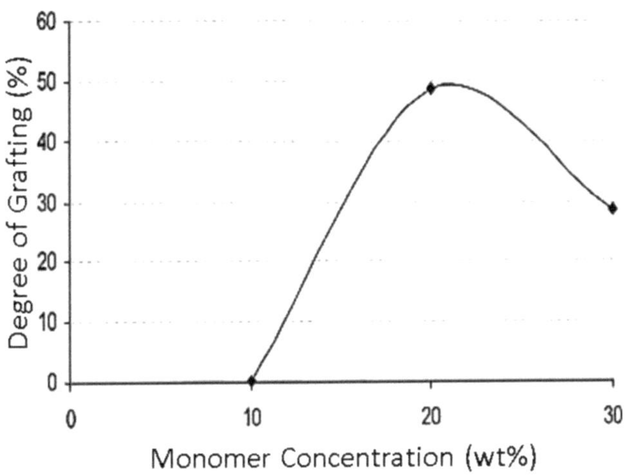

Figure 16. Degree of grafting of SMNR at different total monomer concentration where the styrene-MMA ratio is fixed at 1:1 [125].

5. Conclusions

Compared to NR, NR derivatives exhibit different physical, mechanical and chemical properties; for example, the oxirane groups in ENR enable novel crosslinking system such as

http://dx.doi.org/10.5772/intechopen.69661

ENR-APS-crosslinking system. The *in situ* sol-gel silica-filled ENR-APS network system was reported to have better reinforcing efficiency giving higher mechanical properties. The degree of grafting of vinyl monomers onto NR backbones, carried out via emulsion polymerization in DPNR latex, was found to be more efficient than in HANR latex; the degree of grafting for styrene-grafted NR, when carried out in DPNR latex, was observed to be four times higher than in HANR latex. The vinyl monomer-grafted DPNR such as SNR was found to be stiffer and stronger and can used as impact modifier for PS and as compatibilizer for PP/PS and PP/NR blends.

Author details

Azanam Shah Hashim* and Siew Kooi Ong

*Address all correspondence to: azanam@unikl.edu.my

Universiti Kuala Lumpur, Kuala Lumpur, Malaysia

References

[1] Venkatachalam P, Geetha N, Sangeetha P, Thulaseedharan A. Natural rubber producing plants: An overview. African Journal of Biotechnology. 2013;**12**(12):1297-1310. DOI: 10.5897/AJBX12.016

[2] Brydson JA. Reactivity of diene rubbers—I. In: Brydson JA, editor. Rubber Chemistry. London: Applied Science Publishers Ltd; 1978. pp. 180-184. DOI: 10.1007/978-94-009-9221-4

[3] Phinyocheep P. Chemical modification of natural rubber (NR) for improved performance. In: Kohjiya S, Ikeda Y, editors. Chemistry, Manufacture and Applications of Natural Rubber. United Kingdom: Woodhead Publishing Ltd.; 2014. pp. 68-118. DOI: 10.1533/9780857096913.1.68

[4] Gordon M. Kinetics of cyclization of Hevea latex. Rubber Chemistry and Technology. 1951;**24**(4):940-956. DOI: 10.5254/1.3543123

[5] Fettes EM. Chemical Reactions of Polymers. New York, NY: Interscience Publishers; 1964. pp. 126-127

[6] Riyajan S, Sakdapipanich JT. Cationic cyclization of deproteinized natural rubber latex using sulfuric acid. Kautschuk, Gummi, Kunststoffe. 2006;**59**(3):104-109

[7] Fisher HL. Conversion of rubber into thermoplastic products with properties similar to Gutta-Percha, Balata, and Shellac. I—Methods of preparation and general properties. Industrial and Engineering Chemistry. 1927;**19**(12):1325-1333. DOI: 10.5254/1.3535284

[8] Bruson HA, Sebrell LB, Calvert WC. New chemical reactions of rubber hydrocarbons reactions with metallic halides. Industrial and Engineering Chemistry. 1927;**19**(9):1033-1037. DOI: 10.1021/ie50213a027

[9] van Veersen GJ. The structure of cyclized rubber. Rubber Chemistry and Technology. 1951;**24**(4):957-969. DOI: 10.5254/1.3543124

[10] Riyajan S, Liaw D-J, Tanaka Y, Sakdapipanich JT. Cationic cyclization of purified natural rubber in latex form with a trimethylsilyl triflate as a novel catalyst. Journal of Applied Polymer Science. 2007;**105**(2):664-672. DOI: 10.1002/app.25026

[11] Riyajan S, Tanaka Y, Sakdapipanich JT. Preparation of cyclized deproteinized natural rubber in latex state via a combination of benzotrichloride and sulfuric acid system, and its properties. Rubber Chemistry and Technology. 2007;**80**(2):365-377. DOI: 10.5254/1.3539412

[12] Hourston D, Tabe J. Natural rubber (chemical modification). In: Salamone J, Claypool J, Demby A, editors. Polymeric Material Encyclopaedia. New York, NY: CRC Press Inc.; 1996. pp. 4547-4551

[13] Hashim AS, Ong SK, Jessy RS. A general review of recent developments on chemical modification of natural rubber. Natuurrubber Natural Rubber. 2002;**28**(4):3-9

[14] Ibrahim A, Dahlan M. Thermoplastic natural rubber blends. Progress in Polymer Science. 1998;**23**(4):665-706. DOI: 10.1016/S0079-6700(97)00052-X

[15] Gelling I, Porter M. Chemical modification of natural rubber. In: Roberts AD, editor. Natural Rubber Science and Technology. Oxford [England]; New York: Oxford University Press; 1988. pp. 359-456

[16] Nor HM, Ebdon JR. Telechelic liquid natural rubber: A review. Progress in Polymer Science. 1988;**23**(2):143-177. DOI: 10.1016/S0079-6700(97)00028-2

[17] Gupta S, Kurup M, Devadoss E, Muthiah R, Thomas S. Development and evaluation of a novel binder based on natural rubber and high-energy polyurethane/composite propellants. Journal of Applied Polymer Science. 1985;**30**(3):1095-1112. DOI: 10.1002/app.1985.070300317

[18] Pautrat R, Marteau J. France Patent 7403052: Method for the Preparation of Rubbers with Low Molecular Weights through Degradation of Macromolecular Polyenes, and the Products Thus Obtained. France: Institut FS Caoutcholic Auxil; 1974

[19] Ravindran T, Nayar MRG, Francis JD. A novel method for the preparation of hydroxyl terminated liquid natural rubber. Macromolecular Chemistry Rapid Communications. 1986;**7**(3):159-163. DOI: 10.1002/marc.1986.030070311

[20] Ravindran T, Gopinathan Nayar MR, Francis DJ. Production of hydroxyl-terminated liquid natural rubber—Mechanism of photochemical depolymerization and hydroxylation. Journal of Applied Polymer Science. 1988;**35**(5):1227-1239. DOI: 10.1002/app.1988.070350509

[21] Alimuniar A, Yarmo MA, Rahman MZA, Kohjiya S, Ikeda Y, Yamashita S. Metathesis degradation of natural rubber. Polymer Bulletin. 1990;**23**(1):119-126. DOI: 10.1007/BF00983973

[22] Kawasaki A, Miyamoto Y, Tanaka Y. Proceeding in Regional Conference on Polymeric Materials 1998; 10-11 November 1998; Conference Hall, School of Industrial Technology, USM, Penang, Malaysia. USM, Penang, Malaysia: Universiti Sains Malaysia and Plastics and Rubber Institute Malaysia; 1998. p. 35

[23] Tho NV, Abd Kadir MO, Hashim AS. A comparative study of styrene polymerization in deproteinized and undeproteinized natural rubber latex. Rubber Chemistry and Technology. 2002;**75**(1):111-118. DOI: 10.5254/1.3547663

[24] Ichikawa N, Eng AH, Tanaka Y. Properties of deproteinised natural rubber latex. In: Abdul Aziz bin SAK, editor. Proceeding of International Rubber Technology Conference, Kuala Lumpur; 14-16 June 1993; Kuala Lumpur, Malaysia. Kuala Lumpur, Malaysia: Rubber Research Institute of Malaysia; 1993. pp. 101-110

[25] Makuuchi K. Progress in radiation vulcanization of natural rubber latex. In: Kume T, Maekawa Y, editors. Proceedings of the Takasaki Workshop on Bilateral Cooperation's-Radiation Processing of Natural Polymers; 1-2 November 1999; Takasaki, Japan. Takasaki, Japan: Japan Atomic Energy Research Institute; 1999. pp. 32-37

[26] Hashim AS. A preliminary investigation on *in situ* polymerization of vinyl monomers in DPNR latex. In: Proceedings of the 6th JSPS-VCC Seminar on Integrated Engineering; 27-28 November1996; Kyoto, Japan. Kyoto, Japan; 1996. pp. 83-90

[27] Eng AH, Kawahara S, Tanaka Y. Determination of low nitrogen content of purified natural rubber. Journal of Natural Rubber Research. 1993;**8**(2):109-113

[28] Mariamma George K, Rajammal G, Amma G, Mathew N. Preparation and properties of low protein NR latex using stabilised liquid papain. Journal of Rubber Research. 2007;**10**(3):131-142

[29] Pichayakorn W, Suksaeree J, Boonme P, Taweepreda W, Ritthidej GC. Preparation of deproteinized natural rubber latex and properties of films formed by itself and several adhesive polymer blends. Industrial and Engineering Chemistry Research. 2012;**51**(41):13393-13404. DOI: 10.1021/ie301985y

[30] Pendle TD. Challenges for N R latex products in medical and food-related applications. In: Abdul Aziz bin SAK, editor. The Proceedings of the International Rubber Technology Conference 1993; 14-16 June 1993; Kuala Lumpur. Kuala Lumpur: Rubber Research Institute of Malaysia; 1993. pp. 3-18

[31] Hinchiranan N, Charmondusit K, Prasassarakich P, Rempel G. Hydrogenation of synthetic *cis*-1,4-polyisoprene and natural rubber catalyzed by $[Ir(COD)py(PCy_3)]PF_6$. Journal of Applied Polymer Science. 2006;**100**(5):4219-4233. DOI: 10.1002/app.23710

[32] Mahittikul A, Prasassarakich P, Rempel G. Noncatalytic hydrogenation of natural rubber latex. Journal of Applied Polymer Science. 2007;**103**(5):2885-2895. DOI: 10.1002/app.25449

[33] Sakdapipanich JT, Rojruthai P. Natural rubber: Biosynthesis, structure, properties and application. In: Thomas S, Chan CH, Pothen L, Rajisha K, Maria HJ, editors. Natural

Rubber Materials: Volume 1: Blends and IPNs. 1st ed. Cambridge, United Kingdom: Royal Society of Chemistry; 2014. pp. 28-52. DOI: 10.1039/9781849737647-00028

[34] Ramp FL, De Witt EJ, Trapasso LE. Homogeneous hydrogenation catalyzed by boranes. The Journal of Organic Chemistry. 1962;**27**(12):4672. DOI: 10.1021/jo01059a056

[35] Falk JC. Lithium based coordination catalysts for the hydrogenation of diene and vinyl-aromatic polymers. Die Makromolekulare Chemie. 1972;**160**(1):291-299. DOI: 10.1002/macp.1972.021600123

[36] Rachapudy H, Smith GG, Raju VR, Graessley WW. Properties of amorphous and crystallizable hydrocarbon polymers. III. Studies of the hydrogenation of polybutadiene. Journal of Polymer Science: Polymer Physics Edition. 1979;**17**(7):1211-1222. DOI: 10.1002/pol.1979.180170706

[37] Schulz DN, Tuner SR, Golub MA. Recent advances in the chemical modification of unsaturated polymers. Rubber Chemistry and Technology. 1982;**55**(3):809-859. DOI: 10.5254/1.3535904

[38] Edwards HGM, Farwell DW, Johnson AF, Lewis IR, Webb N, Ward NJ. Spectroscopic studies of an ambient-pressure process for the selective hydrogenation of polybutadienes. Macromolecules. 1992;**25**(2):525-529. DOI: 10.1021/ma00028a005

[39] Hahn SF. An improved method for the diimide hydrogenation of butadiene and isoprene containing polymers. Journal of Polymer Science Part A: Polymer Chemistry. 1992;**30**(3):397-408. DOI: 10.1002/pola.1992.080300307

[40] Gan SN, Subramaniam N, Yahya R. Hydrogenation of natural rubber using nickel 2-ethylhexanoate catalyst in combination with triisobutylaluminum. Journal of Applied Polymer Science. 1996;**59**(1):63-70. DOI: 10.1002/(SICI)1097-4628(19960103)

[41] Singha NK, De PP, Sivaram S. Homogeneous catalytic hydrogenation of natural rubber using $RhCl(PPh_3)_3$. Journal of Applied Polymer Science. 1997;**66**(9):1647-1652. DOI: 10.1002/(SICI)1097-4628(19971128)66:9<1653::AID-APP4>3.0.CO;2-C

[42] Tangthongkul R, Prasassarakich P, Rempel GL. Hydrogenation of natural rubber with $Ru[CH{=}CH(Ph)]Cl(CO)(PCy_3)_2$ as a catalyst. Journal of Applied Polymer Science. 2005;**97**(6):2399-2406. DOI: 10.1002/app.21922

[43] Inoue S-I, Nishio T. Synthesis and properties of hydrogenated natural rubber. Journal of Applied Polymer Science. 2007;**103**(6):3957-3963. DOI: 10.1002/app.25158

[44] Mahittikul A, Prasassarakich P, Rempel GL. Hydrogenation of natural rubber latex in the presence of $[Ir(cod)(PCy_3)(py)]PF_6$. Journal of Molecular Catalysis A: Chemical. 2009;**297**(2):135-141. DOI: 10.1016/j.molcata.2008.09.006

[45] Augustine RL, Warner RW. Heterogeneous catalysis in organic chemistry: Effect of hydrogen presaturation of the catalyst in the hydrogenation of olefins. Journal of Organic Chemistry. 1981;**46**(13):2614-2622. DOI: 10.1021/jo00326a003

[46] Yokota K, Hirabayashi T. Hydrogenation of alternating butadiene–methyl methacrylate copolymer. Polymer Journal. 1981;**13**(8):813-816. DOI: 10.1295/polymj.13.813

[47] Weinstein AH. Elastomeric tetramethylene-ethylethylene-acrylonitrile copolymers. Rubber Chemistry and Technology. 1984;**57**(1):203-215. DOI: 10.5254/1.3535996

[48] Kongparakul S, Ng FTT, Rempel GL. Metathesis hydrogenation of natural rubber latex. Applied Catalysis A: General. 2011;**405**(1-2):129-136. DOI: 10.1016/j.apcata.2011.07.039

[49] Nang TD, Katabe Y, Minoura Y. Diimide reduction of *cis*-1,4-polyisoprene with *p*-toluene-sulphonylhydrazide. Polymer. 1976;**17**(2):117-120. DOI: 10.1016/0032-3861(76)90079-3

[50] Phinyocheep P, Pasiri S, Tavichai O. Diimide hydrogenation of isoprene-styrene diblock copolymers. Journal of Applied Polymer Science. 2003;**87**(1):76-82. DOI: 10.1002/app.11672

[51] Samran J, Phinyocheep P, Daniel P, Derouet D, Buzare J-Y. Spectroscopic study of di-imide hydrogenation of natural rubber. Macromolecular Symposia. 2004;**216**(1):131-144. DOI: 10.1002/masy.200451214

[52] Samran J, Phinyocheep P, Daniel P, Derouet D, Buzaré J-Y. Raman spectroscopic study of non-catalytic hydrogenation of unsaturated rubbers. Journal of Raman Spectroscopy. 2004;**35**(12):1073-1080. DOI: 10.1002/jrs.1256

[53] Samran J, Phinyocheep P, Daniel P, Kittipoom S. Hydrogenation of unsaturated rubbers using diimide as a reducing agent. Journal of Applied Polymer Science. 2005;**95**(1):16-27. DOI: 10.1002/app.20811

[54] Ikeda Y, Phinyocheep P, Kittipoom S, Ruancharoen J, Kokubo Y, Morita Y, Hijikata K, Kohjiya S. Mechanical characteristics of hydrogenated natural rubber vulcanizates. Polymers for Advanced Technologies. 2008;**19**(11):1608-1615. DOI: 10.1002/pat.1176

[55] Harwood HJ, Russel DB, Verthe JJA, Zymonas J. Diimide as a reagent for the hydrogenation of unsaturated polymers. Die Makromolekulare Chemie. 1973;**163**(1):1-12. DOI: 10.1002/macp.1973.021630101

[56] Mango LA, Lenz RW. Hydrogenation of unsaturated polymers with diimide. Die Makromolekulare Chemie. 1973;**163**(1):13-36. DOI: 10.1002/macp.1973.021630102

[57] Bloomfield GF. Rubber, polyisoprenes, and allied compounds. Part VI. The mechanism of halogen-substitution reactions, and the additive halogenation of rubber and of dihydro-myrcene. Rubber Chemistry and Technology. 1944;**17**(4):759-771. DOI: 10.5254/1.3546696

[58] van Amerongen GJ. Chlorinated rubber from latex. Industrial and Engineering Chemistry. 1951;**43**(11):2535-2540. DOI: 10.1021/ie50503a038

[59] Kraus G, Reynolds WB. Chlorination of natural and synthetic polyisoprenes. Journal of the American Chemical Society. 1950;**72**(12):5621-5626. DOI: 10.1021/ja01168a073

[60] Zhong JP, Li SD, Wei YC, Peng Z, Yu HP. Study on preparation of chlorinated natural rubber from latex and its thermal stability. Journal of Applied Polymer Science. 1999;**73**(14):2863-2867. DOI: 10.1002/(SICI)1097-4628(19990929)73:14<2863::AID-APP9>3.0.CO;2-2

[61] Zhong JP, Li SD, Yu H, Wei YC, Peng Z, Qu JA, Guo CK. Thermooxidative decomposition and its kinetics on chlorinated natural rubber from latex. Journal of Applied Polymer Science. 2001;**81**(6):1305-1309. DOI: 10.1002/app.1554

[62] Li SD, Cheung MK, Zhong JP, Yu HP. Effect of atmosphere on the thermal decomposition of chlorinated natural rubber from latex. Journal of Applied Polymer Science. 2001;**82**(10):2590-2598. DOI: 10.1002/app.2110

[63] Cai Y, Li SD, Li CP, Li PW, Wang C, Lv MZ, Xu K. Thermal degradations of chlorinated natural rubber from latex and chlorinated natural rubber from solution. Journal of Applied Polymer Science. 2007;**106**(2):743-748. DOI: 10.1002/app.26531

[64] Radabutra S, Thanawan S, Amornsakchai T. Chlorination and characterization of natural rubber and its adhesion to nitrile rubber. European Polymer Journal. 2009;**45**(7):2017-2022. DOI: 10.1016/j.eurpolymj.2009.04.008

[65] Pummere D, Burkard A. Über kautschuk. Berichte Der Deutschen Chemischen Gesellschaft. 1922;**55**(10):3458-3472. DOI: 10.1002/cber.19220551015

[66] Hashim AS, Kohjiya S. Preparation and properties of epoxidized natural rubber. KGK–Kautschuk Gummi Kunststoffe. 1993;**46**:208-213

[67] Baker CSL, Gelling IR, Newell R. Epoxidized natural rubber. Rubber Chemistry and Technology. 1985;**58**(1):67-85. DOI: 10.5254/1.3536059

[68] Gelling IR. Epoxidised natural rubber. Journal of Natural Rubber Research. 1991;**6**(3):184-205

[69] Tsukahara Y, Yonemura T, Hashim AS, Kohjiya S, Kaeriyama K. Preparation and properties of epoxidized natural rubber/poly (e-caprolactone) self-vulcanizable blends. Journal of Materials Chemistry. 1996;**6**(12):1865-1870. DOI: 10.1039/JM9960601865

[70] Mallick A, Bhattacharya AK, Gupta BR, Tripathy DK, De SK. Effect of HAF carbon black filler on chemorheological behavior of poly(acrylic acid)-epoxidized natural rubber blend. Journal of Applied Polymer Science. 1997;**65**(1):135-142. DOI: 10.1002/(SICI)1097-4628(19970705)65:1<135::AID-APP17>3.0.CO;2-2

[71] Okwu UN, Okieimen FE. Preparation and properties of thioglycolic acid modified epoxidised natural rubber and its blends with natural rubber. European Polymer Journal. 2001;**37**(11):2253-2258. DOI: http://dx.doi.org/10.1016/S0014-3057(01)00112-4

[72] Ismail H, Hairunezam HM. The effect of a compatibilizer on curing characteristics, mechanical properties and oil resistance of styrene butadiene rubber/epoxidized natural rubber blends. European Polymer Journal. 2001;**37**(1):39-44. DOI: http://dx.doi.org/10.1016/S0014-3057(00)00099-9

[73] Ismail H, Suzaimah S, Hairunezam HM. Curing characteristics, mechanical properties and oil resistance of styrene butadiene rubber/epoxidized natural rubber blends. Journal of Elastomers and Plastics. 2002;**34**(2):119-130. DOI: 10.1106/009524402022603

[74] Sahakaro K, Pantupon N. Improving the mechanical properties of ENR/NBR blends via masterbatches with initial accelerator concentration gradients. International Polymer Processing. 2008;**23**(2):140-145. DOI: 10.3139/217.2020

[75] Wang Y, Zhang X, Oh J, Chung K. Effect of carbon black on self-crosslinking network structure of polychloroprene rubber and epoxidized natural rubber blends. Polymer Composites. 2015;**38**(3):463-471. DOI: 10.1002/pc.23604

[76] Hashim AS, Ong SK. Application of epoxidized natural rubber (NR) in pressure sensitive adhesives (PSAs). In: Kohjiya S, Ikeda Y, editors. Chemistry, Manufacture and Applications of Natural Rubber. 1st ed. United Kingdom: Elsevier; 2014. pp. 353-381. DOI: http://dx.doi.org/10.1533/9780857096913.2.353

[77] Hashim AS, Kohjiya S. Preparation and properties of epoxidized natural rubber network crosslinked by ring opening reaction. Polymer Gels and Networks. 1994;**2**(3-4):219-227. DOI: 10.1016/0966-7822(94)90006-X

[78] Hashim AS, Kohjiya S. Curing of epoxidized natural rubber with *p*-phenylenediamine. Journal of Polymer Science Part A: Polymer Chemistry. 1994;**32**(6):1149-1157. DOI: 10.1002/pola.1994.080320617

[79] Hashim AS, Kohjiya S, Ikeda Y. Moisture cure and *in-situ* silica reinforcement of epoxidized natural rubber. Polymer International. 1995;**38**(2):111-117. DOI: 10.1002/pi.1995.210380202

[80] Hashim AS, Akiba M. Vulcanization and crosslinking in elastomers. Progress in Polymer Science. 1997;**22**(3):475-521. DOI: 10.1016/S0079-6700(96)00015-9

[81] Hashim AS, Kawabata N, Kohjiya S. Silica reinforcement of epoxidized natural rubber by the sol-gel method. Journal of Sol-Gel Science and Technology. 1995;**5**(3):211-218. DOI: 10.1007/BF00487018

[82] Derouet D, Radhakrishnan N, Brosse J-C, Boccaccio G. Phosphorus modification of epoxidized liquid natural rubber to improve flame resistance of vulcanized rubbers. Journal of Applied Polymer Science. 1994;**52**(9):1309-1316. DOI: 10.1002/app.1994.070520915

[83] Phinyocheep P, Phetphaisit CW, Derouet D, Campistron I, Brosse JC. Chemical degradation of epoxidized natural rubber using periodic acid: Preparation of epoxidized liquid natural rubber. Journal of Applied Polymer Science. 2005;**95**(1):6-15. DOI: 10.1002/app.20812

[84] Klaichim W, Klinpituksa P, Waehamad W-A. A novel polymeric herbicide based on phenoxyacetic acid derivatives. Songklanakarin Journal of Science and Technology. 2009;**31**(1):57-62

[85] Abdul Razak NES, Mohd Nor H, Wan Ali WK. Synthesis of hydroxyl terminated epoxidized natural rubber as a binder for solid rocket propellant. Applied Mechanics and Materials. 2015;**695**:127-130. DOI: 10.4028/www.scientific.net/AMM.695.127

[86] Azhar NHA, Md Rasid H, Yusoff SFM. In: The 2016 UKM FST Postgraduate Colloquium: Proceedings of the Universiti Kebangsaan Malaysia, Faculty of Science and Technology 2016 Postgraduate Colloquium. Chemical Modifications of Liquid Natural Rubber. In: AIP Conference Proceedings of The 2016 UKM FST Postgraduate Colloquium; 13-14

April 2016; UKM, Bangi. Bangi, Selangor, Malaysia: Faculty of Science and Technology, UKM, Bangi, Selangor, Malaysia; 2016. p. 030024. DOI: http://doi.org/10.1063/1.4966762

[87] Rooshenass P, Yahya R, Gan SN. Comparison of three different degradation methods to produce liquid epoxidized natural rubber. Rubber Chemistry and Technology. 2016;**89**(1):177-198. DOI: http://dx.doi.org/10.5254/RCT.15.84878

[88] Karthni K, Abd. Aziz MA, Nur Athirah M. Development of glass coating using liquid epoxidised natural rubber. International Journal of Engineering and Applied Sciences. 2014;**5**(7):12-16

[89] Abu Bakar MA, Ahmad S, Kuntjoro W. Effect of epoxidized natural rubber on mechanical properties epoxy reinforced kenaf fibre composites. Pertanika Journal of Science & Technology. 2012;**20**(1):129-137

[90] Tan SK, Ahmad S, Chia CH, Mamun A, Heim HP. A comparison study of liquid natural rubber (LNR) and liquid epoxidized natural rubber (LENR) as the toughening agent for epoxy. American Journal of Materials Science. 2013;**3**(3):55-61. DOI: 10.5923/j.materials.20130303.02

[91] Kargarzadeh H, Ahmad I, Abdullah I, Thomas R, Dufresne A, Thomas S, Hassan A. Functionalized liquid natural rubber and liquid epoxidized natural rubber: A promising green toughening agent for polyester. Journal of Applied Polymer Science. 2015;**132**(3) 41292(1-15). DOI: 10.1002/app.41292

[92] Axtell FH, Phinyocheep P, Kriengchieocharn P. The effect of modified natural rubber compatibilizers on polyamide 6/natural rubber blends. Journal of the Science Society of Thailand. 1996;**22**(3):201-216. DOI: 10.2306/scienceasia1513-1874.1996.22.201

[93] Carone E, Jr, Kopcak U, Gonçalves MC, Nunes SP. *In situ* compatibilization of polyamide 6/natural rubber blends with maleic anhydride. Polymer. 2000;**41**(15):5929-5935. DOI: http://dx.doi.org/10.1016/S0032-3861(99)00800-9

[94] Nakason C, Kaesman A, Homsin S, Kiatkamjornwong S. Rheological and curing behavior of reactive blending. I. Maleated natural rubber–cassava starch. Journal of Applied Polymer Science. 2000;**81**(11):2803-2813. DOI: 10.1002/app.1728

[95] Nakason C, Kaesaman A, Supasanthitikul P. The grafting of maleic anhydride onto natural rubber. Polymer Testing. 2004;**23**(1):35-41. DOI: 10.1016/S0142-9418(03)00059-X

[96] Saelao J, Phinyocheep P. Influence of styrene on grafting efficiency of maleic anhydride onto natural rubber. Journal of Applied Polymer Science. 2005;**95**(1):28-38. DOI: 10.1002/app.20810

[97] Bevilacqua EM. Grafting in natural rubber. Journal of Polymer Science Part A: Polymer Chemistry. 1957;**24**(106):292-296. DOI: 10.1002/pol.1957.1202410612

[98] Baker CSL. Modified natural rubber. In: Bhowmick AK, Stephens HL, editors. Handbook of Elastomers. 2nd ed. New York, NY: Marcel Dekker; 2001. pp. 61-131

[99] Lenka S, Nayak PL, Das AP, Mishra SN. Grafting of vinyl monomers onto natural rubber. I. Graft copolymerization of methyl methacrylate onto natural rubber using quinquevalent vanadium ion as the initiator. Journal of Applied Polymer Science. 1985;**30**(1):429-433. DOI: 10.1002/app.1985.070300138

[100] Lenka S, Nayak PL, Mohanty IB, Mishra SN. Graft copolymerization onto rubber. VI. Graft copolymerization of methyl methacrylate onto rubber using acetylacetonate complex of manganese (III). Journal of Applied Polymer Science. 1985;**30**(6):2711-2714. DOI: 10.1002/app.1985.070300636

[101] Lenka S, Nayak PL, Basak A. Graft copolymerization onto natural rubber. XII. Graft copolymerization of methyl methacrylate onto natural rubber using dimethyl aniline–cupric ion redox system. Journal of Polymer Science Part A: Polymer Chemistry. 1986;**24**(11):3139-3144. DOI: 10.1002/pola.1986.080241138

[102] Nayak PL, Basak A. Graft copolymerization onto natural rubber. XV. Graft copolymerization of methyl methacrylate onto natural rubber using a potassium permanganate-ascorbic acid redox system. Journal of Applied Polymer Science. 1986;**32**(3):4271-4275. DOI: 10.1002/app.1986.070320338

[103] Erbil HY. Graft copolymerisation of some hydrophilic vinyl monomers in natural rubber. Journal of Natural Rubber Research. 1986;**1**(4):234-239

[104] Hourston DJ, Romaine J. Modification of natural rubber latex. II. Natural rubber poly(methyl methacrylate) composite latexes synthesized using amine-activated hydroperoxide. Journal of Applied Polymer Science. 1990;**39**(7):1587-1594. DOI: 10.1002/app.1990.070390715

[105] Thiraphattaraphun L, Kiatkamjornwong S, Prasassarakich P, Damronglerd S. Natural rubber-g-methyl methacrylate/poly(methyl methacrylate) blends. Journal of Applied Polymer Science. 2001;**81**(2):428-439. DOI: 10.1002/app.1455

[106] Lee DY, Subramaniam N, Fellows C, Gilbert RG. Structure-property relationships in modified natural rubber latexes grafted with methyl methacrylate and vinyl neo-decanoate. Journal of Polymer Science: Part A: Polymer Chemistry. 2002;**40**(7):809-822. DOI: 10.1002/pola.10165

[107] George B, Maiti SN, Varma IK. Graft copolymerization of methyl methacrylate on to natural rubber: Effect of polymerization conditions on particle morphology. Journal of Elastomers & Plastics. 2006;**38**(4):319-331. DOI: https://doi.org/10.1177/0095244306067129

[108] Nakason C, Pechurai W, Sahakaro K, Kaesaman A. Rheological, thermal, and curing properties of natural rubber-g-poly(methyl methacrylate). Journal of Applied Polymer Science. 2006;**99**(4):1600-1614. DOI: 10.1002/app.22518

[109] Kochthongrasamee T, Prasassarakich P, Kiatkamjornwong S. Effect of redox initiator on graft copolymerization of methyl methacrylate onto natural rubber. Journal of Applied Polymer Science. 2006;**101**(4):2587-2601. DOI: 10.1002/app.23997

[110] Kalkornsurapranee E, Sahakaro K, Kaesaman A, Nakason C. From a laboratory to a pilot scale production of natural rubber grafted with PMMA. Journal of Applied Polymer Science. 2009;**114**(1):587-597. DOI: 10.1002/app.30529

[111] Abu Bakar R, Fauzi MS. Natural rubber-grafted-poly (methyl methacrylate): Influence of coagulating agents on properties and appearances. Journal of Chemistry and Chemical Engineering. 2012;**6**:962-966

[112] Jessy R, Hashim AS. *In situ* polymerization of methyl methacrylate in deproteinized natural rubber latex. In: Proceedings of National Symposium on Polymeric Materials 2000, Penang, Malaysia; 1-2 June 2000; Penang, Malaysia; 2000

[113] Yap GS. Studies on mechanical properties of poly(methyl methacrylate) and poly(methyl methacrylate)-modified natural rubber blend [thesis]. Penang: Universiti Sains Malaysia; 2008

[114] Neliappan V, El-Aasser MS, Klein A, Daniels ES, Roberts JE, Pearson RA. Effect of the core/shell latex particle interphase on the mechanical behaviour of rubber-toughened poly(methyl methacrylate). Journal of Applied Polymer Science. 1997;**65**(3):581-593. DOI: 10.1002/(SICI)1097-4628(19970718)65:3<581::AID-APP18>3.0.CO;2-#

[115] Sperling LH, Klein A, Sambasivam M, Kim KD. Molecular basis of healing and fracture at polymer interfaces. Polymers for Advanced Technologies. 1994;**5**(9):453-472. DOI: 10.1002/pat.1994.220050901

[116] Bloomfield GF, Swift P. The polymerization of vinyl monomers in natural-rubber latex. Rubber Chemistry and Technology. 1956;**29**(4):1119-1126. DOI: http://dx.doi.org/10.5254/1.3542612

[117] Neoh SB, Hashim AS. Highly grafted polystyrene-modified natural rubber as toughener for polystyrene. Journal of Applied Polymer Science. 2004;**93**(4):1660-1665. DOI: 10.1002/app.20576

[118] Hashim AS, Ong SK. Study on polypropylene/natural rubber blend with polystyrene-modified natural rubber as compatibilizer. Polymer International. 2002;**51**(7):611-616. DOI: 10.1002/pi.920

[119] Hashim AS, Khalil HA, Neoh SB, Ong SK. Polystyrene-modified natural rubber as modifier for polypropylene and polystyrene. In: Proceedings of the 3rd International Conference on Recent Advances in Materials, Minerals and Environment 2003; 20-22 October 2003; Bayview Beach Resort, Penang. Penang: School of Materials & Mineral Resources Engineering, USM, Engineering Campus, Nibong Tebal, Penang; 2003. pp. 205-211

[120] Ong SK, Sia CS, Hashim AS. Styrene-modified natural rubber as compatibilizer/modifier in PP/NR and PP/PS blends. In: The Proceeding of Golden Jubilee International Polymer Conference; 16-17 March 2010; Sunway Resort Hotel & Spa, Kuala Lumpur. Kuala Lumpur: Plastics Rubber Institute of Malaysia; 2010

[121] Ong SK, Hashim AS. Effect of polystyrene-modified natural rubber as high molecular weight modifier in polypropylene based binary blends. International Journal of Engineering & Technology. 2011;**11**(4):39-48

[122] Al-Malaika S, Amir EJ. Thermoplastic elastomers I. Effect of processing variables on tensile properties of natural rubber/polypropylene blends. Journal of Natural Rubber Research. 1986;**1**(2):104-121

[123] Elliott DJ. Natural rubber–Polypropylene blends. In: De SK, Bhowmick AK, editors. Thermoplastic Elastomers from Rubber-Plastic Blends. 1st ed. New York, NY: Ellis Horwood; 1990. p. 102

[124] Man SHC, Hashim AS, Md. Akil H. Preparation and characterization of styrene-methyl methacrylate in deproteinized natural rubber latex (SMMADPNR). e-Polymers. 2007;**7**(1):919-928

[125] Man SHC, Hashim AS, Md. Akil H. Properties of styrene-methyl methacrylate grafted DPNR latex at different monomer concentrations. Journal of Applied Polymer Science. 2008;**109**(1):9-15. DOI: 10.1002/app.28047

Chapter 9

Preparation, Characterization, and Preliminary Biocompatibility Evaluation of Carbon Ion-Implanted Silicone Rubber

Xin Zhou, Yiming Zhang, Xiaohua Shi and
Dongli Fan

Additional information is available at the end of the chapter

http://dx.doi.org/10.5772/intechopen.69251

Abstract

Silicone rubber (SR) is a common soft tissue filler material used in plastic surgery. However, it suffers from poor biocompatibility. Previous studies have found that the ion implantation technology can be used to improve the biocompatibility of metal materials. However, it is not clear whether it can improve the biocompatibility of polymer materials. In this study, carbon ion SR was prepared by carbon ion implantation. After that, the characteristics of ion implanted SR were investigated. Then, *Escherichia coli* was utilized to test the antibacterial ability of the carbon ion implanted SR. Besides, the dermal fibroblasts were used to evaluate the cytocompatibility. From the results, carbon ion implantation had no significant effect on the hardness, tensile strength and elongation at break of SR. At the same time, there was no significant change in the surface morphology of SR. But the results show that the surface nano-morphology, surface element composition, hydrophobic and ζ potential of the surface of SR changed significantly. The changes further mediated the lower adhesion of bacteria and enhanced biocompatibility. In conclusion, the carbon ion implantation technology can improve the surface properties of silicone rubber, and further improve its biocompatibility.

Keywords: silicone rubber, ion implantation, polymer, surface modification, biocompatibility

1. Introduction

In the clinical application of plastic surgery, silicone rubber (SR) due to its good physical and chemical stability, physical inertia, easy processing, and corrosion resistance is still the most commonly used clinical soft tissue filling material [1–4]. However, a strong hydrophobic SR surface, leading to its poor biocompatibility, the fibrous connective tissue, forms a capsule around the material [5]. The capsule thickening and contraction over time easily lead to deformation and displacement of implant materials. Besides, certain materials made from silicone rubber, such as catheters, are widely used in medicine but have several limitations, for example, bacteria can readily colonize the surfaces of silicone rubber, facilitating infection and even causing patient death in certain cases [6–9]. It is of significance to carry out suitable surface modification to enhance its biocompatibility and further to improve the clinical effect of soft tissue filling [10].

At present, the surface modification of silicone rubber can be treated by plasma treatment, grafting copolymerization, and biomimetic coating, but there are some defects, such as the steps were cumbersome and substrate temperature and the conditions are not easy to control [11–18]. In this study, the technology of ion implantation which is mainly used in metal and semiconductor materials is introduced into the surface modification of silicone rubber material. This technology has its unique advantages. The substrate temperature can be controlled, and there are a variety of ion sources that can be chosen and a micrographic structure that also can be formed on the surface of the material. At the same time, the ion implantation does not affect the bulk properties of material [19–21].

In this study, graphite was used as carbon source, and carbon ion was used to treat the surface of silicone rubber. Then, scanning electron microscope (SEM), Fourier transformation infrared spectroscopy (FTIR), X-ray diffraction (XRD), X-ray photoelectron spectroscopy (XPS), atomic force microscopy (AFM), water contact angle (WCA), ζ potential, Shore hardness A, elongation at break, and breaking strength were conducted to investigate the characteristics of modified silicone rubber. After that, *E. coli* was utilized to test the antibacterial ability of the carbon ion-implanted silicone rubber. Then, the dermal fibroblasts were used to evaluate the cytocompatibility through CCK-8, immunofluorescence, and WB. Furthermore, evaluation of tissue compatibility was carried out by the hematoxylin and eosin staining, Masson's staining, and immunohistochemistry test after the sample was subcutaneously implanted to SD rats for 7, 30, 90, and 180 days. The long-term goal of this study is to gain a better biomaterial for use by plastic surgeons or other clinical applications.

2. Materials and methods

2.1. Sample preparation

SR sheets with dimensions of 100 × 100 × 1 mm were prepared from a two-component silicone system. Both component A and component B were clinical-quality liquids provided by Chenguang Research Institute of Chemical Industry, China. Three doses of carbon ions were implanted using the ion implanter, respectively. The doses are 1×10^{15} ions/cm^2. After that, SR and C-SR sheets were manufactured into disc-like samples with a diameter of 6 mm using a hole puncher and into square samples with dimensions of 10 × 10 × 1 mm. The disc-like samples were used in in vitro antibacterial adhesion tests, and the square samples were used for in vivo tissue compatibility and in vitro cytocompatibility evaluation. All samples were sterilized with 75% alcohol overnight. In all experiments, SR is served as the control.

2.2. Characterization of carbon ion-implanted silicone rubber

2.2.1. *Scanning electron microscopy*

Clean the sample with ethanol, then place it in a 37° incubator, and let it dry. After that, put the sample into the vacuum pump and gold plating on the surface under the negative pressure state. After cooling, the surface morphology was observed by scanning electron microscope.

2.2.2. *Fourier transformation infrared*

Put clean and dry samples in Fourier transformation infrared spectrometer, and then set the wave number to 4 cm^{-1}; scan range is 400–4000cm^{-1}. The surface composition of the samples was measured.

2.2.3. *X-ray photoelectron spectrometer*

The samples were cleaned and put in the X-ray photoelectron spectrometer. The Al Kα as radiation source, press to collect photoelectron emission from the samples and the photoelectron spectrum. Then, the content of chemical elements on the surface of the material was analyzed.

2.2.4. *X-ray diffraction*

The materials to be tested were cleaned by ethanol and then put into the X-ray diffraction spectrometer and analysis crystal structure of the carbon ion-implanted silicone rubber.

2.2.5. *Atomic force microscope*

The sample are cleaned by ethanol before being scanned by the Environment-Controlled Scanning Probe Microscope (Nanonavi E-Sweep, NSK Ltd., Tokyo, Japan), and each sample was imaged with 5 × 5 μm scanned area. The surfaces were analyzed by measuring the average surface roughness (Ra) of five randomly chosen images per sample from selected areas of 1 × 1 μm under atomic force microscope (AFM) analysis software (NanoNavi II, SII Nano Technology Inc., Tokyo, Japan). Ra is defined as the average absolute deviation of the roughness irregularities from the mean line over one sampling length and gives a good general description of height variations. Three replicas were used.

2.2.6. *Water contact angle*

The contact angle of the prepared samples was measured by DSA100 contact angle measuring instrument; the selected liquid is distilled water. When measuring the contact angle of the sample, six different parts of the sample were selected and measured at least three times.

2.2.7. *ζ Potential*

After the samples are cleaned by ethanol, the surface zeta potentials of materials were measured with a Zeta Potential Analyzer (DelsaNano C, Beckman Coulter, Germany). The measurements were carried out in 1 mmol NaCl electrolyte solution and with standard particles for flat surface cell (Otsuka Electronics Co., Ltd., Japan). Each sample chooses five points, and each point is tested 20 times.

2.2.8. *Shore hardness*

The samples were washed with absolute ethanol and then measured by an A type of the Shore hardness tester; each sample was tested five times, and the average value was obtained.

2.2.9. *Tear strength and elongation at break*

After the samples are cleaned by ethanol, an electronic universal testing machine was used to gain parameters of the tear strength and elongation at break. Each sample was tested five times, and the average value was obtained.

2.3. Evaluation of antibacterial properties

2.3.1. *Bacterial culture preparation*

Gram-negative *E. coli* (ATCC 25922) was employed to bacterial experiments. The strains were streaked on blood agar plates from frozen stocks and grown for 24 h at 37°C in ambient air. The agar plates were then kept at 4°C until further use. For each experiment, one colony from an agar plate was inoculated into 10 ml of tryptone soy broth (TSB) and incubated for 24 h. The bacterial suspension was then added to 0.9% sterile sodium chloride to a final concentration of 1.5×10^6 colony-forming units per ml (CFU/ml), after which a McFarland standard was prepared (in practical terms, OD600 nm = 0.132). The samples were placed on 96-well culture plates and separately incubated in 200 μl of the bacterial suspension at 37°C for 1 or 24 h. After that, the plate colony counting, fluorescence staining, and scanning electron microscopy (SEM) observations were conducted.

2.3.2. *Plate colony counting*

After the sample incubation in bacterial culture for 1 and 24 h, the bacteria on each sample were gently rinsed with phosphate-buffered saline (PBS), respectively, and ultrasonically detached in 1 ml of PBS solution. The bacteria in the PBS were re-cultivated on agar plates for colony counting. The antibacterial rates were determined based on the following relationship: antibacterial rate (%) = (CFU of control−CFU of experimental groups)/CFU of control × 100%. This assay was performed in triplicate.

2.3.3. *Fluorescence staining*

After incubation for 1 and 24 h, the various samples were gently rinsed with PBS before staining the bacteria on the samples. The staining was performed by applying LIVE/DEAD® BacLight™ Bacterial Viability Kit (L7029, Molecular Probes®, OR, USA) for 15 min in darkness and was examined by laser scanning confocal microscopy (LEICA TCS SP5, Germany). The areas of green and red color in the pictures were then analyzed by using Image-Pro Plus version 6.0 (Media Cybernetics, Inc., USA) and then calculating the proportion of red coloration based on the following relationship: red proportion (%) = red area/(green area + red area) × 100%.

2.3.4. *Scanning electron microscopy (SEM) observation*

After bacterial incubation for 1 and 24 h, the samples were rinsed with PBS to remove free bacterial cells and then fixed in 2.5% glutaraldehyde for 3 h at room temperature. The samples were then progressively dehydrated in a series of ethanol solutions (15, 30, 50, 70, 80, 90, 95, and 100%) for 15 min each. After that, the specimens underwent critical-point drying and coating with a thin conductive layer of Au. Finally, the morphology and adhesion of the bacteria on the various samples were determined by SEM (VEGA 2 SEM, TESCAN Inc., Brno, Czech Republic).

2.4. In vitro cytocompatibility evaluation

2.4.1. *Cell culture*

The use of primary dermal fibroblasts derived from discarded human neonatal circumcision specimens and the institutional ethical committee of the Third Military Medical University, People's Republic of China, approved this study. Cells were routinely cultured in Dulbecco's Modified Eagle's Medium (DMEM, Gibco) supplemented with 10% FBS and incubated at 37°C in a humidified atmosphere supplied with 5% CO_2. Trypsin-EDTA (Gibco) was used to detach cells. The loading concentration of cells for all experiments was 5×10^4 cells/ml. Incubation conditions remained unchanged.

2.4.2. *CCK-8*

To study cell proliferation, primary dermal fibroblasts were seeded on the samples and cultured in DMEM supplemented with 10% FBS at 37°C in a humidified 5% CO_2 atmosphere. Primary dermal fibroblast viability was quantified by cell counting kit (CCK)-8 assay after 24 h of incubation. The CCK-8 solution and the medium were mixed at a volume ratio of 1:10. A total of 1 ml of mixed solution was added into each well after removing the medium and reincubated for further 2 h. The optical density (OD) was measured at the wavelength of 450 nm by using microplate reader.

2.4.3. *Immunocytochemical staining*

Immunocytochemical staining was performed as described previously. Primary dermal fibroblasts were cultured on different samples for 24 h and then rinsed with PBS and fixed with 4% paraformaldehyde. Cells were incubated with 1% Triton X-100 for 20 min and blocked for 20 min with PBS containing 5% fetal calf serum and 0.1% bovine serum albumin (BSA; Sigma-Aldrich, St. Louis, MO, USA). Afterward, slides were incubated for 60 min with the Actin Tracker Green antibody (1:100; Beyotime Biotechnology, Shanghai, China). Finally, counterstaining of nuclei was performed with 4′, 6-diamidino-2-phenylindole (DAPI) (Biotium, Hayward, CA, USA) staining. Then, the cells were observed and photographed with a laser confocal scanning microscope (Leica TCS SP5, Germany).

2.4.4. *Western blotting*

The primary dermal fibroblasts were seeded on the samples for 24 h. Next, the cells were washed with ice-cold phosphate-buffered saline (PBS) and lysed with RIPA lysis buffer and protease inhibitor cocktail (Roche Diagnostics, Indianapolis, IN, USA). The lysates were sonicated for 4 s and were separated by centrifugation at 4°C and 14,000g for 2 min. Protein concentration was determined by a BCA protein assay kit. Aliquots of 40 μg of proteins from each treatment were separated by 10% sodium dodecyl sulfate-polyacrylamide gel electrophoresis and transferred onto polyvinylidene difluoride (PVDF) membranes (Millipore, Bedford, UK). After blocking with 10% instant nonfat dry milk for 2 h, the membranes were incubated with primary antibody (dissolved in phosphate-buffered saline Tween 20) overnight at 4°C followed by incubation with secondary antibody for 1 h at room temperature. The immunosignal was detected with the enhanced chemiluminescence detection system (Amersham Biosciences, Piscataway, MD, USA). The detail of primary antibody used in this study was listed in **Table 1**. As a loading control, glyceraldehyde-3-phosphate dehydrogenase (GAPDH) was probed and visualized.

Antibody name	Catalogue no.	Isotype	Dilution rate	Molecular weight (kDa)	Manufacturer
FAK	12636-1-AP	R	1:500	110	Proteintech
Vinculin	sc-7649	G	1:200	117	Santa
Talin	14168-1-AP	R	1:500	230	Proteintech
Zyxin	60254-1-Ig	M	1:500	78	Proteintech
Paxillin	10029-1-Ig	R	1:1000	68	Proteintech
GAPDH	MAB374	M	1:300	36	Millipore

Table 1. The detail of primary antibody used in this study.

2.5. In vivo tissue compatibility evaluation

2.5.1. Animals and surgery

This research was performed in accordance with the *Guide for the Care and Use of Laboratory Animals* published by the US National Institutes of Health (Washington, DC; National Academies Press, 2011), and all of the animal protocols were approved by the Institutional Animal Care and Use Committee of the Third Military Medical University, China. A total of 16 female Sprague-Dawley rats (weighing 160–200 g) were used (2 groups of 8 animals each), and all rats were housed under a 12-h light/dark cycle with free access to water and food. Prior to surgery, all of the rats were anesthetized with 3% pentobarbital sodium (1 ml/1000 g). The skin was swabbed with iodine, and four parallel incisions (10 mm) were performed. The material samples were implanted subcutaneously along the back region. The implants and their surrounding tissue were retrieved from each group by wide excision at 7, 30, 90, and 180 days after implantation and were then fixed in 4% paraformaldehyde solution. After that, HE and Masson's staining and immunohistochemistry were conducted.

2.5.2. HE and Masson's staining

The fixed tissues were sectioned (6 μm thick) and stained using a HE Staining Kit (C0105, Beyotime Inc., Shanghai, China). The thickness of the fibrotic capsule around each implant was determined at five equidistant points for statistical accuracy. Collagen deposition in the tissue around the implants was studied by Masson's trichrome staining, which was performed using a staining kit (MST-8003, Maixin Biological Technology Co., Ltd., Fujian, China). All procedures were performed based on the manufacturer's instructions.

2.5.3. Immunohistochemistry

Immunohistochemistry was performed on 4% paraformaldehyde-fixed cryostat sections of frozen tissue specimens. Endogenous peroxidase and nonspecific antibody binding were blocked with 3% H_2O_2 and 100% methanol, with a ratio of 1:5 and a blocking time of 30 min. Next, 0.02 M PBS was used for antigen retrieval while heating in a water bath, followed by treatment with 5% blocking reagent at 37°C for 30 min. The slides were then incubated at 4°C for 12 h with a primary antibody against CD68, CD4, TNF-α, α-SMA, or elastin (1:25) (Boster Biological Engineering Co., Ltd., Hubei, China). After washing in PBS, a secondary antibody was applied for 30 min. Visualization was achieved by adding 3, 3′-diaminobenzidine chromogen.

2.6. Statistics

All data are expressed as the mean ± standard deviation (SD); statistics were analyzed using SPSS statistical software. One-way ANOVA combined with multiple comparisons performed along with Tukey's multiple comparison tests was utilized to determine the level of significance. In all of the statistical evaluations, $P < 0.05$ was considered significant.

3. Results

3.1. Ion implantation changes the surface roughness and zeta potential of SR

After ion implantation, SEM, AFM, FTIR, XPS, XRD, water contact angle measure instrument, zeta potential detection instrument, Shore A durometer, and an electronic universal testing machine were used to investigate the change in properties of carbon ion silicone rubber.

The SEM results failed to find any significant differences between virgin SR and three C-SRs (**Figure 1**), indicating that carbon ion implantation did not change the macroscale surface of SR.

At the same time, there are no any significant differences, or the difference was very small on the results of FTIR (**Figure 2**), XRD (**Figure 3**), Shore hardness (**Figure 4A**), tear strength (**Figure 4B**),

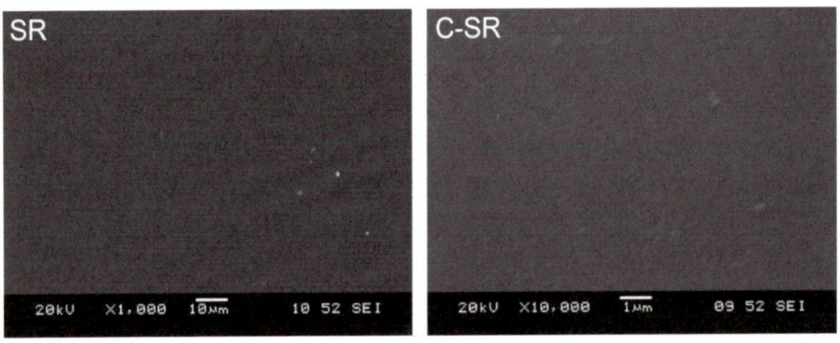

Figure 1. Representative scanning electron microscopic images of virgin silicone rubber and carbon ion-implanted silicone rubber.

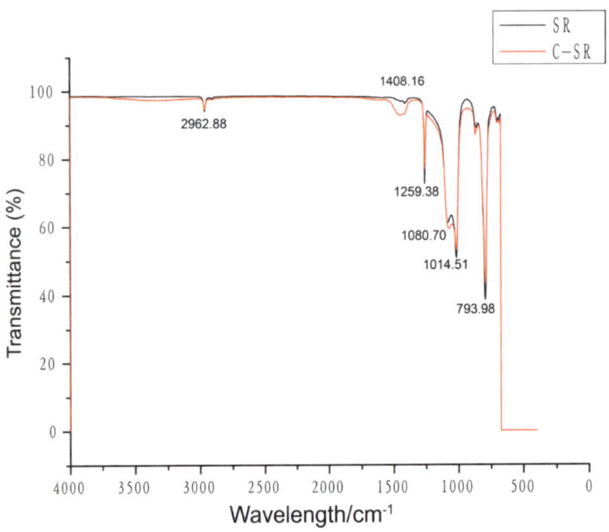

Figure 2. The Fourier transformation infrared spectroscopy results of virgin silicone rubber and carbon ion-implanted silicone rubber.

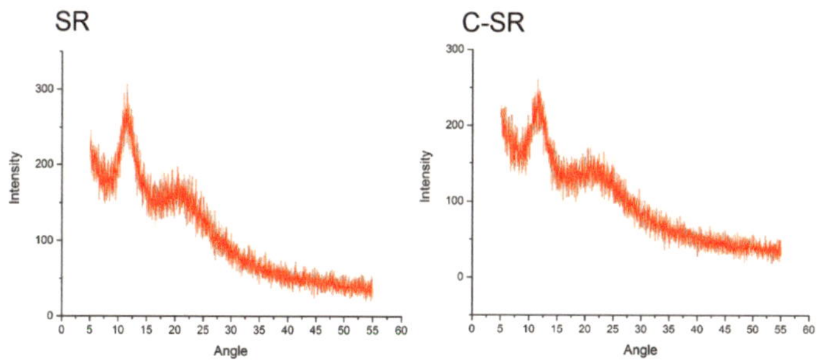

Figure 3. The XRD results of virgin silicone rubber and carbon ion-implanted silicone rubber.

and elongation at break (**Figure 4C**). The FTIR showed that there has been no new peak, noting that there has been no functional group produced. The Shore hardness, tear strength, and elongation at break showed that the ion implantation cannot change the bulk properties.

From the results of water contact angle, we found that carbon ion implantation significantly decreased the water contact angle of SR (**Figure 5**).

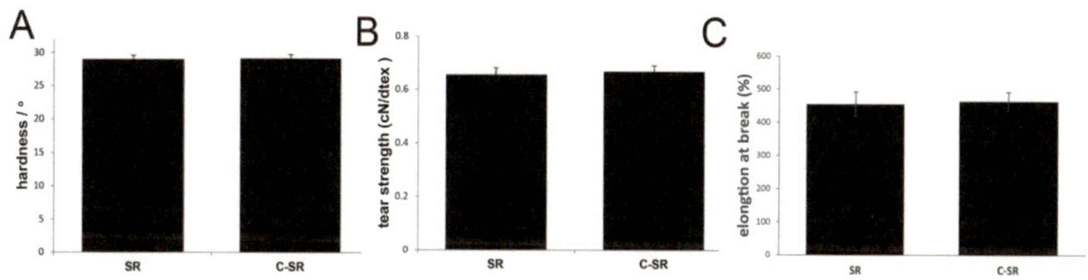

Figure 4. The mechanical properties of virgin silicone rubber and carbon ion-implanted silicone rubber. (A) Shore hardness. (B) Tear strength. (C) Elongation at break.

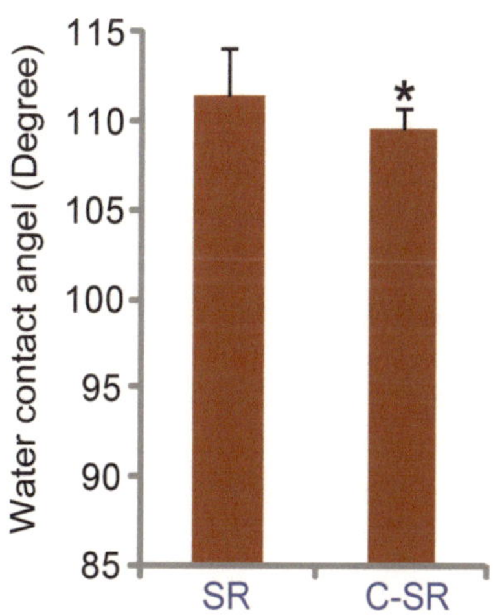

Figure 5. Water contact angle of virgin silicone rubber and carbon ion-implanted silicone rubber. (*P < 0.05 compared with silicone rubber).

Besides, the XPS results showed that carbon ion implantation significantly changed the surface silicone oxygen rate and chemical element distribution of SR (**Figure 6**) (**Table 2**); note that with the ion implantation, the carbon content in the material increased, while the Si content decreased, suggesting that implanted carbon atom may replace the Si of virgin SR, interrupting the original Si-O assemble, so the surface-free energy increases, thereby theoretically decreasing material's water contact angle. Furthermore, AFM images revealed that the surfaces of C-SRs were composed of larger irregular peaks and deeper valleys, while virgin SR exhibited a relatively smooth and more homogeneous surface (**Figure 7A**). The surface roughness of the CSR, which underwent carbon ion implantation, was the highest (**Figure 7B**).

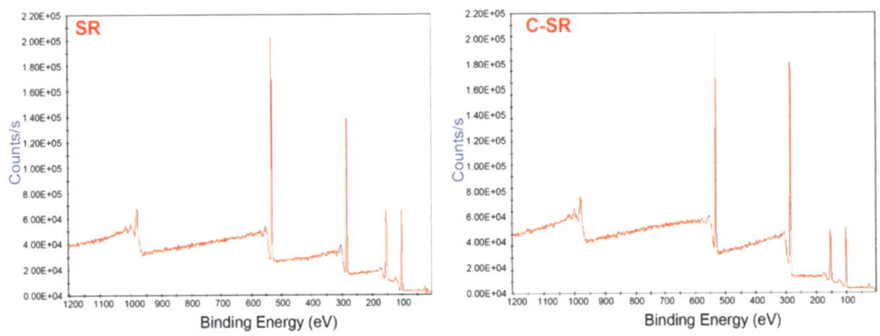

Figure 6. XPS results of virgin silicone rubber and carbon ion-implanted silicone rubber.

Group	Si 2p	C 1s	O 1s
SR	28.82	47.57	23.65
C-SR	18.94	58.40	22.67

Table 2. Chemical composition (in at.%) from the XPS analysis.

In addition, all samples exhibited negative zeta potentials and reflect that the surfaces of all samples were negatively charged. The absolute value of the zeta potential increased with the ion dose (**Figure 8**). Considering the influence of surface roughness on contact angle, we propose that ion implantation can change the surface roughness of the material and increase the surface potential of the material.

3.2. Ion implantation inhibits bacterial adhesion on SR

Preventing bacterial adhesion and biofilm formation by improving the surface antibacterial adhesion property of the silicone rubber is critical for eliminating various types of infections. After ion implantation, we use Gram-negative *E. coli* (American Type Culture Collection 25922) to evaluate the ability to resist bacteria adhesion. From the result, after 1 h of incubation, the rate of *E. coli* adherence on the carbon ion silicone rubber increased to approximately 11% (**Figure 9**) ($P < 0.05$). After 24 h of incubation, the rates of bacterial adherence were slightly lower, but did not significantly decrease compared with that after 1 h of incubation ($P > 0.05$).

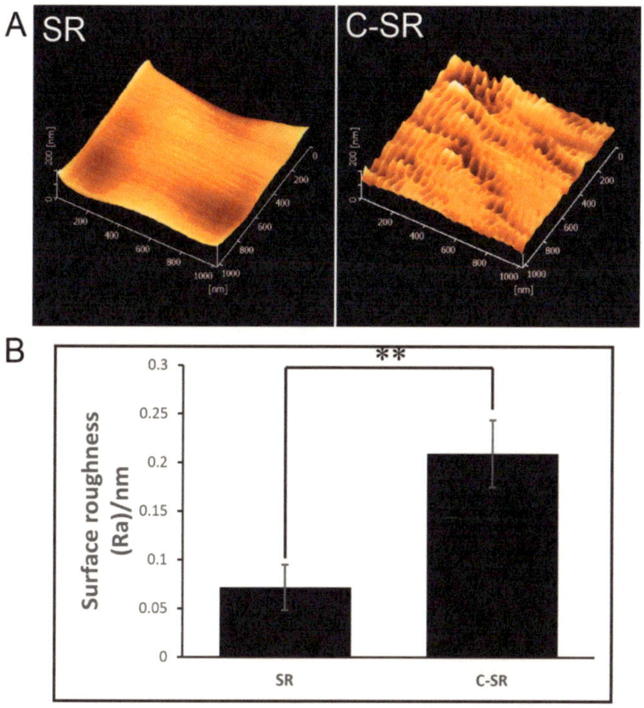

Figure 7. AFM results of virgin silicone rubber and carbon ion-implanted silicone rubber. (A) Representative atomic force microscope images. (B) Surface roughness. (**P < 0.01 compared with silicone rubber).

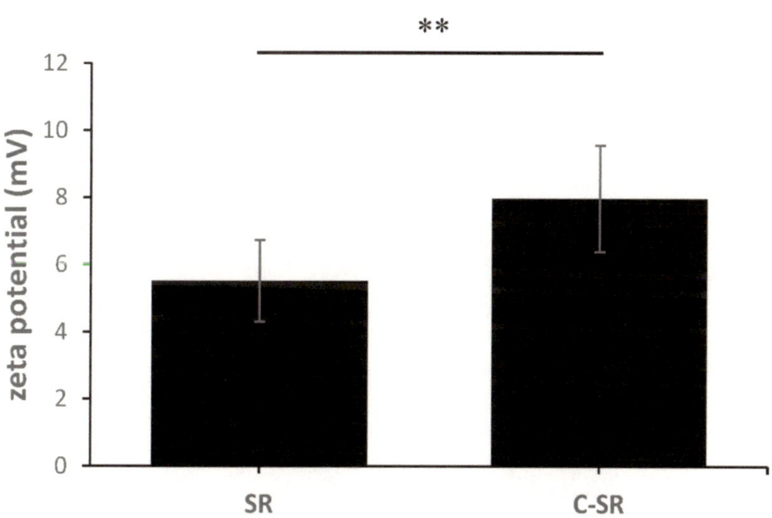

Figure 8. Zeta potential of virgin silicone rubber and carbon ion-implanted silicone rubber. (**P < 0.01 compared with silicone rubber).

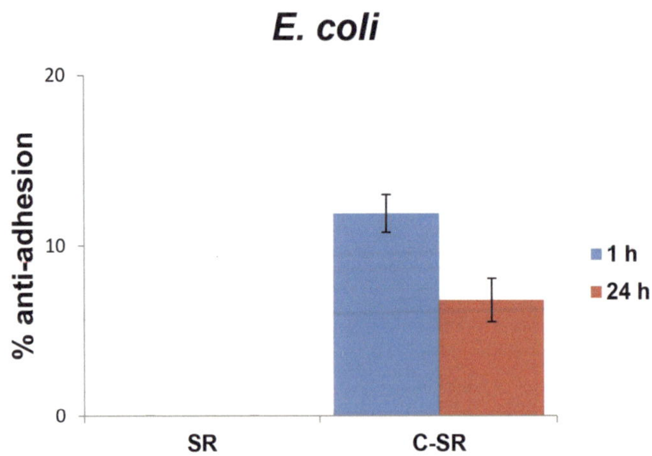

Figure 9. The antiadhesion rates (percentage) of virgin silicone rubber and carbon ion-implanted silicone rubber. After all samples were cultured in bacterial suspension for 1 and 24 h, bacteria on the surface of all samples were recultured on the plate, and bacterial colonies were subsequently counted. According to the number of colonies, the antiadhesion rates (percentage) for *E. coli* were counted. The data are presented as the mean ± SD (n = 3); $^{*}P < 0.05$ compared with silicone rubber.

The ability of carbon ion silicone rubber to prevent viable bacteria colonization was also verified by fluorescence staining. The results showed that the amount of bacterial adhesion to the surface of carbon ion silicone rubber was reduced compared with the virgin silicone rubber (**Figure 10**).

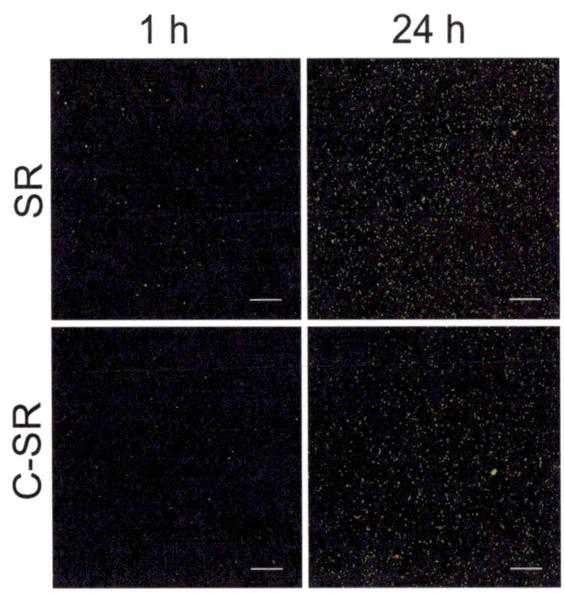

Figure 10. Representative images of fluorescence staining observation of virgin silicone rubber and carbon ion-implanted silicone rubber. Representative images showing bacteria viability on SR and C-SR after 24 h of incubation, as indicated by staining with a LIVE/DEAD BacLight Bacterial Viability Kit (Thermo Fisher Scientific, Waltham, Mass.). The live bacteria appear green, whereas the dead bacteria are red (original magnification, ×200).

Scanning electron microscopy was performed to examine the attached bacteria. The results showed that bacteria were observed on surfaces of all samples, but there were differences in quantity (**Figure 11**).

3.3. Ion implantation enhances the cell proliferation and adhesion on SR

Cell adhesion and proliferation were evaluated by the immunofluorescence, WB, and CCK-8 experiments. After cell culture on samples for 24 h, the cell adhesion was reflected by the immunofluorescence and focal adhesion kinase (FAK), and adhesion-associated proteins vinculin, talin, zyxin, and paxillin were also detected by WB. As shown in **Figure 12**, primary dermal fibroblasts cultured on C-SRs have a higher adherence rate and higher cell areas.

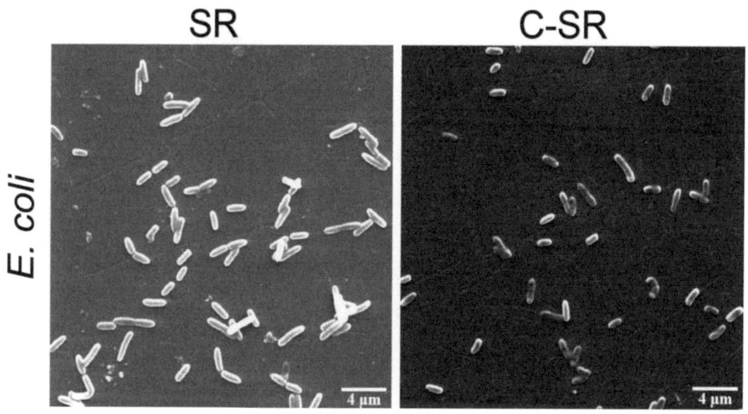

Figure 11. Representative scanning electron microscopic images of the bacteria on SR and C-SR after incubation for 24 h.

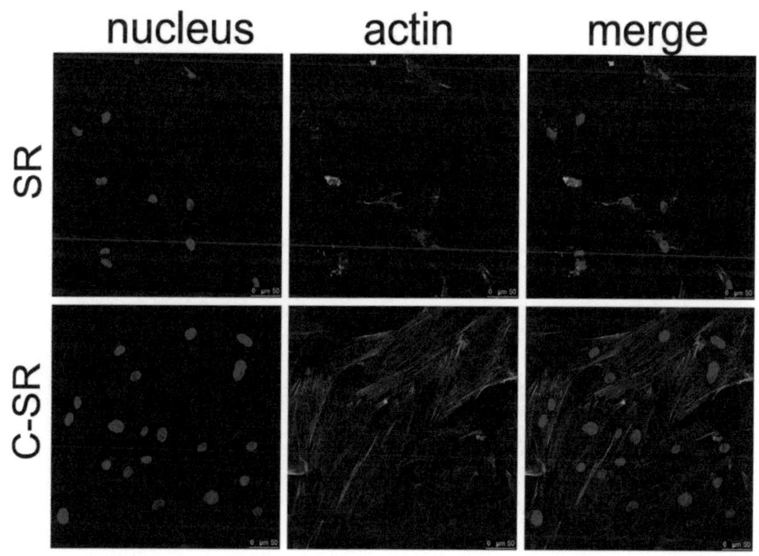

Figure 12. Representative cytoskeleton images of the fibroblasts on SR and C-SR after incubation for 24 h.

The WB results also showed that the primary dermal fibroblasts cultured on C-SR expressed higher levels of FAK and adhesion-associated proteins vinculin, talin, zyxin, and paxillin (**Figure 13**).

Besides, the cell proliferation was higher on C-SR (**Figure 14**). These results together indicate that carbon ion implantation could provide a better environment for cell adhesion and proliferation.

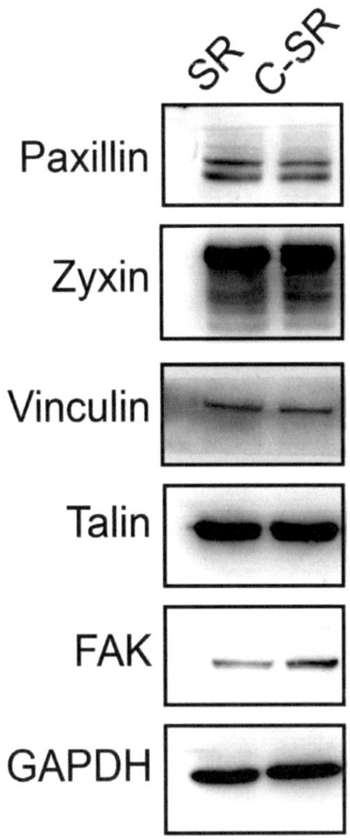

Figure 13. The WB results of FAK and adhesion-associated proteins vinculin, talin, zyxin, and paxillin.

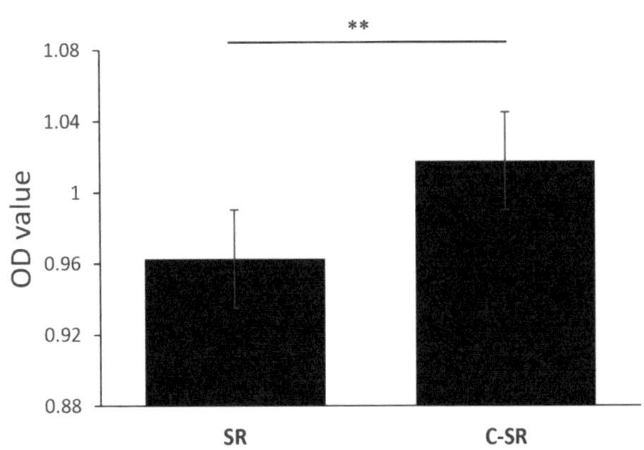

Figure 14. Cell proliferation on SR and C-SR after incubation for 24 h.

http://dx.doi.org/10.5772/intechopen.69251

3.4. Ion implantation enhances the tissue compatibility of SR

After ion implantation, the host responses were evaluated by surveying inflammation and fibber capsule formation that developed after subcutaneous implantation in Sprague-Dawley rats for 7, 30, 90, and 180 days. The thickness values of tissue capsules around the implants were identified from hematoxylin and eosin-stained sections of the peri-implant soft tissues and were analyzed as one of the physiologic responses to implantation. At 7 days after implantation, silicone rubber had the thinnest tissue capsules, and carbon ion silicone rubber had thicker (**Figure 15**) ($P > 0.05$) and weaker tissue capsules. Interestingly, the thickness decreased with longer implantation (**Figure 15**). At 180 days after implantation, silicone rubber and C-SR had the thickest and the thinnest tissue capsules, respectively (**Figure 15**).

In addition, collagen deposition was revealed using Masson's trichrome staining. Our results show that collagen gradually became sparser over time and with increasing carbon ion doses. Carbon ion silicone rubber had obviously lower collagen deposition than silicone rubber (**Figure 16**) ($P < 0.05$).

To gain insight into inflammatory foreign body responses and capsule contracture to the samples, the major biomarkers CD68, CD4, tumor necrosis factor-α, elastin, and α-smooth muscle actin were detected using immunohistochemistry. The results showed that all samples present lower expression of CD68, with no significant differences. The distribution of CD4 in the inflammatory infiltrate, which was induced by the samples, was investigated to further understand the local immunomodulation against these types of materials. The results show that there were many positive staining areas of CD4 in silicone rubber after 90 days, but positive staining in the carbon ion silicone rubber decreased with time. After 90 days, CD4 significantly decreased compared with silicone rubber.

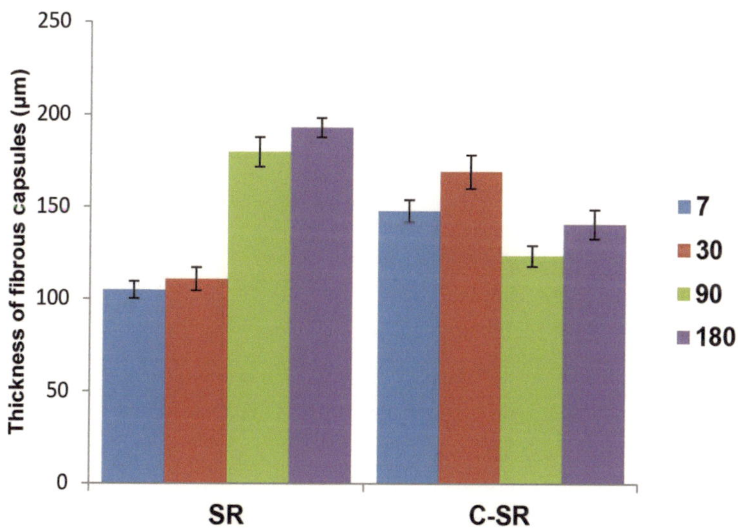

Figure 15. The capsule thicknesses around the implants.

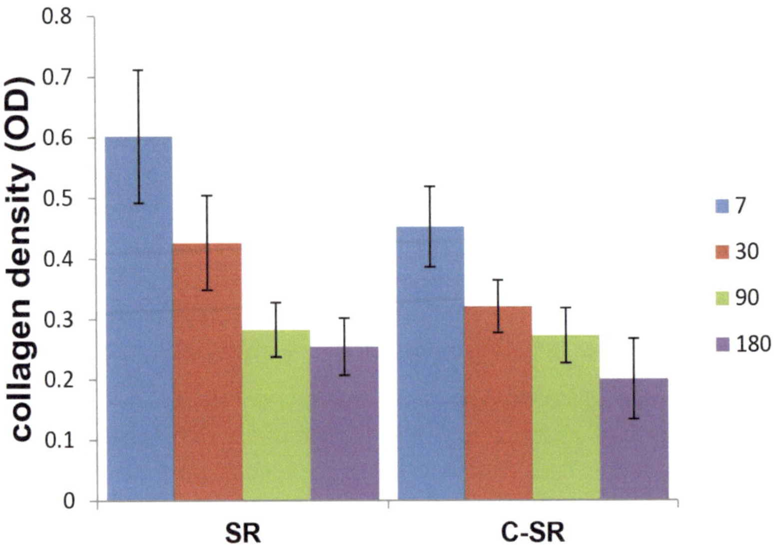

Figure 16. The collagen density around the implants.

In addition, the expression results of pro-inflammatory cytokine tumor necrosis factor-α by macrophage cells show that silicone rubber had an obviously positive staining area (**Table 3**).

Furthermore, the positive staining areas of α-smooth muscle actin and elastin have no difference; the positive staining area of α-smooth muscle actin appeared predominantly in silicone rubber than in carbon ion silicone rubber (**Table 4**). Elastin was intensely expressed in silicone rubber, particularly after 30 days (**Table 5**).

	7 days	30 days	90 days	180 days
SR	++	++	+++	+++
C1-SR	++	++	+	+

Table 3. Semiquantitative evaluation of TNF-α in peri-implant tissue.

	7 days	30 days	90 days	180 days
SR	+	++	++	++
C1-SR	+	+	+	+

Table 4. Semiquantitative evaluation of α-SMA in peri-implant tissue.

	7 days	30 days	90 days	180 days
SR	++	+++	+++	++
C1-SR	+	++	++	+

Table 5. Semiquantitative evaluation of elastin in peri-implant tissue.

4. Discussion

SR is the main method for the treatment of deformities and defects of body surface tissues and organs due to congenital or trauma. However, the SR material has poor cytocompatibility; cells are not easy to adhere to the surface of the material. It easy to form the fiber capsule on the surface, at the same time, deformation, displacement, and even cause skin infection and ulceration. By modifying the surface of SR and obtaining the modified SR material with good biocompatibility which it is expected to further improve the clinical therapeutic effect [22–25]. In this study, C ion implantation was used to modify SR. The surface properties of the materials were analyzed by a series of characterization techniques. The bacterial adherence and cell compatibility of C-SR were also observed. At the same time, a preliminary study on the histocompatibility was also carried out.

After C ion implantation, the surface properties of the material changed. The surface properties of materials can affect the biocompatibility of materials. In the further study, we observed the biocompatibility of the modified materials. CCK-8 test showed that the proliferation of human dermal fibroblasts on C-SR surface was better than that of SR. The adhesion of cells on the surface of materials mainly depends on the formation of adhesion complex. In this study, we found that the adhesion complex on C-SR had higher expression. These results indicated that C-SR was more favorable for cell adhesion and growth than SR; C-SR has excellent biocompatibility.

In this study, the surface properties and cell compatibility of modified C-SR were observed and analyzed. It was found that the surface properties of the modified materials changed somewhat compared with that of SR after carbon ion implantation. It is mainly because of the formation of new morphologies on the surface of SR after ion implantation; the change of surface properties may affect the hydrophobicity of material surface and improve the adsorption ability of the material to extracellular matrix proteins, so as to promote cell adhesion and proliferation on the surface of the material. Further studies also showed that C-SR was more favorable for cell adhesion and proliferation than SR. This suggests that if the modified material is implanted in the body, it will have a good affinity with the surrounding tissue, can reduce the degree of formation of the surrounding material, and can also reduce the deformation and displacement of the implant material and other complications. It can be concluded from the results that C ion implantation technology can improve the biocompatibility of polymer materials, and C-SR has great application prospect.

5. Conclusion

Our study evaluates the in vitro antibacterial properties and the in vivo host response to carbon ion-implanted silicone rubber. The results of our study indicate that the carbon ion silicone rubbers have good biocompatibility and lower bacterial adhesion and lower foreign body reaction with relatively thin fibrous capsules. All results show that ion implantation should be considered for further investigation and application, and carbon ion-implanted silicone rubber might be a better biomaterial for decreasing silicone rubber-initiated complications.

Acknowledgements

This work was funded by grants from the National Natural Science Foundation of China (81571918, 81401610) and a grant for Scientific Personnel Innovation from Chongqing (cstc2013kjrc-qnrc10003).

Author details

Xin Zhou, Yiming Zhang, Xiaohua Shi and Dongli Fan*

*Address all correspondence to: fdltmmu@sina.com

Department of Plastic and Cosmetic Surgery, Xinqiao Hospital, Third Military Medical University, Chongqing, China

References

[1] Liu P, Chen Q, Yuan B, Chen M, Wu S, Lin S, SHEN J. Facile surface modification of silicone rubber with zwitterionic polymers for improving blood compatibility. Materials Science & Engineering C-Materials for Biological Applications. 2013;**33**(7):3865-3874. DOI: 10.1016/j.msec.2013.05.025

[2] Lugowski SJ, Smith DC, Bonek H, Lugowski J, Peters W, Semple J. Analysis of silicon in human tissues with special reference to silicone breast implants. Journal of Trace Elements in Medicine and Biology. 2000;**14**(1):31-42. DOI: 10.1016/S0946-672X(00)80021-8

[3] Erlich MA, Parhiscar A. Nasal dorsal augmentation with silicone implants. Facial Plastic Surgery. 2003;**19**(4):325-330. DOI: 10.1055/s-2004-815652

[4] Cazacu M, Racles C, Vlad A, Antohe M, Forna N. Silicone-based composite for relining of removable dental prosthesis. Journal of Composite Materials. 2009;**43**(10):2045-2055. DOI: 10.1177/0021998309340447

[5] Anderson JM, Rodriguez A, Chang DT. Foreign body reaction to biomaterials. Seminars in Immunology. 2008;**20**(2):86-100. DOI: 10.1016/j.smim.2007.11.004

[6] Vasilev K, Cook J, Griesser HJ. Antibacterial surfaces for biomedical devices. Expert Review of Medical Devices. 2009;**6**(5):553-567. DOI: 10.1586/erd.09.36

[7] Bazaka K, Jacob MV, Crawford RJ, Ivanova EP. Plasma-assisted surface modification of organic biopolymers to prevent bacterial attachment. Acta Biomaterialia. 2011;**7**(5):2015-2028. DOI: 10.1016/j.actbio.2010.12.024

[8] Bergmann PA, Tamouridis G, Lohmeyer JA, Mauss KL, Becker B, Knobloch J, Mail Nder P, Siemers F. The effect of a bacterial contamination on the formation of capsular contracture with polyurethane breast implants in comparison with textured silicone implants: An animal study. Journal of Plastic Reconstructive and Aesthetic Surgery. 2014;**67**(10):1364-1370. DOI: 10.1016/j.bjps.2014.05.040

[9] Tang CY, Chen D-z, Chan KYY, Chu KM, Ng PC, Yue TM. Fabrication of antibacterial silicone composite by an antibacterial agent deposition, solution casting and crosslinking technique. Polymer International. 2011;**60**(10):1461-1466. DOI: 10.1002/pi.3102

[10] Rohrich RJ. Advances in breast augmentation. Plastic and Reconstructive Surgery. 2006;**118**(7):1S-2S. DOI: 10.1097/01.prs.0000247292.96867.06

[11] Xue L, Zhang Y, Zuo Y, Diao S, Zhang J, Feng S. Preparation and characterization of novel UV-curing silicone rubber via thiol-ene reaction. Materials Letters. 2013;**106**(1):425-427. DOI: 10.1016/j.matlet.2013.05.084

[12] van der Houwen EB, Kuiper LH, Burgerhof JG, van der Laan BF, Verkerke GJ. Functional buckling behavior of silicone rubber shells for biomedical use. Journal of the Mechanical Behavior of Biomedical Materials. 2013;**28**(1):47-54. DOI: 10.1016/j.jmbbm.2013.07.002

[13] Abbasi F, Mirzadeh H, Katbab AA. Modification of polysiloxane polymers for biomedical applications: A review. Polymer International. 2001;**50**(12):1279-1287. DOI: 10.1002/pi.783

[14] Magennis EP, Hook AL, Williams P, Alexander MR. Making silicone rubber highly resistant to bacterial attachment using thiol-ene grafting. ACS Applied Materials & Interfaces. 2016;**8**(45):30780-30787. DOI: 10.1021/acsami.6b10986

[15] Ziraki S, Zebarjad SM, Hadianfard MJ. A study on the role of polypropylene fibers and silica nanoparticles on the compression properties of silicone rubber composites as a material of finger joint implant. International Journal of Polymeric Materials and Polymeric Biomaterials. 2017;**66**(1):48-52. DOI: 10.1080/00914037.2016.1180623

[16] Rahman CV, Ben-David D, Dhillon A, Kuhn G, Gould TWA, Muller R, Rose F, Shakesheff KM, Livne E. Controlled release of BMP-2 from a sintered polymer scaffold enhances bone repair in a mouse calvarial defect model. Journal of Tissue Engineering and Regenerative Medicine. 2014;**8**(1):59-66. DOI: 10.1002/term.1497

[17] Ling T, Lin J, Tu JJ, Liu SQ, Weng WJ, Cheng K, Wang HM, Du PY, Han GR. Mineralized collagen coatings formed by electrochemical deposition. Journal of Materials Science-Materials in Medicine. 2013;**24**(12):2709-2718. DOI: 10.1007/s10856-013-5028-9

[18] Meretoja VV, Rossi S, Peltola T, Pelliniemi LJ, Narhi TO. Adhesion and proliferation of human fibroblasts on sol-gel coated titania. Journal of Biomedical Materials Research Part A. 2010;**95A**(1):269-275. DOI: 10.1002/jbm.a.32829

[19] Tsuji H, Izukawa M, Ikeguchi R, Kakinoki R, Sato H, Gotoh Y, Ishikawa J. Surface treatment of silicone rubber by carbon negative-ion implantation for nerve regeneration. Applied Surface Science. 2004;**235**(1):182-187. DOI: 10.1016/j.apsusc.2004.05.121

[20] Tsuji H, Sommani P, Hattori M, Yamada T, Sato H, Gotoh Y, Ishikawa J. Negative ion implantation for pattering mesenchymal-stem cell adhesion on silicone rubber and differentiation into nerve cells with keeping their adhesion pattern. Surface and Coatings Technology. 2009;**2003**(17-18):2562-2565. DOI: 10.1016/j.surfcoat.2009.02.076

[21] Tsuji H, Sommani P, Kitamura T, Hattori M, Sato H, Gotoh Y, Ishikawa J. Nerve-cell attachment properties of polystyrene and silicone rubber modified by carbon negative-ion implantation. Surface and Coatings Technology. 2007;**201**(19-20):8123-8126. DOI: 10.1016/j.surfcoat.2006.01.074

[22] Baba K, Hatada R, Flege S, Ensinger W. Preparation and properties of Ag-containing diamond-like carbon films by magnetron plasma source ion implantation. Advances in Materials Science and Engineering. 2012;**2012**(1):536853. DOI: 10.1155/2012/536853

[23] Dudognon J, Vayer M, Pineau A, Erre R. Mo and Ag ion implantation in austenitic, ferritic and duplex stainless steels: A comparative study. Surface and Coatings Technology. 2008;**203**(1-2):180-185. DOI: 10.1016/j.surfcoat.2008.08.069

[24] Fang J, Zhao JH, Sun Y, Ma HY, Yu XL, Ma Y, Ni YX, Zheng L, Zhou YM. Biocompatibility and antibacterial properties of zinc-ion implantation on titanium. Journal of Hard Tissue Biology. 2014;**23**(1):35-43

[25] Hou XG, Wang XM, Luan WJ, Li DJ, Dong L, Ma J. Study of antibacterial, hydrophilic and nanomechanical properties of TiOx films modified by Ag + beam implantation. Surface and Coatings Technology. 2013;**229**:71-75. DOI: 10.1016/j.surfcoat.2012.04.092

Chapter 10

Devulcanization of Elastomers and Applications

Fabiula Danielli Bastos de Sousa

Additional information is available at the end of the chapter

http://dx.doi.org/10.5772/intechopen.68585

Abstract

In no other phase of human development was produced such amount of waste as currently. Its composition and quantity are directly related to the way the population lives, socioeconomic condition and the ease of access to consumer goods. The irregular disposition of such waste may cause harmful environmental impacts. One of the most dangerous solid wastes is the vulcanized rubber, which, besides having high natural degradation time, has many chemical additives on its formulation and the possibility to store rain water when disposed in landfills and may become a breeding place for vectors. So, recycling comes against this problem and devulcanization is a way of recycling that restores the fluidity of the rubber. One of the applications of this devulcanized rubber is in the production of polymeric blends. Devulcanization of rubbers, the application of this material in polymeric blends based on thermoplastic/recycled rubber, and the parameters involved during the processing of these materials will be addressed in this work.

Keywords: sustainability, recycling, devulcanization, revulcanization, polymer blends

1. Introduction

Elastomers consist in a class of materials widely used today in many fields of application. Their typical properties such as high levels of elasticity and damping are what make this class so important. For this, the elastomers must, first, go through a complex process known as vulcanization [1]. Chemically, it is the process in which the molecular chains of rubber, independently, are joined by chemical bonds forming primary cross-linkings, which lead to the formation of a three-dimensional network in the material. This structural organization allows to maintain or increase the elastic properties and to reduce the plastic behavior of the material. The elastomer becomes insoluble and with greater mechanical resistance, different from nonvulcanized elastomer.

However, while the vulcanization provides improvements in the properties of the elastomers and with it the possibility of a wide use as consumer goods, it brings difficulties for recycling after the use, once the vulcanized polymer becomes a thermosetting, making it impossible for subsequent molding into another product.

Presently, a key expression is the sustainable development, which refers to the responsible search for economic and material development without damaging the human being and the environment, using natural resources wisely so that future generations will not be harmed. Among the related subjects are the waste problem, and one of the possible solutions is the recycling of recyclable materials (especially postconsumer polymeric materials) [2].

Among the most harmful polymeric materials include those that contain heavy metals fillers, plasticizers and vulcanized elastomers. The latter ones, for being nonreprocessable due to the presence of cross-linkings, may cause serious public health problems, since they may be accumulators of rain water when disposed in landfills (especially tires), becoming a place able to the proliferation of vectors, such as the *aedes aegypti* mosquito, transmitter of dengue, chikungunya and zika (and day by day the scenario is getting worse, since new diseases transmitted by the same vector are being discovered) [2]. Adding, vulcanized elastomers are materials which require long periods of time to degrade naturally due to their structure of cross-linkings, presence of stabilizers and other additives in their formulation [3], making them infusible and difficult to be reprocessed. Besides, these are considered high value-added materials, due to the range of chemical additives on the formulation, making their discard a waste. According to Imbernon and Norvez [4], "Because of the scarcity and increasing prices of natural resources, and of the growing environmental awareness, waste management has become a crucial issue in today's society."

Presently, the high growth of consumption and inadequate disposal of polymeric materials have increased problems related to solid urban residues. The ideal would be that 100% of all the polymeric materials used around the world was recycled, assisting in sustainable development. However, the real scenario is quite different for bumping into numerous difficulties, which generates serious environmental, public health, economic and governmental problems [2]. In this way, devulcanization is against this serious global problem, seeking a viable reuse way of postconsumer vulcanized elastomers.

This work aims to discuss about devulcanization of vulcanized elastomers and some possible applications of this material, with focus on the preparation of polymeric blends containing recycled elastomers (based on thermoplastic and recycled rubber), and the factors that affect the properties of the final material, with the purpose of being a useful literature serving in obtaining final materials applicable in consumer goods.

2. Devulcanization of elastomers

A very prominent form of recycling is the devulcanization, which is the process of total or partial cleavage of the cross-linkings formed during the initial vulcanization [5–8]. Despite returning to the material its flow capacity, the higher the devulcanization degree, the greater the breakage of the main polymer chain links. The effect of this degradation is the significant reduction of stiffness and other mechanical properties when the material is revulcanized (second vulcanization process). Therefore, for the choice of the parameters of the devulcanization process, must be taken into account the balance between the processability and mechanical properties of the final material [9].

The literature presents several works that discuss the different methods used to devulcanize rubbers, as mechanical and chemical mechanical method [10–12], microwaves method [9, 13–19], ultrasound method [20–22], chemical method [23, 24], microbial method [25, 26], and still other methods as bioreactor and spraying for solid-state shear [27, 28]. Subsequently, when the goal is to use the recycled elastomer in the production of a blend, the role of devulcanization is to increase the interaction between the raw and recycled material, reducing the degradation of the properties of the finished product, and making it possible to increase the amount of recycled elastomer in the compound raw phase/recycled phase [29].

A well-established way of recycling vulcanized elastomers is through the production of polymeric blends, that is, physical mixes of two or more polymers that can be miscible or not. As two or more properties of the polymers can be combined, the blends have been studied widely with the aim of improving the physical properties compared to neat polymers, that is, obtain materials with additional properties, and minimal loss of original properties [30], as well as being more economically viable to unite two existing polymers to synthesize another nonexistent [31], for the creation of a new molecule.

A plethora of polymeric blends composed of elastomers can be obtained. There are basically two types of polymeric blends composed in at least one of the phases an elastomer: blends composed of two or more types of elastomers (elastomeric blends) and blends composed of a thermoplastic phase and the other elastomeric. These can be of two types: when there is a high concentration of elastomer (thermoplastic elastomers—TPEs) and when there is low concentration of elastomer (toughened plastic). In all cases, the goal is to obtain materials with desired properties, additional to the properties of the neat materials.

In addition to restore the fluidity of the rubber, devulcanization is able to chemically change the structure of the material [19]. All these changes certainly affect its revulcanization. So, revulcanization is still more complex than the vulcanization itself, since other parameters influence it. And especially in the production of dynamically revulcanized blends (blends based on thermoplastic and recycled rubber in high concentrations, in which the last phase was revulcanized during processing), beyond this, others related to processing get place, to be discussed ahead.

According to Karger-Kocsis et al. [32], among the vast number of possibilities concerning blends containing ground tire rubber (GRT), value-added application can especially be expected in thermoplastic elastomers, and rubber combinations. In the literature, several studies about elastomeric blends with satisfactory results are easily found [16, 33]. Despite the great advances obtained by this type of material, a great difficulty, or even a disadvantage, is that the rejects produced during processing are not easily recycled and reworked and may cause environmental problems due to incorrect disposal of this waste. On the other hand, rejects of TPE blends can be easily reworked, which presents some remarkable advantages, being the most important one the ability of being processed as a thermoplastic, presenting the performance of a vulcanized elastomer. However, even today with all the progress achieved in the field of science and technology, the use of recycled elastomer in this type of blend remains a major challenge. Even so, this class of polymeric blends represents a major current trend for the use of recycled elastomers and will be addressed in more details in the following section.

3. Thermoplastic elastomers

Thermoplastic elastomers contain high concentrations of elastomeric phase (usually above 50%), combining the processability of thermoplastics and the functional performance of vulcanized rubbers at room temperature [34–37]. There are three distinct classes of thermoplastic elastomers, namely: block copolymer, thermoplastic/dynamically vulcanized elastomer blends—thermoplastic vulcanizates (TPVs) and ionic thermoplastic.

The unique combination of properties allows the processing of TPEs in conventional equipment used for processing of thermoplastics in processes such as injection and blow molding, film production and extruded profiles, keeping the elastomeric properties. Such behavior is attributed to their structures that contain both flexible and elastic fields of high extensibility with low glass transition temperature (T_g), and rigid low extensibility areas with a T_g and/or crystalline melting temperature (T_m) high [38].

The key advantages of TPEs include [38]: (i) their ability to become fluid with heat and then hardening with cooling gives manufacturers the possibility to produce articles with rubber behavior using equipments commonly used in the processing of thermoplastics. (ii) Little or no mixing of additives necessary for the production of TPEs. The majority is ready for manufacturing. Rubbers, however, require the mixture of all the additives. (iii) TPEs, once prepared, do not need vulcanization stage. Their processing consists of fewer stages than the processing for obtaining a vulcanized rubber. (iv) Scraps produced in the production process can be reprocessed. Scraps generated in processing of vulcanized rubbers, however, have their potential reuse limited, and the cost of its production is higher, due to the loss of material and disposal cost of scraps. (v) Thermoplastic processing consumes less total energy by having a more efficient processing and smaller time cycles.

A particular type of TPE blend, thermoplastic vulcanizate blend, known as TPV, is largely adopted when using devulcanized elastomers, due to its typical features and properties. It will be described in more details in the following sections.

4. Thermoplastic vulcanizate blends

TPV is a type of TPE produced via dynamic vulcanization of the elastomeric phase of an immiscible blend of thermoplastic in molten state and elastomer under high shear rates [39–43]. TPVs are materials widely used in automotive [12, 44] and electronics industries, civil construction, wiring and cables, biomedical products [12, 45–47], among others. Due to the high applicability of this kind of blend, the use of recycled rubber can be useful and worth being studied.

Dynamic vulcanization is the vulcanization of the elastomeric phase in a molten mixture with other polymer(s) [48]. The process produces a cross-linked polymer dispersion in a continuous polymer matrix phase not cross-linked [38, 40, 49–52]. The continuity of the thermoplastic phase provides the thermo-plasticity and mechanical resistance necessary to blends [53], while the dynamically vulcanized rubber particles give elasticity, flexibility and stability [36, 46, 54].The process can be described as follows: after enough fusion-blend of thermoplastic and rubber, vulcanization agents are added. The vulcanization of rubber phase occurs with a continuation of the mixture. After the output of the mixer, the cold blend can be chopped, extruded, injected, molded, pelletized, etc [43, 52].

The literature presents a vast number of works showing the differences in the properties of polymer blends resulting from dynamic vulcanization, among them: improvement in mechanical properties [55], greater thermal stability [56], minor swelling of the extruded [43], better reprocessability [46, 57], increase in the service temperature [49, 55], greater weather resistance [12] among others, depending on the analyzed system. Several papers also feature improvements in mechanical properties as a result of the dynamic vulcanization, but through the use of compatibilizing agents [36, 48, 58–60] as a result of greater refinement of morphology [34, 61], in general. Among the improvements, it is also found reduced permanent elongation, increased fatigue resistance, greater stability of morphology and better chemical resistance.

TPV containing recycled elastomers, a special issue nowadays as a possible solution to the problem of solid urban residues, especially vulcanized rubbers, will be addressed in the following section.

5. Thermoplastic vulcanizate blends containing recycled elastomers

The reuse, recycling and recovery of waste of cross-linked rubbers are of great scientific and technological interest. As discussed previously, there is great difficulty in recycling, as they are infusible and insoluble materials, which have difficult processing [57] due to their structure of cross-linkings. In this context, many efforts have been made regarding the preparation and characterization of polymer blends containing GTR and various thermoplastics, as an alternative to recycling [62, 63].

The properties of these materials depend on the concentration of the recycled material, as well as the adhesion among phases [64, 65]. According to Zhang et al. [66], the adhesion between the GTR and the polymer matrix is usually very weak due to the three-dimensional structure of the cross-linkings, in the case of blends, in which the GTR is just ground. Cañavate et al. [63] report that the lack of adhesion among phases is due to the large particles of GTR, their superficial characteristics and structure of cross-linkings, hindering their adsorption by molecules of the thermoplastic matrix, being that the use of only ground GTR into blends makes the processing a difficult step [67]. For Kumar et al. [68], for the production of TPVs containing recycled rubber, the addition of a raw rubber or the devulcanization (at least partial) of recycled material is prerequisites. The devulcanization improves the compatibility between GTR and the matrix [4]. However, despite all the difficulties presented by Cespedes et al., "The use of GTR is an excellent option for reducing the cost of TPVs, and GTR is an environmentally friendly alternative because of its upcycling applications" [69].

In order to improve adhesion and interaction among phases, many authors have used compatibilization techniques [48, 66, 70], devulcanization of elastomeric phase [12, 20, 57, 66, 71–74], addition of a third elastomeric phase or replacing part of recycled rubber for a raw one [53, 63, 68, 69], functionalization [75], filler addition [47, 70, 74], among others, beyond the dynamic vulcanization, which notoriously increases the adhesion and interaction among phases of the blends [20, 34, 55, 57, 68, 76]. Additionally, the dynamic vulcanization in blends containing recycled material gives them greater added value [37].

The next section will present the stages involved during the evolution of morphology of thermoplastic vulcanizate blends and some important parameters able to influence the development of the final morphology of these blends.

6. Evolution of the morphology of thermoplastic vulcanizate blends during processing and important parameters

The final morphology of a blend is achieved during its processing, so it is a crucial stage in getting the final desired properties, since they are consequence of its morphology.

Many factors can change the morphology of polymer blends during processing such as temperature, residence time (processing on extruders), intensity of the mixture (speed of the extruder and setting of the screw), composition of the blend, viscosities and elasticities ratio, and interfacial tension among the phases [77]. In this way, the final morphology of immiscible polymer blends depends on the properties of the individual components, as well as processing conditions [78, 79].

Regarding the parameters related to the processing, the literature presents several works in which the processing variables are changed and analyzed [44, 55, 57, 72, 80–89].

As an important example of processing variable, temperature should be close to the T_m of the thermoplastic phase (or a little greater) [44] and be able to activate the vulcanization reaction of the elastomeric phase [46]. It is known that the behavior of vulcanization of the elastomeric phase varies according to the adopted temperature, besides that it should not be high enough so that, combined with the high shear rates involved in the process (especially in twin screw extruders), promotes high level of degradation in both phases of the blend. However, no matter the studied parameter, all of them are very important, since they are able to directly affect the final morphology of blends and, then, their final properties.

Figure 1 shows in a schematic way the transformation of the morphology of thermoplastic vulcanizate blends during processing. According to the schema, initially a rubber-thermoplastic blend is formed, with co-continuous morphology and rubber phase non–cross-linked (stage A). In the next stage, rubber phase becomes stretched and strongly deformed due to the beginning of dynamic vulcanization. A rubber-thermoplastic blend is formed, with cocontinuous morphology as well (stage B). Due to dynamic vulcanization, rubber phase is able to break up in the following stage (stage C). So, a TPV is formed, the rubber phase is cross-linked and is dispersed in thermoplastic phase. In the lasts stages (stages D and E), rubber particles have the distribution improved on the matrix phase.

On the whole, the blend completely changes its morphology, from co-continuous to dispersed phase. However, so that the blend presents typical final properties for noble uses, the processing must be carefully analyzed and optimized. In the case of using an extruder, the analysis becomes more complex because of the large number of variables involved, but at the same time, it becomes a big advantage in the improvement of the technique. A good example of possible parameter change in an extruder is the number of feeders.

When producing a thermoplastic vulcanizate blend in an extruder containing two feeders, it is possible to introduce independently each phase through each one of the feeders. In the case of blends in which the elastomeric phase was previously mixed to vulcanization additives, this can be added in the second feeder, whereas, in the case of elastomeric phase not be previously mixed to additives, the two phases of the blend can be added together in the same feeder, while the additives can be added in the second one.

The change of some processing parameters was deeply studied by de Sousa et al. [61]. The authors produced dynamically revulcanized blends based on 60 wt% of devulcanized GTR (GTR5.5) and 40 wt% of high density polyethylene (HDPE), by using a twin screw extruder. Processing parameters such as screw speed and feeding mode were varied. In the production of blends using only one feeder, both HDPE and devulcanized rubber were added together. For blends produced with two feeders, the HDPE phase was added in the first feeder and the GTR phase in the second one. Vulcanization additives were previously added to the GTR phase. The authors proved the importance of knowing previously the rheological properties of the rubber phase, as well as matching these properties to the processing conditions.

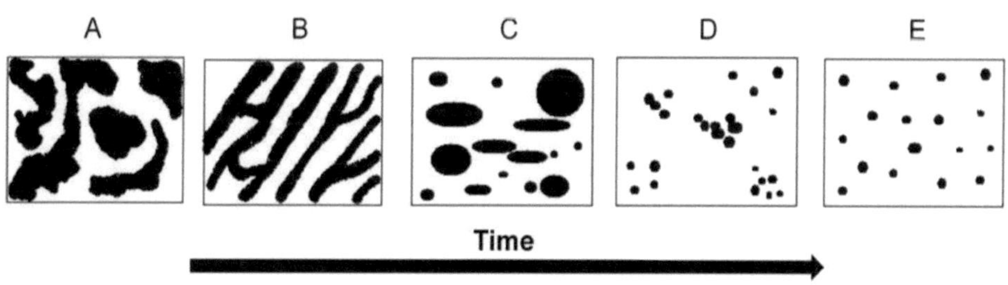

Figure 1. Schematic morphology transformation during the dynamic vulcanization of polymeric blends. The black part represents the elastomeric phase, and the white represents the thermoplastic phase.

According to de Sousa et al. [61], overall, the effects of screw speed on the mechanical properties were not significant. Furthermore, the mechanical properties of the blends were far below those of classical TPVs, probably because of the poor compatibility and adhesion between GTR5.5 and HDPE. In general, the blends produced by using the second feeding mode obtained a finer morphology.

Figure 2 shows some vulcanization parameters involved in the processing of dynamically revulcanized blends, as well as the scheme of a probable evolution of morphology during processing. The screw profile is for feeding mode 2, since it produced finer morphology and consequently slightly higher mechanical properties.

According to **Figure 2**, at point 1 of the extruder, there is only physical mixing among the phases and there is no revulcanization of the GTR5.5. At the beginning of the second high shear zone (point 2), the revulcanization reaction gets place (the residence time of the rubber from its introduction in the extruder to this point is about the same of ts_1 (scorch time of the reaction)) and, around this point, the blend presents a cocontinuous morphology, in which the elastomeric phase is stretched in the flow direction [61].

Due to the high elongational flow in this zone (use of mixing or kneading blocks), the rubber along vulcanization process can deform (due precisely to the three-dimensional structure formation of cross-linkings and consequent increase in viscosity of the blend [76]) enough for

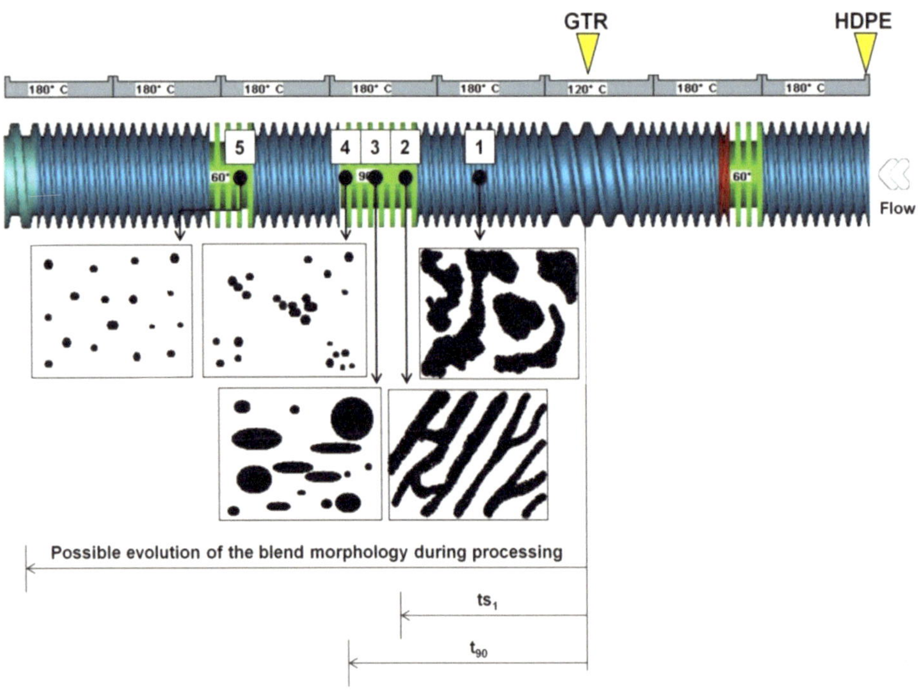

Figure 2. Screw profile relative to feeding mode 2 used in the preparation of the blends, showing the schema of the possible evolution of the morphology and rheology of the elastomeric phase involved during processing. Reprinted with permission from Ref. [61]. Copyright 2017, John Wiley and Sons.

break of their particles in other smaller than the ones before [60] and phase inversion may occur. At this stage, the overall viscosity is increased (the elastomeric phase with high elasticity is stretched and breaks into smaller particles due to high shear rates, intense elongational flow and high elasticity generated by cross-linkings (point 3), resulting in high mechanical stresses [61]. The elastomeric phase deforms until it reaches a critical tension, when it breaks up into small particles [90]. The biggest changes of morphology occur in the first high-shear region, in which both phases are together [39, 91–93].

At point 4, still under the effect of high shear rates, rubber particles break into smaller particles, and at point 5, there is a better distribution in the thermoplastic matrix. The end of the second shear zone (point 4) refers approximately to the optimum cure time (t_{90}) of the elastomeric phase which, in the case of GTR5.5, is 44 s. The residence time of rubber from its introduction in the extruder to the end point of the reaction must be equivalent to t_{90}, and this point should be in a high shear zone of the extruder for breakage of rubber particles in micrometric dimensions [61].

In the point 5, cross-linked rubber particles, at this moment, have a very high viscosity and elasticity, and it occurs only the distribution of the particles in the matrix, improving macroscopic homogeneity, being necessary the use of mixing elements. It must be pointed out that the dispersion process occurs instantly with the vulcanization reaction, and both processes are influencing each other. The rapid increase in the cross-linkings of rubber also leads to an increase in the surface tension of the elastomeric phase. Cross-linkings and the high superficial tension will reduce the driving force for the coalescence and, therefore, the characteristic of the particles of vulcanized rubber phase will be preserved, even in a new mixture after the completion of the cross-linking process [67, 90].

The increase of interfacial tension as a result of cross-linkings and the high elasticity of the particles produce the relaxation of the deformed structures; in an ideal case, spherical particles are formed. Due to the high viscosities ratio, elastomeric cross-linked particles can only be distributed, and no longer dispersed on a new mixing cycle [90]. In agreement, some results [94] revealed that the developed microstructure is highly affected by the type of the melt compounding process, as well as the feeding mode.

It is important to address here that, for both screw profiles used, the first mixing zone served to melt the HDPE, the second one to dynamically revulcanize the GTR and the last one to improve the distribution of the rubber particles in the HDPE. In the case of the blends produced through feeding mode 1, the high shear rate in the first mixing zone could bring about premature revulcanization of the rubber phase, since the components were added together. As the length of this zone and the corresponding residence time were short, the time for the reaction to go to completion was longer than the residence time on the zone, which probably happened in the second transport zone. Thus, the rubber domains were not satisfactorily well dispersed and distributed in the HDPE matrix. However, in the case of the blends produced through feeding mode 2, the residence time of the rubber phase inside the extruder from its introduction to the end of the second mixing zone was closer to the optimum cure time of the GTR5.5 at 180°C (43.8 s). Therefore, the mixing zone was long enough for the revulcanization reaction to go to completion and the dispersed rubber domains to have its size reduced in the HDPE matrix [61].

The blend produced by using the second feeding mode and at 250 rpm presented higher finer morphology and consequently better mechanical properties, despite the fact that the compatibility and adhesion between the phases were poor. The finer morphology is due to good match between processing conditions and rheological properties of the GTR5.5. The residence times of the GTR5.5 inside the extruder from its introduction to the respective points shown in **Figure 2** were approximately 50 and 30 s, respectively, which were very close to the values of $t_{90}$and ts_1 (44 and 27 s, respectively) [61].

At this point, the presentation of some equations can be useful to understand all the modification occurred to the blends during the processing. The elongation and breakage of polymeric particles suspended in another polymer under shearing flow were first studied by Taylor [95]. According to the author, two dimensionless parameters that enable the prediction of morphology in the molten state are the number of Capillarity [Eq. (1)] and the viscosities ratio [Eq. (2)]:

$$C_a = \frac{\sigma R}{\alpha} \tag{1}$$

where σ represents the shear stress, R is the radius of the particle or drop, and α is the interfacial tension between the phases of the blend.

$$p = \frac{\eta_d}{\eta_m} \tag{2}$$

where η_d and η_m are the viscosities of the dispersed phase and the matrix phase, respectively.

If the value of C_a is small, interfacial forces dominate and the particles acquire the shape of ellipsoids. Above a critical value $C_{a\,crit}$, the particles become unstable and break down [96].

When two immiscible polymers are mixed to form a blend, in accordance with the principle of minimum dissipation of energy, it is expected that the most viscous polymer forms the dispersed phase and the less viscous forms the matrix phase. When this principle is satisfied, it means that the viscosities ratio [Eq. (2)] was the predominant factor in determining the state of dispersion of the blend [97].The viscosities ratio strongly affects the development of morphology during the reaction [52].

Additionally, in order to study the development of the phase morphology and predict the inversion phase region in immiscible polymer blends, Avgeropoulus et al. [98] developed an empirical model based on torque ratio in internal mixer and volumetric fraction of each phase [Eq. (3)]. Jordhamo et al. [99] proposed a similar equation [Eq. (4)], based on the viscosities ratio:

$$\frac{\varnothing_1}{\varnothing_2} \cdot \frac{T_2}{T_1} = X \tag{3}$$

$$\frac{\varnothing_1}{\varnothing_2} \cdot \frac{\eta_2}{\eta_1} = Y \tag{4}$$

which results in the following variants morphologies:

X, Y > 1—Phase 1 is continuous or matrix and phase 2 is dispersed.

X, Y < 1—Phase 2 is continuous or matrix and phase 1 is dispersed.

X, Y = 1—Two phases are continuous or region of phase inversion.

http://dx.doi.org/10.5772/intechopen.68585

Where $ø_1$, $ø_2$ are volumetric fractions, T_1, T_2 are measures of torques at a same temperature, and η_1, η_2 are viscosity values for phases 1 and 2, respectively [50].

According to Zhang et al. [57], soon after the end of the dynamic vulcanization, vulcanized rubber particles present a high surface tension, which agglomerate them. Therefore, it is necessary the application of high shear rates, which are generated due to the presence of mixing blocks in processing performed in extruder, as in point 5 (**Figure 2**). Also, Yao et al. [100] deeply studied the morphology evolution of bromo-isobutylene-isoprene rubber (BIIR)/polypropylene (PP) TPV blends. It was depicted that the dynamic vulcanization increases the compatibility among the phases, demonstrated by the increase in interfacial phase thickness and the decrease in interfacial tension. During the processing, single nanoparticles of elastomeric phase are being formed, and their agglomeration is getting lesser as dynamic vulcanization advances. Thus, Sararoudi et al. [101] concluded that the extent of agglomerations among the vulcanized rubber particles in the twin screw extruder not only depends on the rubber content, but also are controlled by a common agglomeration and disagglomeration mechanism which is, in turn, governed by the screw speed.

For the case of blends in which all phases are added together in the extruder in one feeder, studies show that as soon as the complete melting of the thermoplastic phase is reached, the blend reaches quickly its final morphology [39, 85] due to generated interfaces among phases. According to Covas et al. [102], the increase in the interfacial area raised soon after the melting of the thermoplastic, which induced chemical conversion and the evolution of morphology. Therefore, the choice of the parameters of the processing, the number of feeders to be used depending on the screw profile adopted is a factor of great importance. Van Duin and Machado [39] studied the dynamic vulcanization reaction of ethylene-propylene-diene rubber (EPDM)/HDPE blend through the withdrawal of peer-to-peer samples on twin screw extruder during processing. According to the authors, the cross-linking of EPDM phase began when the HDPE still was not fully melted, and the final morphology of blends was reached very quickly. The phase inversion occurred due to formation of cross-linkings. Machado and van Duin [103] analyzed the properties of EPDM/HDPE TPV blends and found that, the higher the content of EPDM, the greater the viscous dissipation, the higher the melting and, consequently, the greater the rate of cross-linkings.

The type of equipment can also alter the particle size distribution of rubber, in the case of morphology of dispersed phase. Studies show that blends produced on extruder tend to have smaller particle sizes compared to blends produced in internal mixer [20, 71, 104–106] due to the higher shear rate during processing in extruders and intensive flow field [105]. The elastomeric phase of the blends produced on extruder can also present greater cross-linking density, as verified by Sengupta and Noordermeer [104]. However, the distribution of sizes of particles is more uniform in the blends produced in internal mixer due to longer residence time and greater total shear stress, promoting the breakdown of particles [104]. According to Shahbikian et al. [105], that produced EPDM/PP blends in internal mixer and extruder, even with the shortest residence time for processing in extruders, the cure reaction occurred quickly, resulting in EPDM particles of sizes more heterogeneous and with greater cross-linking density. By combining the effects of time, temperature and shearing, the matrix phase acquires elasticity, and it is extruded into sheets. At the same time, there is a break of these sheets due to elongational and shear forces generated in the mixing equipment. So, there is a dynamic balance between the process of breakdown of phases and coalescence.

Among other factors mentioned, the final morphology is the result of processes of coalescence and breakage of the elastomeric phase particles (in the case of blends with the morphology of dispersed phases) during processing. In the case of TPV blends containing devulcanized rubber, the devulcanization acts on the process of break, while the dynamic revulcanization acts on the reduction of the coalescence process of particles [73].

In short, the process of devulcanization makes the rubber fluid, aiding in the process of breaking what, consequently, helps in reducing the size of the particles, increases the contact area among the phases and increases the transmission of tensions. Dynamic revulcanization helps in stabilizing the morphology by inhibition of coalescence process among the particles of the dispersed phase [73].According to Goharpey et al. [107], dynamic vulcanization can prevent the coalescence of the rubber particles from the early stage of the dynamic vulcanization.

As a conclusion, the stage of processing (dynamic vulcanization) is of extreme importance, and all aspects involved should be carefully analyzed and optimized, as they may change the final morphology of blends and, with it, completely alter their final properties [108].

7. Conclusion

The irregular disposition of solid urban residues, especially vulcanized elastomers like tires, may bring together dangerous environmental impacts. In this way, a very outstanding form of vulcanized elastomers recycling is the devulcanization. Along with a variety of possible uses, the devulcanized elastomer can be used in the formation of polymeric blends. Among a vast number of polymeric blends composed of elastomers, TPV is largely adopted when using devulcanized elastomers, due to its typical features and properties. However, parameters of processing as well as the devulcanization process itself must be considered and carefully analyzed, since they are able to dictate the final morphology of the polymeric blend and, consequently, its properties as a whole.

TPV is considered a "green" polymer, since its recyclability promotes the environmental protection and resource saving. When it is composed by a recycled elastomeric phase, it aids the petroleum sources saving, saves raw materials and energy, not harming the environment, and still being a source of income for many families who survive from the collection of recyclable materials. The production of blends composed of recycled rubber is still a major challenge for the academic community, since its final properties need to justify all energy expenditures necessary for production, in addition to being economically viable. However, attention should be given to the subject, since it is a possible solution to the problem of final disposal of urban solid waste. On the other hand, it is observed that the subject is still little explored in literature, possibly due to difficulties encountered in the production and final properties obtained.

Author details

Fabiula Danielli Bastos de Sousa

Address all correspondence to: fabiuladesousa@gmail.com

Technology Development Center – CDTec, Universidade Federal de Pelotas, Pelotas, RS, Brazil

References

[1] Arrillaga A, Zaldua AM, Atxurra RM, Farid AS. Techniques used for determining cure kinetics of rubber compound. European Polymer Journal. 2007;**43**(11):4783-4799

[2] de Sousa FDB. Vulcanization of Natural Rubber: Past, present and future perspectives. In: Hamilton JL, editor. Natural Rubber: Properties, Behavior and Applications. New York: Nova Science Publisher; 2016. pp. 47-88

[3] Roche N, Ichchou MN, Salvia M, Chettah A. Dynamic damping properties of thermoplastic elastomers based on EVA and recycled ground tire rubber. Journal of Elastomers and Plastics. 2011;**43**(4):317-340

[4] Imbernon L, Norvez S. From landfilling to vitrimer chemistry in rubber life cycle. European Polymer Journal. 2016;**82**:347-376

[5] Sienkiewicz M, Kucinska-Lipka J, Janik H, Balas A. Progress in used tyres management in the European Union: A review. Waste Management. 2012;**32**(10):1742-1751

[6] Oh JS, Isayev AI. Continuous ultrasonic devulcanization of unfilled butadiene rubber. Journal of Applied Polymer Science. 2004;**93**(3):1166-1174

[7] Kleps T, Piaskiewicz M, Parasiewicz W. The use of thermogravimetry in the study of rubber devulcanization. Journal of Thermal Analysis and Calorimetry. 2000;**60**(1):271-277

[8] Warner WC. Methods of devulcanization. Rubber Chemistry and Technology. 1994;**67**(3):559-566

[9] Scuracchio CH, Waki DA, Bretas RES. Caracterização térmica e reológica de borracha de pneu desvulcanizada por microondas. Polímeros: Ciência e Tecnologia. 2006;**16**(1):46-52

[10] Zhang XX, Lu ZX, Tian D, Li H, Lu CH. Mechanochemical devulcanization of ground tire rubber and its application in acoustic absorbent polyurethane foamed composites. Journal of Applied Polymer Science. 2013;**127**(5):4006-4014

[11] Yazdani H, Ghasemi I, Karrabi M, Azizi H, Bakhshandeh GR. Continuous devulcanization of waste tires by using a co-rotating twin screw extruder: Effects of screw configuration, temperature profile, and devulcanization agent concentration. Journal of Vinyl & Additive Technology. 2013;**19**(1):65-72

[12] Hassan MM, Badway NA, Elnaggar MY, Hegazy EA. Effects of peroxide and gamma radiation on properties of devulcanized rubber/polypropylene/ethylene propylene diene monomer formulation. Journal of Applied Polymer Science. 2014;**131**(16)

[13] Bani A, Polacco G, Gallone G. Microwave-induced devulcanization for poly(ethylene-propylene-diene) recycling. Journal of Applied Polymer Science. 2011;**120**(5):2904-2911

[14] Scuracchio CH, Waki DA, da Silva M. Thermal analysis of ground tire rubber devulcanized by microwaves. Journal of Thermal Analysis and Calorimetry. 2007;**87**(3):893-897

[15] Hirayama D, Saron C. Chemical modifications in styrene-butadiene rubber after microwave devulcanization. Industrial & Engineering Chemistry Research. 2012;**51**(10): 3975-3980

[16] Zanchet A, Carli LN, Giovanela M, Crespo JS, Scuracchio CH, Nunes RCR. Characterization of microwave-devulcanized composites of ground SBR scraps. Journal of Elastomers and Plastics. 2009;**41**(6):497-507

[17] Zanchet A, Carli LN, Giovanela M, Brandalise RN, Crespo JS. Use of styrene butadiene rubber industrial waste devulcanized by microwave in rubber composites for automotive application. Materials & Design. 2012;**39**:437-443

[18] de Sousa FD, Scuracchio CH. The Role of carbon black on devulcanization of natural rubber by microwaves. Materials Research-Ibero-American Journal of Materials. 2015;**18**(4):791-797

[19] Garcia PS, de Sousa FDB, de Lima JA, Cruz SA, Scuracchio CH. Devulcanization of ground tire rubber: Physical and chemical changes after different microwave exposure times. Express Polymer Letters. 2015;**9**(11):1015-1026

[20] Hong CK, Isayev AI. Plastic/rubber blends of ultrasonically devulcanized GRT with HDPE. Journal of Elastomers and Plastics. 2010;**33**(1):47-71

[21] Zhu SH, Penlidis A, Tzoganakis C, Ginzel E. Ultrasonic properties and morphology of devulcanized rubber blends. Journal of Applied Polymer Science. 2012;**124**(3):2062-2070

[22] Sun X, Isayev AI. Ultrasound devulcanization: Comparison of synthetic isoprene and natural rubbers. Journal of Materials Science. 2007;**42**(17):7520-7529

[23] De D, Das A, Dey B, Debnath SC, Roy BC. Reclaiming of ground rubber tire (GRT) by a novel reclaiming agent. European Polymer Journal. 2006;**42**(4):917-927

[24] Sonnier R, Leroy E, Clerc L, Bergeret A, Lopez-Cuesta JM. Polyethylene/ground tyre rubber blends: Influence of particle morphology and oxidation on mechanical properties. Polymer Testing. 2007;**26**(2):274-281

[25] Li YH, Zhao SH, Wang YQ. Microbial desulfurization of ground tire rubber by sp.: A novel technology for crumb rubber composites. Journal of Polymers and the Environment. 2012;**20**(2):372-380

[26] Li YH, Zhao SH, Wang YQ. Microbial desulfurization of ground tire rubber by Thiobacillus ferrooxidans. Polymer Degradation and Stability. 2011;**96**(9):1662-1668

[27] Chaubey T, Arastoopour H. Studying the pulverization mechanism of rubber with a modified design of the solid-state shear extrusion process. Journal of Applied Polymer Science. 2011;**119**(2):1075-1083

[28] Shahidi N, Arastoopour H, Ivanov G. Pulverization of rubber using modified solid state shear extrusion process (SSSE). Journal of Applied Polymer Science. 2006;**102**(1):119-127

[29] Levin VY, Kim SH, Isayev AI, Massey J, vonMeerwall E. Ultrasound devulcanization of sulfur vulcanized SBR: Crosslink density and molecular mobility. Rubber Chemistry and Technology. 1996;**69**(1):104-114

[30] da Costa HM, Ramos VD, da Silva WS, Sirqueira AS. Analysis and optimization of polypropylene (PP) ethylene-propylene-diene monomer (EPDM)/scrap rubber tire (SRT) mixtures using RSM methodology. Polymer Testing. 2010;**29**(5):572-578

[31] Bhadane PA, Cheng J, Ellul MD, Favis BD. Decoupling of reactions in reactive polymer blending for nanoscale morphology control. Journal of Polymer Science Part B-Polymer Physics. 2012;**50**(23):1619-1629

[32] Karger-Kocsis J, Meszaros L, Barany T. Ground tyre rubber (GTR) in thermoplastics, thermosets, and rubbers. Journal of Materials Science. 2013;**48**(1):1-38

[33] Hirayama D, Scuracchio CH, Saron C. Microwaves devulcanization of SBR containing carbon black. Journal of Research Updates in Polymer Science. 2016;**5**(2):52-59

[34] Nevatia P, Banerjee TS, Dutta B, Jha A, Naskar AK, Bhowmick AK. Thermoplastic elastomers from reclaimed rubber and waste plastics. Journal of Applied Polymer Science. 2002;**83**(9):2035-2042

[35] Ismail H, Suryadiansyah. Thermoplastic elastomers based on polypropylene/natural rubber and polypropylene/recycle rubber blends. Polymer Testing. 2002;**21**(4):389-395

[36] Chatterjee T, Wiessner S, Naskar K, Heinrich G. Novel thermoplastic vulcanizates (TPVs) based on silicone rubber and polyamide exploring peroxide cross-linking. Express Polymer Letters. 2014;**8**(4):220-231

[37] Fainleib A, Grigoryeva O, Starostenko O, Danilenko I, Bardash L. Reactive compatibilization of recycled low density polyethylene/butadiene rubber blends during dynamic vulcanization. Macromolecular Symposia. 2003;**202**:117-126

[38] White JR, De SK. Rubber Technologist's Handbook. Reino Unido: Rapra Technology; 2001

[39] van Duin M, Machado AV. EPDM-based thermoplastic vulcanisates: Crosslinking chemistry and dynamic vulcanisation along the extruder axis. Polymer Degradation and Stability. 2005;**90**(2):340-345

[40] Passador FR, Pessan LA, Rodolfo A. PVC/NBR blends by reactive processing I: In situ dynamic vulcanization process. Polímeros: Ciência e Tecnologia. 2007;**17**(2):80-84

[41] Babu RR, Singha NK, Naskar K. Melt viscoelastic properties of peroxide cured polypropylene-ethylene octene copolymer thermoplastic vulcanizates. Polymer Engineering and Science. 2010;**50**(3):455-467

[42] Antunes CF, van Duin M, Machado AV. Morphology and phase inversion of EPDM/PP blends - Effect of viscosity and elasticity. Polymer Testing. 2011;**30**(8):907-915

[43] Coran AY. Vulcanization: Conventional and dynamic. Rubber Chemistry and Technology. 1995;**68**:351-375

[44] Yeh JT, Lin SC. Optimized processing conditions for the preparation of dynamically vulcanized EPDM/PP thermoplastic elastomers containing PP resins of various melt indexes. Journal of Applied Polymer Science. 2009;**114**(5):2806-2815

[45] Babu RR, Singha NK, Naskar K. Dynamically vulcanized blends of polypropylene and ethylene-octene copolymer: Comparison of different peroxides on mechanical, thermal, and morphological characteristics. Journal of Applied Polymer Science. 2009; **113**(3):1836-1852

[46] Tang YC, Lu K, Cao XJ, Li YJ. Nanostructured thermoplastic vulcanizates by selectively cross-linking a thermoplastic blend with similar chemical structures. Industrial & Engineering Chemistry Research. 2013;**52**(35):12613-12621

[47] Wu HG, Ning NY, Zhang LQ, Tian HC, Wu YP, Tian M. Effect of additives on the morphology evolution of EPDM/PP TPVs during dynamic vulcanization in a twin-screw extruder. Journal of Polymer Research. 2013;**20**(10)

[48] Lu X, Wang WW, Yu L. Waste ground rubber tire powder/thermoplastic vulcanizate blends: Preparation, characterization, and compatibility. Journal of Applied Polymer Science. 2014;**131**(3)

[49] Mark JE, Erman B, Eirich FR. Science and Technology of Rubber. United States: Elsevier; 2005

[50] Babu RR, Singha NK, Naskar K. Phase morphology and melt rheological behavior of uncrosslinked and dynamically crosslinked polyolefin blends: Role of macromolecular structure. Polymer Bulletin. 2011;**66**(1):95-118

[51] Nakason C, Wannavilai P, Kaesaman A. Effect of vulcanization system on properties of thermoplastic vulcanizates based on epoxidized natural rubber/polypropylene blends. Polymer Testing. 2006;**25**(1):34-41

[52] Antunes CF, Machado AV, van Duin M. Morphology development and phase inversion during dynamic vulcanisation of EPDM/PP blends. European Polymer Journal. 2011;**47**(7):1447-1459

[53] Jacob C, De PP, Bhowmick AK, De SK. Recycling of EPDM waste. II. Replacement of virgin rubber by ground EPDM vulcanizate in EPDM/PP thermoplastic elastomeric composition. Journal of Applied Polymer Science. 2001;**82**(13):3304-3312

[54] Joseph S, Oommen Z, Thomas S. Melt elasticity and extrudate characteristics of polystyrene/polybutadiene blends. Materials Letters. 2002;**53**(4-5):268-276

[55] George J, Varughese KT, Thomas S. Dynamically vulcanised thermoplastic elastomer blends of polyethylene and nitrile rubber. Polymer. 2000;**41**(4):1507-1517

[56] George S, Varughese KT, Thomas S. Thermal and crystallisation behaviour of isotactic polypropylene/nitrile rubber blends. Polymer. 2000;**41**(14):5485-5503

[57] Zhang XX, Lu CH, Liang M. Preparation of thermoplastic vulcanizates based on waste crosslinked polyethylene and ground tire rubber through dynamic vulcanization. Journal of Applied Polymer Science. 2011;**122**(3):2110-2120

[58] Chen J, Chen JW, Chen HM, Yang JH, Chen C, Wang Y. Effect of compatibilizer and clay on morphology and fracture resistance of immiscible high density polyethylene/polyamide 6 blend. Composites Part B-Engineering. 2013;**54**:422-430

[59] Li S, Lang FZ, Wang ZB. Zinc dimethacrylate-reinforced thermoplastic vulcanizates based on ethylene-vinyl acetate copolymer/chlorinated polyethylene rubber/nitrile butadiene rubber blends. Polymer-Plastics Technology and Engineering. 2013;**52**(7):683-689

[60] Li S, Liu T, Wang LJ, Wang ZB. Dynamically Vulcanized nitrile butadiene rubber/ethylene-vinyl acetate copolymer blends compatibilized by chlorinated polyethylene. Journal of Macromolecular Science Part B-Physics. 2013;**52**(1):13-21

[61] de Sousa FDB, Scuracchio CH, Hu GH, Hoppe S. Effects of processing parameters on the properties ofmicrowave-devulcanized ground tire rubber/polyethylenedynamically revulcanized blends.Journal of Applied Polymer Science. 2016;**133**(23)

[62] Bianchi O, Fiorio R, Martins JN, Zattera AJ, Scuracchio CH, Canto LB. Crosslinking kinetics of blends of ethylene vinyl acetate and ground tire rubber. Journal of Elastomers and Plastics. 2009;**41**(2):175-189

[63] Cañavate J, Casas P, Colom X, Nogues F. Formulations for thermoplastic vulcanizates based on high density polyethylene, ethylene-propylene-diene monomer, and ground tyre rubber. Journal of Composite Materials. 2011;**45**(11):1189-1200

[64] Anandhan S, De PP, Bhowmick AK, De SK, Bandyopadhyay S. Thermoplastic elastomeric blend of nitrile rubber and poly(styrene-co-acrylonitrile). II. Replacement of nitrile rubber by its vulcanizate powder. Journal of Applied Polymer Science. 2003;**90**(9):2348-2357

[65] Anandhan S, Bhowmick AK. Thermoplastic vulcanizates from post consumer computer plastics/nitrile rubber blends by dynamic vulcanization. Journal of Material Cycles and Waste Management. 2013;**15**(3):300-309

[66] Zhang SL, Zhang ZX, Kim JK. Study on thermoplastic elastomers (TPEs) of waste polypropylene/waste ground rubber tire powder. Journal of Macromolecular Science Part B-Physics. 2011;**50**(4):762-768

[67] Magioli M, Sirqueira AS, Soares BG. The effect of dynamic vulcanization on the mechanical, dynamic mechanical and fatigue properties of TPV based on polypropylene and ground tire rubber. Polymer Testing. 2010;**29**(7):840-848

[68] Kumar CR, Fuhrmann I, Karger-Kocsis J. LDPE-based thermoplastic elastomers containing ground tire rubber with and without dynamic curing. Polymer Degradation and Stability. 2002;**76**(1):137-144

[69] Cespedes RIN, Gamez JFH, Velázquez MGN, Belmontes FA, de León RED, Fernández OSR, Orta CAA, Hernández EH. Thermoplastic elastomers based on high-density polyethylene, ethylene-propylene-diene terpolymer, and ground tire rubber dynamically vulcanized with dicumyl peroxide. Journal of Applied Polymer Science. 2014;**131**(4)

[70] Razmjooei F, Naderi G, Bakhshandeh G. Preparation of dynamically vulcanized thermoplastic elastomer nanocomposites based on LLDPE/reclaimed rubber. Journal of Applied Polymer Science. 2012;**124**(6):4864-4873

[71] Luo T, Isayev AI. Rubber/plastic blends based on devulcanized ground tire rubber. Journal of Elastomers and Plastics. 1998;**30**(2):133-160

[72] Scaffaro R, Dintcheva NT, Nocilla MA, La Mantia FP. Formulation, characterization and optimization of the processing condition of blends of recycled polyethylene and ground tyre rubber: Mechanical and rheological analysis. Polymer Degradation and Stability. 2005;**90**(2):281-287

[73] de Sousa FDB, Gouveia JR, de Camargo Filho PMF, Vidotti SE, Scuracchio CH, Amurin LG, Valera TS. Blends of ground tire rubber devulcanized by microwaves/HDPE - Part A: Influence of devulcanization process. Polímeros: Ciência e Tecnologia. 2015;**25**(3):256-264

[74] de Sousa FDB, Gouveia JR, de Camargo Filho PMF, Vidotti SE, Scuracchio CH, Amurin LG, Valera TS. Blends Ground tire rubber devulcanized by microwaves/HDPE - Part B: Influence of clay addition. Polímeros: Ciência e Tecnologia. 2015;**25**(4):382-391

[75] Tolstov A, Grigoryeva O, Fainleib A, Danilenko I, Spanoudaki A, Pissis P, Grenet J. Reactive compatibilization of polyethylene/ground tire rubber inhomogeneous blends via interactions of pre-functionalized polymers in interface. Macromolecular Symposia. 2007;**254**:226-232

[76] Jose J, Nag A, Nando GB. Processing and characterization of recycled polypropylene and acrylonitrile butadiene rubber blends. Journal of Polymers and the Environment. 2010;**18**(3):155-166

[77] Thitithammawong A, Nakason C, Sabakaro K, Noordermeer JWM. Thermoplastic vulcanizates based on epoxidized natural rubber/polypropylene blends: Selection of optimal

peroxide type and concentration in relation to mixing conditions. European Polymer Journal. 2007;**43**(9):4008-4018

[78] Paul DR, Bucknall CB. Polymer Blends: Formulation. United States: Wiley-Interscience; 2000

[79] Ultracki LA. Polymer Alloys and Blends. New York: Hanser Publishers; 1989

[80] Mousa A, Ishiaku US, Ishak ZAM. Rheological properties of dynamically vulcanized poly(vinyl chloride)/epoxidized natural rubber thermoplastic elastomers: Effect of processing variables. Polymer Testing. 2000;**19**(2):193-204

[81] John B, Varughese KT, Oommen Z, Thomas S. Melt rheology of HDPE/EVA blends: The effects of blend ratio, compatibilization, and dynamic vulcanization. Polymer Engineering and Science. 2010;**50**(4):665-676

[82] Goharpey F, Foudazi R, Nazockdast H, Katbab AA. Determination of twin-screw extruder operational conditions for the preparation of thermoplastic vulcanizates on the basis of batch-mixer results. Journal of Applied Polymer Science. 2008;**107**(6):3840-3847

[83] Machado AV, Covas JA, van Duin M. Evolution of morphology and of chemical conversion along the screw in a corotating twin-screw extruder. Journal of Applied Polymer Science. 1999;**71**(1):135-141

[84] Joseph S, Thomas S. Morphology, morphology development and mechanical properties of polystyrene/polybutadiene blends. European Polymer Journal. 2003;**39**(1):115-125

[85] Yquel V, Machado AV, Covas JA, Flat JJ. Contribution of the melting stage to the evolution of the morphology and chemical conversion of immiscible polyamide/polyethylene blends in twin-screw extruders. Journal of Applied Polymer Science. 2009;**114**(3): 1768-1776

[86] Kim JK, Lee SH, Hwang SH. Study on the thermoplastic vulcanizate using ultrasonically treated rubber powder. Journal of Applied Polymer Science. 2003;**90**(9):2503-2507

[87] Kalkornsurapranee E, Nakason C, Kummerlowe C, Vennemann N. Development and preparation of high-performance thermoplastic vulcanizates based on blends of natural rubber and thermoplastic polyurethanes. Journal of Applied Polymer Science. 2013;**128**(4):2358-2367

[88] Pesneau I, Champagne MF, Huneault MA. PP/EMA TPV: Dynamic cross-linking through an alcoholysis reaction. Polymer Engineering and Science. 2002;**42**(10):2016-2031

[89] Verbois A, Cassagnau P, Michel A, Guillet J, Raveyre C. New thermoplastic vulcanizate, composed of polypropylene and ethylene-vinyl acetate copolymer crosslinked by tetrapropoxysilane: Evolution of the blend morphology with respect to the crosslinking reaction conversion. Polymer International. 2004;**53**(5):523-535

[90] Harrats C, Thomas S, Groeninckx G. Micro and nanostructured multiphase polymer blend systems: Phase morphology and interfaces.United States: Taylor & Francis; 2006

[91] Sundararaj U, Macosko CW, Rolando RJ, Chan HT. Morphology development in polymer blends. Polymer Engineering and Science. 1992;**32**(24):1814-1823

[92] Machado AV, Covas JA, Van Duin M. Chemical and morphological evolution of PA-6/Epm/Epm-g-MA blends in a twin screw extruder. Journal of Polymer Science Part a-Polymer Chemistry. 1999;**37**(9):1311-1320

[93] van Duin M, Machado AV, Covas J. A look inside the extruder: Evolution of chemistry, morphology and rheology along the extruder axis during reactive processing and blending. Macromolecular Symposia. 2001;**170**:29-39

[94] Shafieizadegan-Esfahani ARA, Addollahi MM, Katbab AA. Effects of compounding procedure on morphology development, melt rheology, and mechanical properties of nanoclay reinforced dynamically vulcanized EPDM/polypropylene thermoplastic vulcanizates. Polymer Engineering and Science. 2016;**56**(8):914-921

[95] Taylor G. The formation of emulsions in definable fields of flow. Proceedings of the Royal Society of London Series A. 1934;**146**:501-523

[96] Harrats C. Multiphase polymer-based materials. United States: CRC Press; 2009

[97] Lee JK, Han CD. Evolution of polymer blend morphology during compounding in a twin-screw extruder. Polymer. 2000;**41**(5):1799-1815

[98] Avgeropoulos GN, Wessert FC, Biddinson PH, Boehm GGA. Heterogeneous blends of polymers: Rheology and morphology. Rubber Chemistry and Technology. 1976;**49**:93-104

[99] Jordhamo GM, Manson JA, Sperling LH. Phase continuity and inversion in polymer blends and simultaneous interpenetrating networks. Polymer Engineering and Science. 1986;**26**:517-524

[100] Yao P, Wu H, Ning N, Zhang L, Tian H, Wu Y, Hu G, Chan TW, Tian M. Properties and unique morphological evolution of dynamically vulcanizedbromo-isobutylene-isoprene rubber/polypropylene thermoplastic elastomer. RSC Advances. 2016;**6**:11151-11160

[101] Sararoudi SS, H. Nazockdast H, Katbab AA. Study on parameters affecting the morphology development of dynamically vulcanized thermoplastic elastomers based on EPDM/PP in a co-rotating twin screw extruder. Rubber Chemistry and Technology. 2004;**77**(5):847-855

[102] Covas JA, Carneiro OS, Maia JM, Filipe SA, Machado AV. Evolution of chemistry, morphology and rheology of various polymer systems along a twin-screw extruder. Canadian Journal of Chemical Engineering. 2002;**80**(6):1065-1074

[103] Machado A, van Duin M. Dynamic vulcanisation of EPDM/PE-based thermoplastic vulcanisates studied along the extruder axis. Polymer. 2005;**46**(17):6575-6586

[104] Sengupta P, Noordermeer JWM. Effects of composition and processing conditions on morphology and properties of thermoplastic elastomer blends of SEBS-PP-oil and dynamically vulcanized EPDM-PP-oil. Journal of Elastomers and Plastics. 2004;**36**(4):307-331

[105] Shahbikian S, Carreau PJ, Heuzey MC, Ellul MD, Cheng J, Shirodkar P, Nadella HP. Morphology development of EPDM/PP uncross-linked/dynamically cross-linked blends. Polymer Engineering and Science. 2012;**52**(2):309-322

[106] Xiao HW, Huang SQ, Jiang T, Cheng SY. Miscibility of blends of ethylene-propylene-diene terpolymer and polypropylene. Journal of Applied Polymer Science. 2002;**83**(2):315-322

[107] Goharpey F, Katbab AA, Nazockdast H. Mechanism of morphology development in dynamically cured EPDM/PP TPEs. I. Effects of state of cure. Journal of applied polymer science. 2001;**81**(10):2531-2544

[108] Barick AK, Jung JY, Choi MC, Chang YW. Thermoplastic vulcanizate nanocomposites based on thermoplastic polyurethane and millable polyurethane blends reinforced with organoclay prepared by melt intercalation technique: Optimization of processing parameters via statistical methods. Journal of Applied Polymer Science. 2013;**129**(3):1405-1416

Chapter 11

Dielectric Elastomer Sensors

Na Ni and Ling Zhang

Additional information is available at the end of the chapter

http://dx.doi.org/10.5772/intechopen.68995

Abstract

Dielectric elastomers (DEs) represent a class of electroactive polymers (EAPs) that exhibit a significant electromechanical effect, which has made them very attractive over the last several decades for use as soft actuators, sensors and generators. Based on the principle of a plane-parallel capacitor, dielectric elastomer sensors consist of a flexible and stretchable dielectric polymer sandwiched between two compliant electrodes. With the development of elastic polymers and stretchable conductors, flexible and sensitive dielectric elastomer tactile sensors, similar to human skin, have been used for measuring mechanical deformations, such as pressure, strain, shear and torsion. For high sensitivity and fast response, air gaps and microstructural dielectric layers are employed in pressure sensors or multiaxial force sensors. Multimodal dielectric elastomer sensors have been reported that can detect mechanical deformation but can also sense temperature, humidity, as well as chemical and biological stimulation in human-activity monitoring and personal healthcare. Hence, dielectric elastomer sensors have great potential for applications in soft robotics, wearable devices, medical diagnostic and structural health monitoring, because of their large deformation, low cost, ease of fabrication and ease of integration into monitored structures.

Keywords: dielectric elastomers, soft capacitance sensors, electronic skin, flexible mechanical sensors, robot sensors, tactile sensors

1. Introduction

In recent years, the soft sensing systems have attracted considerable attention because of their potential applications in assistive soft robots, healthcare and entertainment [1]. In contrast to traditional rigid sensors, their advantageous compliant properties enable the soft sensors to safely monitor soft movements or interactions with humans. Many of these devices can be used to measure strain, pressure, force, light, humidity and temperature similar to the human skin, which have advantageous properties such as flexibility, stretchability, highly sensitivity and technological compatibility with a large area [2, 3].

Various transduction methods for fabricating flexible sensors have been developed [4], including piezoresistivity, capacitance, piezoelectricity, optics and wireless antennas. Current applications require highly sensitive, flexible, stretchable and low-cost devices, where capacitive-based sensors exhibit better potential for use, because of their high strain sensitivity, static force measurement and low power consumption [5]. These capacitive tactile sensors measure the magnitude of mechanical forces or strain by converting mechanical solicitations into an electrical signal. Dielectric elastomer (DE) sensors are one type of the capacitive-based sensors, which are flexible, soft and stretchable for measuring deformations, forces and pressures.

The dielectric elastomers (DEs) are a type of field-activated polymers that belong to the family of electroactive polymers (EAPs) [6]. The typical structure of a dielectric elastomer (DE) is a dielectric material sandwiched between two electrodes, which can produce a large strain response and high electromechanical efficiency from an electric field. Based on the electromechanical effect, DEs can be used as soft actuators, sensors and generators.

Recently, several reviews have been published in the literature, which have detailed the properties and chemistry of dielectric elastomers [6, 7]. However, these reviews focused primarily on the development of dielectric elastomer actuators and generators. A detailed overview of the recent progress in dielectric elastomer sensors has not been reported, so this current review focuses on the recent research in dielectric elastomer sensors. In the first section, several sensing principles of dielectric elastomer sensors are introduced. In the second and third section, the materials of the dielectric and compliant electrodes are described. In the fourth section, a detailed overview is given of the recent progress regarding applications of dielectric elastomers in electronic skin, structural health monitoring, tissue elasticity measurements, self-sensing actuators and robotic technologies. Wearable dielectric elastomer sensor systems are also reviewed based on multiple physical sensors. Conclusions and future developments in practical applications of dielectric elastomer sensors are discussed in the final section.

2. Principle of dielectric elastomer sensors

The capacitance of a parallel plate capacitor can be written as:

$$C = \varepsilon_0 \varepsilon_r \frac{A}{d} \tag{1}$$

where C is the capacitance, $\varepsilon_0 = 8.854 \times 10^{-12} F/m.$ is the permittivity of the vacuum, ε_r is the relative permittivity, A is the electrode area and d is the dielectric distance [8]. A capacitive sensor can be designed to provide elastic deformation by sandwiching an elastomer film between two compliant conductive electrodes to form a dielectric elastomer sensor. The deformation of the dielectric elastomer produces a change in capacitance of the sensor. A simple sensing method is shown in **Figure 1**. When the sensor is subjected to external tension or

http://dx.doi.org/10.5772/intechopen.68995

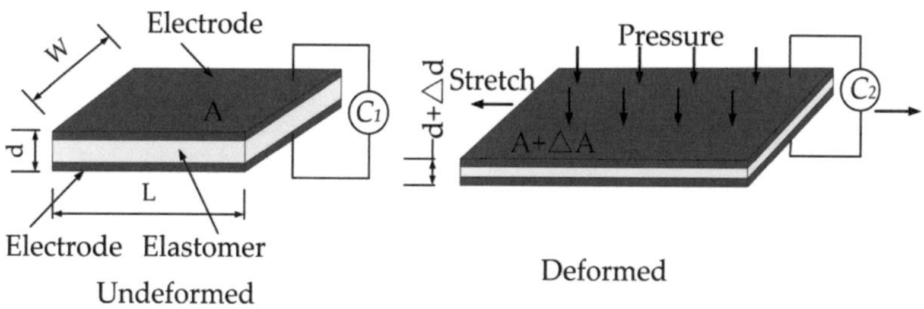

Figure 1. Simple measurement principle of a dielectric elastomer sensor.

compression, the surface of the sensor film expands, while the displacement between the two electrodes decreases, which causes an increase in capacitance.

According to the model of an ideal dielectric elastomer, assuming that the volume and permittivity remain constant, when a dielectric elastomer sensor is stretched in its plane, the change in capacitance can be derived as follows:

$$L_{strech} = \lambda_1 L \tag{2}$$

$$W_{strech} = \lambda_2 W \tag{3}$$

$$d_{strech} = \lambda_3 d = \frac{1}{\lambda_1 \lambda_2} d \tag{4}$$

where λ_1, λ_2, λ_3 are multiples of L, w, d, which are the initial dimensions of the dielectric elastomer as shown in **Figure 1** [9]. Combining the above equations with the Eq. (1), the capacitance C of the dielectric elastomer can be written as:

$$C = C_0 (\lambda_1 \lambda_2)^2 \tag{5}$$

where C_0 is the initial capacitance. When a uniaxial stretch is applied, the length of dielectric elastomer is λ times its initial length and the width and the thickness become $\frac{1}{\sqrt{\lambda}}$ times their initial values, and the capacitance value of the deformed dielectric elastomer is then:

$$C = C_0 \lambda \tag{6}$$

Hence, the capacitance change ΔC. can be expressed as:

$$\Delta C = C_0\ \varepsilon \tag{7}$$

where ε is the strain along the stretched axis of the dielectric elastomer. Eq. (7) implies that the capacitance change is linear with the strain of dielectric elastomer sensor.

Another sensing method of the dielectric elastomer sensors can be found in the relative position change of two electrodes, which produces a capacitance change as shown in **Figure 2**. This sensing method is used for measuring shear forces.

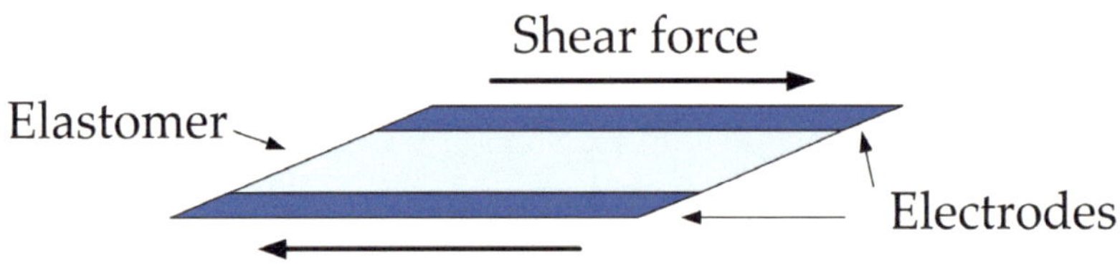

Figure 2. Measurement principle of a dielectric elastomer sensor when shear force is subjected.

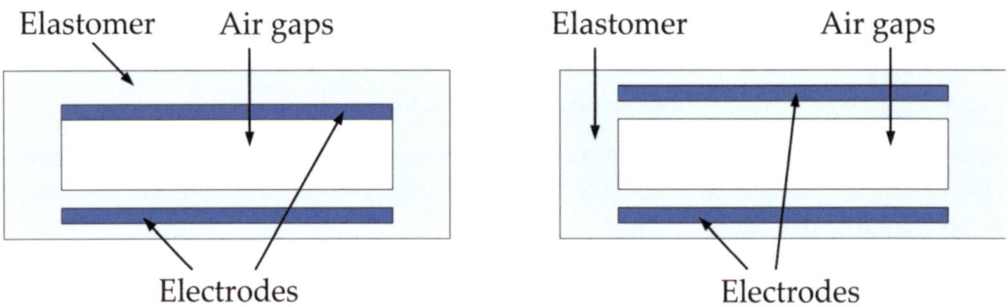

Figure 3. Prototype of a dielectric elastomer sensor with an air gap. Left: one elastic layer between two electrodes and right: two elastic layers between two electrodes.

To improve the sensitivity of dielectric elastomer sensors, **Figure 3** shows a complex sensing method that employs a multilayer dielectric including an air gap and dielectric films. By constructing the air gap, the device is able to sensitively detect multiaxis force and pressure [10, 11]. In addition, the micro/nanostructured dielectric layer can improve the sensitivity of dielectric elastomer sensors as well.

3. Materials of dielectric elastomer sensors

For the large deformation requirements of dielectric elastomer sensors, many classes of dielectric materials have been investigated, including acrylates, silicones, polyurethanes (PU), rubbers, latex rubbers, acrylonitrile butadiene rubbers, olefinic, polymer foams, fluorinated and styrenic copolymers [7]. Acrylates and silicon rubbers such as VHB, polydimethylsiloxane (PDMS) and Ecoflex have been widely used in fabricating flexible sensing devices due to their commercial availability and good performance [9, 12, 13]. The relative dielectric constant and modulus are important properties in the performance of dielectric elastomer sensors. The typical properties of a number of candidate polymers are listed in **Table 1** [7, 14–17].

A commercial acrylate adhesive film, VHB, produced by the 3M Company, is commercially available and exhibits large deformation and transparency. It can be stretched more than six times its initial length [9], and the strain is nearly linear with stress to as much as three times the initial length of the film [7]. VHB acrylates have a dielectric constant that is higher than silicon rubbers.

http://dx.doi.org/10.5772/intechopen.68995

Elastomer material	Relative dielectric constant	Young's modulus
VHB 4910 (acrylate)	4.7	1–2 MPa
Ecoflex 0050 (platinum-catalyzed silicon)	2.65	83 kPa
Ecoflex 0050 (platinum-catalyzed silicon)	2.65	69 kPa
Ecoflex 0010 (platinum-catalyzed silicon)	2.65	55 kPa
Dow Corning, Sylgard 184 (polydimethylsiloxane silicon rubber)	2.75	1.84 MPa

Table 1. Properties of several dielectric elastomers [7, 14–17].

Silicones have also been studied for use as dielectric elastomers [18]. Many silicone materials are produced commercially, such as Dow Cording Sylgard184. The modulus of these materials typically ranges from 0.1 to 2 MPa and their dielectric constants are generally around 3 [7]. Their elongation at break is less than that of VHB adhesive film. Dow Corning Sylgard184 is a type of polydimethylsiloxane (PDMS) that is generally used for fabricating dielectric elastomer sensors as a dielectric [19] and a substrate [10]. The advantages of PDMS include variable mechanical properties, transparency and stability over a wide range of temperatures [20]. For bonding electric materials to its surface, defining potential adhesive and non-adhesive regions can be accomplished by exposure to UV irradiation [20]. Another commercial silicone material, Ecoflex rubbers are also widely used for flexible capacitance sensors [21]. Ecoflex rubbers are platinum-catalyzed silicones that are soft, strong and "stretchy" [22]. Compared to Ecoflex, PDMS is hard and brittle as shown in **Figure 4** [23].

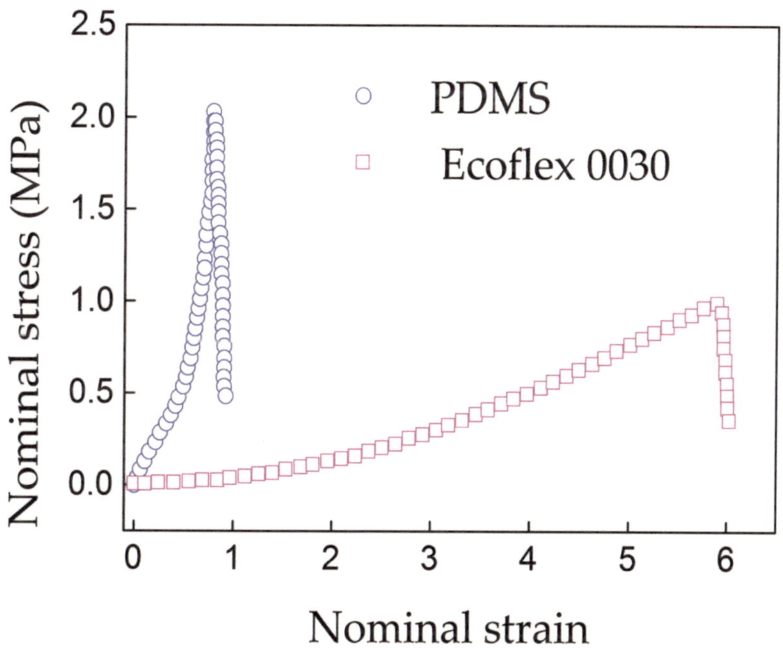

Figure 4. Uniaxial tensile test of Ecoflex 0030 and PDMS [23].

To improve the sensitivity of dielectric elastomer sensors, it is important to develop dielectric elastomers that have a relatively high dielectric constant. There are generally three routes for enhancing the dielectric constant of an elastomer, which include addition of high permittivity inorganic particles [24, 25], addition of conductive fillers [26] and chemical design [27]. Chemical design involves polymer chain modification. Titanium dioxide (TiO_2) [28] and barium titanate ($BaTiO_3$) [29] have reportedly served as high permittivity inorganic particles for improving the dielectric constants of elastomers. Conductive fillers enhance the dielectric constant of elastomers by increasing the effective electrode area or facilitating electronic polarization, such as carbon nanotubes (CNTs) [30, 31], metal particles [26] and conductive polymers [32].

4. Compliant and stretchable electrodes

Much like the dielectrics, good stretching ability and flexibility in the electrodes are necessary to enable the dielectric elastomer sensors to be used in soft robots, healthcare and entertainment. They must have the ability to exhibit large deformation (bend, fold, twist, compress, and stretch) while maintaining a high level of conductive performance, integration and reliability [33]. Of the work performed on stretchable and flexible electrodes, two main types of conductors, electrical conductors and ionic conductors, have been identified as the most desirable for modification of the electrodes based on their conductance.

4.1. Electrical conductors

Sensors employing electrical conductors (e.g., carbon grease, metal films and carbon nanotubes) as electrodes develop signals based on the movement of electrons in the material. The electronic conductors used as compliant electrodes in sensors mainly include carbon-based electrodes [11, 34], metallic thin films [35], composites of conducting materials and elastomers (or rubber fiber mats) [36], conducting films (indium-tin oxide (ITO)-coated poly (ethylene terephthalate (PET)) [37] and liquid conductors [38].

Carbon-based electrodes are widely used for dielectric elastomer sensors because they are compliant, easy to fabricate and exert a low impact on the stiffness of the dielectric [7]. Carbon-based electrodes are usually fabricated from carbon black, graphite [39], carbon nanotubes (CNTs) [34] and single-walled carbon nanotube (SWNT) [1, 40] as loose particles or mixed with the matrix to form a viscous media (carbon grease) [11] or mixed into an elastomer (conductive elastomer) [41]. Goulbourne's experiments showed that the performance of carbon grease was better than that of graphite power, graphite spray or silver grease [42]. Lipomi et al. [34] developed a skin-like pressure and strain sensor composed of carbon nanotubes (CNTs) as compliant electrodes. The advantage of these sensors was that they were transparent compared to the electrodes of sensors fabricated using other carbon-based electrodes such as carbon grease.

In other compliant electrodes composed of a single material (e.g., Au [43], Al [19] and Pt [44], AgNWs [45, 46]), metallic thin films are generally deposited on elastomers by sputtering and using an E-beam evaporator [7]. These electrodes are more suitable for use as pressure sensors [19, 43] or normal and shear force sensors [44], compared to strain sensors [40]. This is because they are thin, highly conductive and can be patterned [7], but their hardness affects the stretching of the capacitance sheets.

Conductive elastomers (composites of conducting materials and an elastomer) offer the advantages of integration into the dielectric layer, robustness and durability. For example, Laflamme et al. [41, 47] reported on a soft, large surface area capacitor network employed in structural health monitoring that was composed of carbon black dispersed in poly-styrene-co-ethylene-co-butyleneco-styrene (SEBS) solution. Experimental testing demonstrated the improved robustness and stable capacitance of the device after it was cut or punched [47]. Cai et al. [36] designed an excellent, durable capacitance strain sensor based on CNTs-doped Dragon skin 30 (Smooth-on, Inc., USA) that could measure strains of up to 300% over thousands of cycles. An elastomeric capacitive sensor array was constructed with conductive PDMS (CNT-PDMS) using micro-contact printing techniques [10]. These devices were stable under various elastic deformations. In additional to these carbon-based electrodes, CNT fabrics were reported that employed a capacitor geometry grid as multimodal skin sensors [2]. In addition to CNTs, elastic conductors composed of silver nanowires (AgNWs) and elastomers have been used for improving mechanical robustness of sensors that can detect stains of up to 50% [48, 49].

Other electrodes used in capacitive sensors include conductive textiles composed of conductive particles and fabrics. A commercial, stretchable conductive textile, termed Electrolyca (Mindsets Ltd, UK), was composed of woven silver with nylon elastic fibers employed as electrodes in a flexible capacitance-sensing element integrated in a measured structure [4]. Another conductive textile, Zelt fabric (Mindsets Ltd, UK), made of soft copper-/tin-coated woven fabrics, was assembled into a dielectric film for use in soft tactile sensors [5]. Park et al. [50] fabricated stretchable, thin-film electrodes using silver nanoparticles and rubber fiber mats. The electrodes retained high conductivity at 140% strain and were compatible with various substrates and could be deployed over large areas.

A conductive film has been reported, which was composed of indium-tin oxide (ITO)-coated poly (ethylene terephthalate) (PET) film and exhibited good performance when its bending radius was greater than 8 mm [51]. This ITO/PET film was suitable for use as a pressure sensor with microstructured elastomeric dielectric film [37, 51, 52].

In addition to the use of compliant electrical conductors, conductive liquids have also been employed based on liquid, metal-filled channels [38, 53, 54] or microliquid metal droplets [55]. This technical approach is beneficial to use in an array of soft tactile sensors.

4.2. Ionic conductors

There are instances when compliant electrodes must meet the stretch requirements of a sensor as well as the biocompatibility and transparency requirements [9]. Sensors employing ionic conductors can develop signals using these ions. Ionic conductors have been studied for use in applications that demand greater stretch and greater transparency than electronic conductors, such as polyacrylamide hydrogels [56–58] and ionogels composed of ionic liquid and polyacrylic acid [59]. Recent work has demonstrated that these materials can be used as the electrodes of transparent sensors and actuators for artificial muscles, skin, axons and kinesthetic sensing [60–64]. A transparent, capacitive sensor termed ionic skin was first developed using a dielectric elastomer covered with ionic electrodes, which were composed of a polyacrylamide hydrogel with NaCl [9]. To solve the problem of water loss from the hydrogel, high-retention hydrogel conductors [58] were applied in a highly stretchable and transparent actuator [60]. This was then formed into a highly stretchable electroluminescence [65] and an ionic cable that exhibited high-speed and long-distance signal transmission [66]. The high-retention hydrogel conductors were very stretchable, with strains that exceeded 1000% as shown in **Figure 5**. The hydrogel conductor was also very tough. We poked a knife at the hydrogel conductor (**Figure 6a**). **Figure 6b** showed it remained intact after being poked [67].

For use in sensors, a good compliant electrode should exhibit low hysteresis of resistance versus strain, a well-controlled surface resistance and limited degradation with load cycling [7].

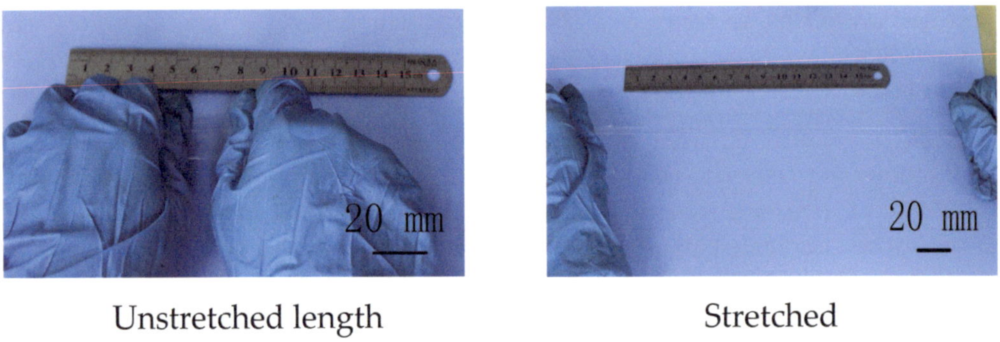

Figure 5. The hydrogel conductor from an original state to a stretched state.

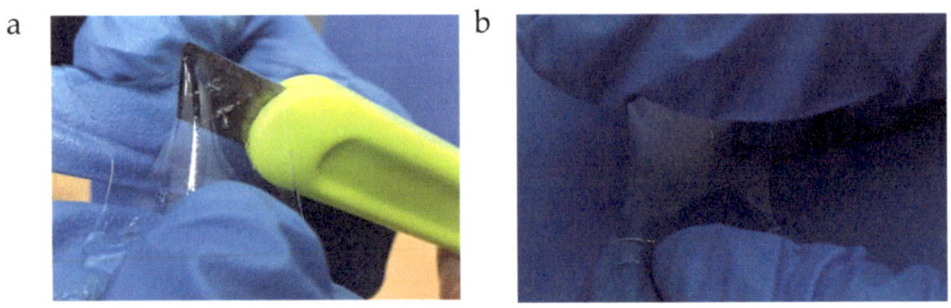

Figure 6. (a) Poking at the hydrogel conductor with a knife and (b) the ionic conductor remaining intact after being poked.

5. Applications of dielectric elastomer sensors

Compared to traditional rigid sensors (resistance strain gages, piezoelectric ceramics sensors, etc.), dielectric elastomer sensors exhibit advantages, including low cost, large deformation (the strain even exceeds 100%) and fast response time, and are easily integrated into monitored structures [5, 50, 52]. They have been used extensively as physical sensors, such as pressure sensors [52], strain sensors [1, 36, 68], normal and shear force sensors [5, 34, 43, 53]. In addition, for simulating multifunctional human skin, multimodal sensors have been made for measuring pressure, temperature and humidity [2, 69]. These outstanding features make these sensors potentially useful in soft robots, human-machine interfaces, human-activity monitoring, personal healthcare and structural health monitoring (**Figure 7**).

5.1. Flexible and stretchable sensors as capacitive electronic skin

Recently, flexible and stretchable strain sensors have been developed that can be worn as electronic skin (e-skin) for soft robots, wearable devices and human motion monitoring in medicine, including diagnosis development, rehabilitation assistance and activity monitoring [3]. These flexible and stretchable capacitive strain sensors consisted of a dielectric polymer film sandwiched by two compliant electrical conductors. Cohen et al. [68] proposed a highly elastic strain sensor composed of dielectric elastomers sandwiched between CNT percolation electrodes. These sensors can be stretched to 100% of their original size over thousands of cycles with a 3% variability, which has resulted in the demonstration of useful applications in a robotics context for transduce joint angles. A highly stretchable and transparent capacitive strain sensor based on CNT elastic electrodes has been proposed [36]. The highly sensitive strain sensors can measure strains up to 300% with excellent durability, stability and fast response, which have potential applications in wearable smart electronics as demonstrated by experiments in glove and respiration monitoring. Yang et al. [1] made a recoverable motion sensor based on the surface-modified $CaCu_3Ti_4O_{12}$ (S-CCTO) nanoparticles involving a self-healing

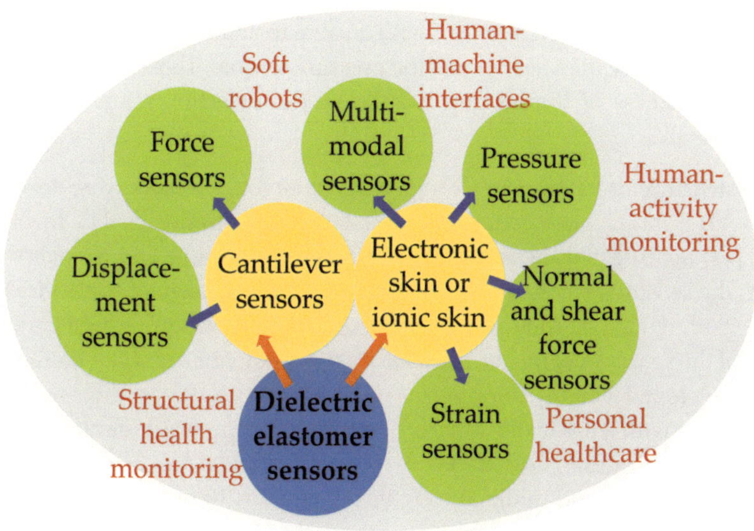

Figure 7. Applications of the dielectric elastomer sensors in soft robots, human-machine interfaces, human-activity monitoring, personal healthcare and structural health monitoring.

polymer matrix. The composites had a high dielectric permittivity of 93 at 100 Hz. These authors' work showed the benefits of the electrical and mechanical self-healing properties of their motion sensors.

Similar to the concept of human skin, the flexible and stretchable dielectric elastomer sensors can be used as soft strain gauges but can also be used for measurements of pressure and three-axis forces. Pressure sensors have been fabricated by sandwiching a soft dielectric between two sets of flexible electrodes, such as metal thin films [43], CNTs [34], AgNW-based [45], liquid electrodes [53]. Dielectric elastomer sensors were designed as skin-like tactile sensors, which could measure pressure produced by the human-body activity. These pressures included a low-pressure regime (<10 kPa) produced by intrabody pressure (intraocular pressure and intracranial pressure), a medium-pressure regime (<100 kPa) generated by wave, vibration or pulses (blood pressure, respiration, phonation, heart, radial artery, jugular venous) and a high-pressure regime (>100 kPa) produced in the foot [52, 68]. Extremely robust pressure sensors were fabricated using flexible polyurethane foams and Au thin films that could measure normal pressures from 1 to 100 kPa for applications as artificial skin and in wearable robotics [43]. Transparent and stretchable pressure and strain sensor arrays have been developed to detect pressures of up to 1 MPa using nanomaterial electrodes. Potential applications include prosthetic limbs, bandages, robotics and touch screens [34, 43]. Li et al. [53] proposed highly deformable tactile sensing arrays based on low modulus platinum-catalyzed silicone polymer (EcoFlex00-30) with embedded liquid metal microfluidic (eutectic gallium indium) arrays. The wearable tactile sensors could be stretched greater than 400% and could conform to curved objects or soft biological tissues.

However, the incompressibility and viscoelasticity of rubbers limit their sensitivity for use in pressure sensors [52]. By constructing an air gap and compressible dielectric layers between two electrodes, the performance of the pressure sensors can be improved [70]. Zhang et al. [21, 71] designed a compressive soft sensor with an air gap and theoretically demonstrated that the air gap could improve the sensitivity of the pressure sensors. These authors' results showed that when the thickness of the air gap was large, the detection range was large but the resolution was small.

Based on the structure containing an air gap, three-axial force sensors were designed using a 2D overlap with a larger top electrode on four bottom electrodes [5, 11]. In this structure, Zhang et al. [11] investigated the effect of the geometry (rectangle and circular sector) of the electrodes on the sensitivity, linearly, hysteresis and detection range of the device using simulations and experimentation. Their results showed that the rectangular strategy improved the performance of the sensor over that of a circular design (four circular sector bottom electrodes). Viry et al. [5] developed a highly compressible and sensitive capacitive three-axial force sensor that was fabricated with a top textile electrode and four bottom textile electrodes sandwiching a fluorosilicone film and an air gap. This device reportedly could detect pressures of up to 190 kPa (estimated up to 400 kPa) and minimal weights of less than 10 mg. Hence, the application of the sensor was extensive, including biomimetic touch, heartbeat monitoring and foot pressure-distribution monitoring.

The micro/nanostructured dielectrics have been widely developed to improve the sensitivity and response and/or relaxation time of the pressure sensors [72]. Mannsfeld et al. [52] proposed a type of organic thin film that consisted of a regularly structured dielectric layer with organic field-effect transistors (OFETs) for use as pressure sensors. The design of this device resulted in faster response and relaxation times (<<1s), and higher sensitivity (0.55 kPa^{-1}) than the unstructured film (0.02 kPa). Using a device with microstructured rubber dielectric layers, Schwartz et al. [37] further developed highly flexible transistor devices with high sensitivity (8.4 kPa^{-1}), high stability and low power consumption that were used as an electronic skin for human-machine interfaces [37]. Various microstructures have been studied for use as the dielectric layer for flexible capacitive pressure sensors, including pyramids [19], microhairs [73] and microspheres [74]. Tee et al. [19] found that the pyramidal structures with relative small sidewall angles were the optimum shapes for rendering improved sensitivity, compared to structures with relative large sidewall angles and square cross-sectional structures. In addition, when the structures were spaced further apart, the sensitivity increased. The authors also demonstrated that these pressure sensors could be used for blood pulse monitoring and next-generation force sensing track pads with high sensitivity, easy manufacturing and low cost [19]. The design of microhair structures for flexible pressure sensors can enhance the signal-to-noise ratio and enable signal detection of the deep-lying internal jugular venous pulses for healthcare monitoring [73]. Li et al. [74] proposed a pressure sensor composed of polystyrene microspheres (dielectric layer) and Au electrodes with a surface micropattern similar to lotus leaves. The devices facilitated measurements over a wide dynamic response range (0–50N) in addition to high sensitivity (0.815 kPa^{-1}) and a fast response time. This device was suitable for use over a wide pressure-measuring range for applications in wearable healthcare devices, patient rehabilitation and biomedical prostheses. In other devices, a dielectric layer of the pressure sensor respectively employed porous PDMS [51], porous silicone foam [75, 76] or graphene oxide (GO) foam [77]. In particular, Chen et al. [51] developed a microstuctured elastomeric dielectric film with air voids (porous PDMS) integrated with ITO/PET electrodes, which was highly sensitive and fast, capable of measuring pressures from 1 to 250 kPa and durable under loads larger than 1 MPa. Experiments demonstrated that these advantages of this device readily let it for use as a smart insole and a wrist pulse monitoring sensor.

5.2. Ionic skin

Ionic conductors have attracted attention in capacitive tactile sensing applications, due to their high transparency and conductivity under large deformation [61–63]. Nie et al. [61] fabricated a capacitive pressure sensor utilizing an ionic gel matrix. This transparent film sensor had a high sensitivity (3.1 nF/kPa) because the electrical double layer produced an ultra-high unit-area capacitance. The sensor also exhibited excellent mechanical and thermal stability and a rapid response. The applicability of this sensor as a wearable device was successfully demonstrated so that it was later incorporated into the consumer electronic devices including a smart watch, augmented reality glasses and a custom fingertip-mounted tactile sensor. Sun et al. [9] developed a highly stretchable and transparent capacitive sensing sheet called ionic skin, which was composed of a dielectric and two ionic conductors. These sensors can detect strains over a wide range (from 1 to 500%) and pressures as low as 1 kPa. The working principle of the sensor was reported in detail in the literature [9]. A hybrid ionic-electronic circuit was formed by connecting the ionic conductors to electronic conductors for transmitting electrical signals. When a low voltage was applied between the two electrodes, electrochemical reactions did not occur, and no electrons or ions crossed the interface between the electrode and the ionic conductor. This allowed electrical double layers to form at the interface, similar to a capacitor C_{EDL}, which was in series with a capacitor C_m formed by the ionic conductors and the dielectric (**Figure 8**). As a result, the capacitance C measured between the two electrodes can be written as $C = C_m/(2C_m/C_{EDL} + 1)$. The capacitance of the dielectric was much less than the electrical double layer since the separation (nanometer) of the charges in the electrode and in the ionic conductor was much less than the distance between the charges separated by the dielectric with a thickness. Consequently, the measured capacitance C was dominated by the capacitance of the dielectric C_m. According to this principle, a transparent capacitive sensing film was fabricated in our laboratory for structural health monitoring.

5.3. Flexible and large-surface dielectric elastomer sensors applied in structural health monitoring

Detecting and locating damage to bridges, roads, buildings and other structures is necessary to ensure a long lifespan of the national infrastructure. Detecting defects and providing warnings in time can prevent the collapse of structures. Soft film sensors employing functional materials have been proposed for monitoring the condition of structures. These sensors include resistance-based, piezoelectric, antenna, vacuum and capacitance-based strain sensors. Some of these technologies are unsuitable for use on a large scale, because their complex manufacturing processes make them too expensive.

In the recent developments in structural health monitoring, Laflamme et al. [78] first proposed a sensing technique for damage localization using a layer of commercial dielectric polymer (DEAP, Danfoss PolyPower) on the surface of a monitored structure with a same reference

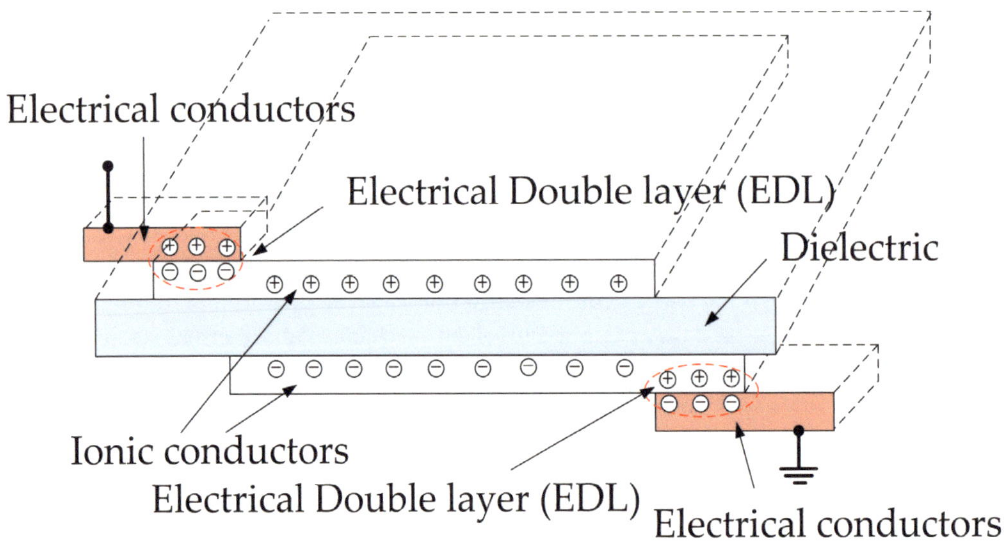

Figure 8. A schematic diagram of additional electric charges accumulated when a voltage is applied between the electrical conductors [9].

capacitance. The next step entailed fabricating soft elastomeric capacitors (SECs) that were inexpensive and useable on large surface areas [79]. The robust and static characterization of the sensors was demonstrated through testing [47]. Static characterization, dynamic monitoring, localization of fatigue cracks and the distribution of thin film sensor arrays on structures were studied using the referenced SECs [80–83]. The electrodes of these sensors were composed of SEBS containing carbon black particles, and the sensors were not transparent. The damaged parts of the subject structure are invisible when the sensors were placed on the surface.

We have developed a transparent, capacitive sensor that is inexpensive and can easily detect cracks in large-scale infrastructures [67]. The sensor was attached to a monitored surface by a bonding layer (**Figure 9**). The sensor consisted of a transparent dielectric elastomer sandwiched between two transparent and stretchable ionic electrodes. The transparent capacitive sensor could be used to monitor large areas of a structure to detect potential damage. The robustness of this sensor was demonstrated by tests (**Figure 10a** and **b**), particularly on the over-reinforced beam. Tests conducted on the over-reinforced concrete beam demonstrated that the sensor was capable of detecting small cracks (**Figure 10c** and **d**).

5.4. Cantilever sensors (force, torque and displacement) for robotic technologies

In the previously described applications of dielectric elastomer sensors, most of these sensors were in the shape of a flat sheet. This is not suitable for measuring other mechanical parameters such as concentrated force and displacement. The cantilever beam sensors are a class of force sensors that are used extensively for weight measurement, environmental monitoring, biological monitoring and gas detection. Most of these devices are rigid sensors that are unsuited for soft robotics applications. There have been few reports in the literature about soft displacement or force dielectric elastomer sensors based on the flexible beam. Lucarotti et al. [4] proposed a strategy to sense-bending angle and force in a soft body based on beam configuration, which was integrated with two capacitive sensing elements for distinguishing between convex and concave bending. In their analysis of the device, they imposed a bending and/or an external force to a soft beam. The results confirmed the superiority of this sensing strategy for use in soft robotics.

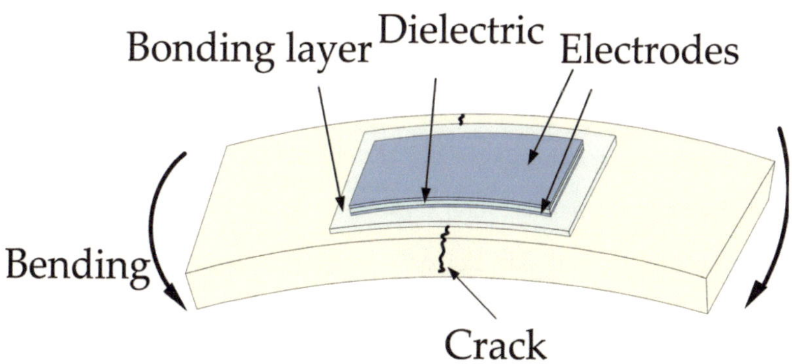

Figure 9. A schematic of a dielectric covered with ionic conductors deployed over the surface of a monitored structure by using a bonding layer [67].

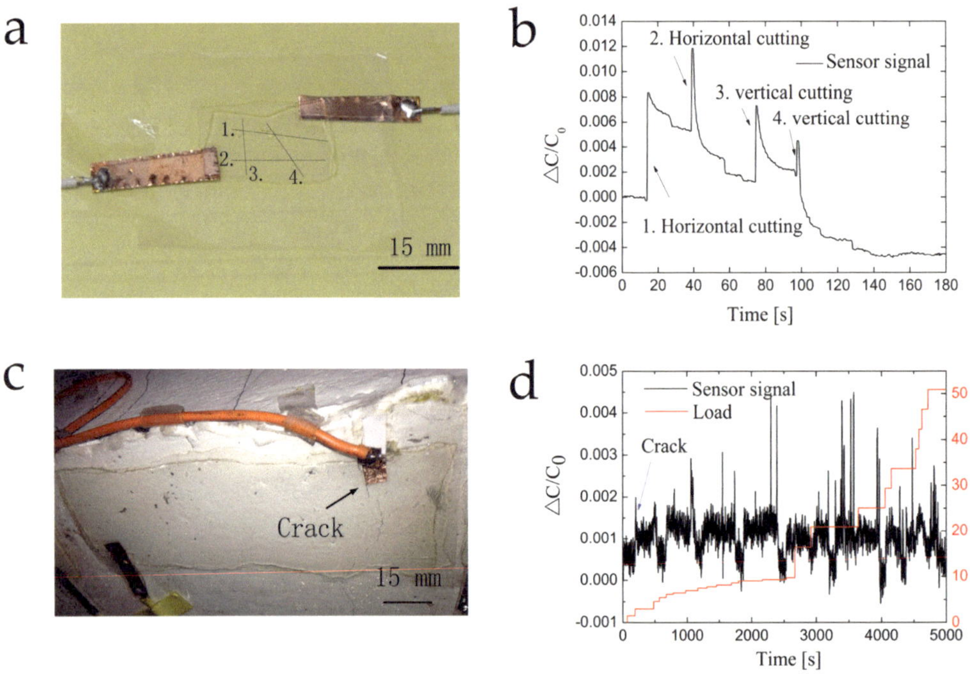

Figure 10. (a) An illustrative photo of four cuts on the transparent capacitive sensor, (b) relative change in capacitance when cuts were made, (c) a photograph of cracks formed under the sensor and d) relative change in capacitance of the sensor and load against time [67].

We have proposed a dielectric elastomer cantilever beam sensor for measuring force or displacement [84]. A dielectric membrane sandwiched between compliant electrodes was pasted to the surface of a soft uniform strength cantilever beam by the bonding layer, which made the sensing device as shown in **Figure 11**. The concentrated force at the free end induced a change in the capacitance of the dielectric membrane. Therefore, the applied force could be determined from the change in the capacitance of the sensor. The beam of uniform strength was used to determine that the deformation of dielectric membrane was homogeneous. The results of the experiments showed the change in capacitance almost increased linearly with the increasing of the force from 0 N to 2 mN, which was applied to the end of the beam. As the results indicate, the dielectric elastomer cantilever beam force sensor appeared to work well. Because the force was linear with the deflection at the free end of the cantilever beam, the proposed device can be used as a displacement sensor.

5.5. Multimodality of dielectric elastomer sensors

The tactile sense of the human skin not only includes physical responses such as strain, pressure, shear and torsion but also temperature and humidity. To mimic the human skin, the electronic skin must be designed to sense multimodality consisting of physical, biological and chemical stimuli. The integration of flexible and stretchable multiple sensors into wearable platforms has been developed for applications in human-activity monitoring, human-machine interfaces and personal healthcare [2, 3, 69, 85]. To assess these parameters, multimodal

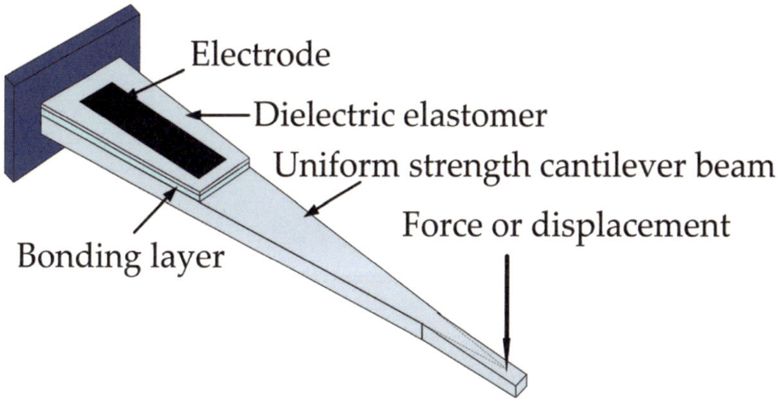

Figure 11. A sketch of the dielectric elastomer cantilever beam force sensor [84].

dielectric elastomer sensors have been tested. Kim et al. [2] described a highly sensitive and multimodal capacitive sensor composed of Ecoflex and carbon nanotube microyarns. The sensor was capable of measuring pressures as low as 0.4 Pa, temperature or humidity gradients and biological variables with different dipole moments in a single pixel. The measurements of mechanical deformation, humidity and biological variables were realized by the change in capacitance. The measurement of temperature was accomplished by determining a change in resistance. Ho et al. [69] developed a transparent and stretchable multimodal electronic skin sensor matrix. This sensor could measure pressure, thermal and humidity simultaneously using independent electrical signals. Graphene (chemical vapor deposition) electrodes sandwiched PDMS, which formed capacitive sensors for measuring pressure and strain. The graphene oxide (GO) and reduced graphene oxide (rGO) were used as impedance humidity sensor and resistive thermal sensor, respectively. The pressure, humidity and thermal sensors were integrated into a layer-by-layer geometry. In addition, excellent sensitivity of the sensors was demonstrated by monitoring moving hot air, breathing and finger touching.

In addition to the previously mentioned applications of dielectric elastomer sensors, other applications have also been developed. For example, a soft capacitive tactile sensor consisting of two sensing elements with different stiffness was reported for measuring tissue elasticity [86]. The results of tests using this device showed that the sensor could detect tissue elasticity from 0.1 to 0.5 MPa. Dielectric elastomer sensors have also been used as self-sensing actuators, such as pneumatic actuators [62], dielectric elastomer actuators [87], coaxial actuators [88] and McKibben actuators [39].

6. Conclusions and outlook

In this chapter, the primary sensing principles and materials used in flexible and stretchable dielectric elastomer sensors have been described along with the efforts to develop dielectric elastomer sensing devices for applications in soft robotics, structure health monitoring, electronic or ionic skin for human-activity and personal healthcare. In addition, developments in our own laboratory on dielectric elastomers were presented, including cantilever force (displacement) sensors for soft robotics and transparent sensors used in structural health monitoring.

Highly sensitive, flexible and stretchable dielectric elastomer mechanical sensors have been developed for soft robots and personal healthcare; however, a tunable measurement range is needed for different positions on the human body or on robots. For practical applications of dielectric elastomer sensors, the performance of the sensors must be further improved to increase their durability and minimize drift and interference. The multiple dielectric elastomer sensors need be developed to improve their sensitivity in the measurement of temperature, humidity, chemical and biological stimuli. For future development, a sensor-integrated wearable platform with low-power consumption will be needed, which is equipped with a wearable power generator with high output efficiency as well as a power-storage device with large capacity [3].

Author details

Na Ni and Ling Zhang*

*Address all correspondence to: zhangl@mail.xjtu.edu.cn

State Key Laboratory for Strength and Vibration of Mechanical Structures, Xi'an Jiaotong University, Xi'an, China

References

[1] Yang Y, Zhu B, Yin D, Wei J, Wang Z, Xiong R, Shi J, Liu Z, Lei Q. Flexible self-healing nanocomposites for recoverable motion sensor. Nano Energy. 2015;**17**:1–9. DOI: 10.1016/j.nanoen.2015.07.023

[2] Kim SY, Park S, Park HW, Park DH, Jeong Y, Kim DH. Highly sensitive and multimodal all-carbon skin sensors capable of simultaneously detecting tactile and biological stimuli. Advanced Materials. 2015;**27**(28):4178–4185. DOI: 10.1002/adma.201501408

[3] Trung TQ, Lee NE. Flexible and stretchable physical sensor integrated platforms for wearable human-activity monitoring and personal healthcare. Advanced Materials. 2016; DOI: 10.1002/adma.201504244

[4] Lucarotti C, Totaro M, Sadeghi A, Mazzolai B, Beccai L. Revealing bending and force in a soft body through a plant root inspired approach. Scientific Reports. 2015;**5**:8788. DOI: 10.1038/srep08788

[5] Viry L, Levi A, Totaro M, Mondini A, Mattoli V, Mazzolai B, Beccai L. Flexible three-axial force sensor for soft and highly sensitive artificial touch. Advanced Materials. 2014;**26**(17):2659–2664. DOI: 10.1002/adma.201305064

[6] Chiba S. Dielectric elastomers. In: Asaka K, Okuzaki H, editors. Soft Actuators. Japan: Springer; 2014. pp. 183–195. DOI: 10.1007/978-4-431-54767-9_13

[7] Biggs J, Danielmeier K, Hitzbleck J, Krause J, Kridl T, Nowak S, Orselli E, Quan X, Schapeler D, Sutherland W, Wagner J. Electroactive polymers: Developments of and perspectives for dielectric elastomers. Angewandte Chemie International Edition. 2013;**52**(36):9409–9421. DOI: 10.1002/anie.201301918

[8] Puers R. Capacitive sensors: When and how to use them. Sensors and Actuators A: Physical. 1993;**37**:93–105. DOI: 10.1016/0924-4247(93)80019-D

[9] Sun JY, Keplinger C, Whitesides GM, Suo Z. Ionic skin. Advanced Materials. 2014;**26**(45):7608–7614. DOI: 10.1002/adma.201403441

[10] Woo SJ, Kong JH, Kim DG, Kim JM. A thin all-elastomeric capacitive pressure sensor array based on micro-contact printed elastic conductors. Journal of Materials Chemistry C. 2014;**2**(22):4415–4422. DOI: 10.1039/C4TC00392F

[11] Zhang H, Wang MY. Multi-axis soft sensors based on dielectric elastomer. Soft Robotics. 2016;**3**(1):3–12. DOI: 10.1089/soro.2015.0017

[12] Lipson H. Challenges and opportunities for design, simulation, and fabrication of soft robots. Soft Robotics. 2014;**1**(1):21–27. DOI: 10.1089/soro.2013.0007

[13] Rosenthal M, Bonwit N, Duncheon C, Heim J. Applications of dielectric elastomer EPAM sensors. In: Bar-Cohen Y, editor. The 14th International Symposium on Smart Structures and Materials & Nondestructive Evaluation and Health Monitoring; 2007 April 6; San Diego, California. International Society for Optics and Photonics; 2007. p. 65241F. DOI: 10.1117/12.715084

[14] Diller E, Zhuang J, Zhan Lum G, Edwards MR, Sitti M. Continuously distributed magnetization profile for millimeter-scale elastomeric undulatory swimming. Applied Physics Letters. 2014;**104**(17):174101. DOI: 10.1063/1.4874306

[15] Roberts P, Damian DD, Shan W, Lu T, Majidi C. Soft-matter capacitive sensor for measuring shear and pressure deformation. In: IEEE International Conference on InRobotics and Automation (ICRA); 6 May 2013; Karlsruhe, Germany. IEEE; 2013. pp. 3529-3534. DOI: 10.1109/ICRA.2013.6631071

[16] Wohl CJ, Palmieri FL, Hopkins JW, Jackson AM, Connell JW, Lin Y, Cisotto AA. Flexible micropost arrays for shear stress measurement. NASA/TP-2015-218986. Hampton, VA: NASA Langley Research Center; December 2015. https://ntrs.nasa.gov/search.jsp?R=20160001622

[17] Wang Q, Tahir M, Zhang L, Zhao X. Electro-creasing instability in deformed polymers: Experiment and theory. Soft Matter. 2011;**7**(14):6583–6589. DOI: 10.1039/C1SM05645J

[18] Pelrine RE, Kornbluh RD, Joseph JP. Electrostriction of polymer dielectrics with compliant electrodes as a means of actuation. Sensors and Actuators A: Physical. 1998;**64**(1):77–85. DOI: 10.1016/S0924-4247(97)01657-9

[19] Tee BC, Chortos A, Dunn RR, Schwartz G, Eason E, Bao Z. Tunable flexible pressure sensors using microstructured elastomer geometries for intuitive electronics. Advanced Functional Materials. 2014;**24**(34):5427–5434. DOI: 10.1002/adfm.201400712

[20] Sun Y, Rogers JA. Structural forms of single crystal semiconductor nanoribbons for high-performance stretchable electronics. Journal of Materials Chemistry. 2007;**17**(9):832–840. DOI: 10.1039/B614793C

[21] Zhang H, Wang MY, Li J, Zhu J. A soft compressive sensor using dielectric elastomers. Smart Materials and Structures. 2016;**25**(3):035045. DOI: 10.1088/0964-1726/25/3/035045

[22] Smooth-on, Inc. Ecoflex Series [Internet]. 2017. Available from: https://www.smooth-on.com/product-line/ecoflex/

[23] Zhang J, Tang J, Hong J, Lu T, Wang H. The design and analysis of pneumatic rubber actuator of soft robotic fish. In: Zhang X, Liu H, Chen Z, Wang N, editors. Intelligent Robotics and Applications. ICIRA 2014. Lecture Notes in Computer Science; Cham, Switzerland: Springer; 2014. pp. 320–327. DOI: 10.1007/978-3-319-13966-1_32

[24] Kossi A, Bossis G, Persello J. Electro-active elastomer composites based on doped titanium dioxide. Journal of Materials Chemistry C. 2015;**3**(7):1546–1556. DOI: 10.1039/C4TC02535K

[25] Maiolino P, Galantini F, Mastrogiovanni F, Gallone G, Cannata G, Carpi F. Soft dielectrics for capacitive sensing in robot skins: Performance of different elastomer types. Sensors and Actuators A: Physical. 2015;**226**:37–47. DOI: 10.1016/j.sna.2015.02.010

[26] Hu W, Zhang SN, Niu X, Liu C, Pei Q. An aluminum nanoparticle–acrylate copolymer nanocomposite as a dielectric elastomer with a high dielectric constant. Journal of Materials Chemistry C. 2014;**2**(9):1658–1666. DOI: 10.1039/c3tc31929f

[27] Kussmaul B, Risse S, Kofod G, Waché R, Wegener M, McCarthy DN, Krüger H, Gerhard R. Enhancement of dielectric permittivity and electromechanical response in silicone elastomers: Molecular grafting of organic dipoles to the macromolecular network. Advanced Functional Materials. 2011;**21**(23):4589–4594. DOI: 10.1002/adfm.201100884

[28] Stoyanov H, Kollosche M, Risse S, McCarthy DN, Kofod G. Elastic block copolymer nanocomposites with controlled interfacial interactions for artificial muscles with direct voltage control. Soft Matter. 2011;**7**(1):194–202. DOI: 10.1039/c0sm00715c

[29] Szabo JP, Hiltz JA, Cameron CG, Underhill RS, Massey J, White B, Leidner J. Elastomeric composites with high dielectric constant for use in Maxwell stress actuators. In: Bar-Cohen Y, editor. Smart Structures and Materials; 2003 Jul 28; San Diego, CA. International Society for Optics and Photonics; 2003. pp. 180–190. DOI: 10.1117/12.484377

[30] Shehzad K, Hakro AA, Zeng Y, Yao SH, Xiao-Hong Y, Mumtaz M, Nadeem K, Khisro NS, Dang ZM. Two percolation thresholds and remarkably high dielectric permittivity in pristine carbon nanotube/elastomer composites. Applied Nanoscience. 2015;**5**(8):969–974. DOI: 10.1007/s13204-015-0403-0

[31] Galantini F, Bianchi S, Castelvetro V, Gallone G. Functionalized carbon nanotubes as a filler for dielectric elastomer composites with improved actuation performance. Smart Materials and Structures. 2013;**22**(5):055025. DOI: 10.1088/0964-1726/22/5/055025

[32] Stoyanov H, Kollosche M, McCarthy DN, Kofod G. Molecular composites with enhanced energy density for electroactive polymers. Journal of Materials Chemistry. 2010;**20**(35):7558–7564. DOI: 10.1039/C0JM00519C

[33] Cheng T, Zhang Y, Lai WY, Huang W. Stretchable thin-film electrodes for flexible electronics with high deformability and stretchability. Advanced Materials. 2015;**27**(22):3349–3376. DOI: 10.1002/adma.201405864

[34] Lipomi DJ, Vosgueritchian M, Tee BC, Hellstrom SL, Lee JA, Fox CH, Bao Z. Skin-like pressure and strain sensors based on transparent elastic films of carbon nanotubes. Nature Nanotechnology. 2011;**6**(12):788–792. DOI: 10.1038/nnano.2011.184

[35] Cheng MY, Lin CL, Lai YT, Yang YJ. A polymer-based capacitive sensing array for normal and shear force measurement. Sensors. 2010;**10**(11):10211–10225. DOI: 10.3390/s101110211

[36] Cai L, Song L, Luan P, Zhang Q, Zhang N, Gao Q, Zhao D, Zhang X, Tu M, Yang F, Zhou W. Super-stretchable, transparent carbon nanotube-based capacitive strain sensors for human motion detection. Scientific Reports. 2013;**3**:3048. DOI: 10.1038/srep03048

[37] Schwartz G, Tee BC, Mei J, Appleton AL, Kim DH, Wang H, Bao Z. Flexible polymer transistors with high pressure sensitivity for application in electronic skin and health monitoring. Nature Communications. 2013;**4**:1859. DOI: 10.1038/ncomms2832

[38] Noda K, Matsumoto K, Shimoyama I. Stretchable tri-axis force sensor using conductive liquid. Sensors and Actuators A: Physical. 2014;**215**:123–129. DOI: 10.1016/j.sna.2013.09.031

[39] Goulbourne NC, Son S, Fox JW. Self-sensing McKibben actuators using dielectric elastomer sensors. In: Bar-Cohen Y, editor. The 14th International Symposium on: Smart Structures and Materials & Nondestructive Evaluation and Health Monitoring; 6 April 2007; San Diego, California. International Society for Optics and Photonics; 2007. p. 652414. DOI: 10.1117/12.716274

[40] Yamada T, Hayamizu Y, Yamamoto Y, Yomogida Y, Izadi-Najafabadi A, Futaba DN, Hata K. A stretchable carbon nanotube strain sensor for human-motion detection. Nature Nanotechnology. 2011;**6**(5):296–301. DOI: 10.1038/NNANO.2011.36

[41] Laflamme S, Saleem HS, Vasan BK, Geiger RL, Chen D, Kessler MR, Rajan K. Soft elastomeric capacitor network for strain sensing over large surfaces. IEEE/ASME Transactions on Mechatronics. 2013;**18**(6):1647–1654. DOI: 10.1109/TMECH.2013.2283365

[42] Fox JW, Goulbourne NC. A study on the effect of flexible electrodes and passive layers on the performance of dielectric elastomer membranes. In: ASME International Mechanical Engineering Congress and Exposition; 1 Jan 2006; Chicago, IL, USA. American Society of Mechanical Engineers; 2006. pp. 425-433. DOI: 10.1115/IMECE2006-15888

[43] Vandeparre H, Watson D, Lacour SP. Extremely robust and conformable capacitive pressure sensors based on flexible polyurethane foams and stretchable metallization. Applied Physics Letters. 2013;**103**(20):204103. DOI: 10.1063/1.4832416

[44] Dobrzynska JA, Gijs MA. Polymer-based flexible capacitive sensor for three-axial force measurements. Journal of Micromechanics and Microengineering. 2012;**23**(1):015009. DOI: 10.1088/0960-1317/23/1/015009

[45] Hu W, Niu X, Zhao R, Pei Q. Elastomeric transparent capacitive sensors based on an interpenetrating composite of silver nanowires and polyurethane. Applied Physics Letters. 2013;**102**(8):38. DOI: 10.1063/1.4794143

[46] Quan Y, Wei X, Xiao L, Wu T, Pang H, Liu T, Huang W, Wu S, Li S, Chen Z. Highly sensitive and stable flexible pressure sensors with micro-structured electrodes. Journal of Alloys and Compounds. 2017;**699**:824–831. DOI: 10.1016/j.jallcom.2016.12.414

[47] Laflamme S, Kollosche M, Connor JJ, Kofod G. Robust flexible capacitive surface sensor for structural health monitoring applications. Journal of Engineering Mechanics. 2012;**139** (7):879–885. DOI: 10.1061 10.1061/(ASCE)EM.1943-7889.0000530

[48] Xu F, Zhu Y. Highly conductive and stretchable silver nanowire conductors. Advanced Materials. 2012;**24**(37):5117–5122. DOI: 10.1002/adma.201201886

[49] Yao S, Zhu Y. Wearable multifunctional sensors using printed stretchable conductors made of silver nanowires. Nanoscale. 2014;**6**(4):2345–2352. DOI: 10.1039/C3NR05496A

[50] Park M, Im J, Shin M, Min Y, Park J, Cho H, Park S, Shim MB, Jeon S, Chung DY, Bae J. Highly stretchable electric circuits from a composite material of silver nanoparticles and elastomeric fibres. Nature Nanotechnology. 2012;**7**(12):803–809. DOI: 10.1038/ nnano.20 12. 206

[51] Chen S, Zhuo B, Guo X. Large area one-step facile processing of microstructured elastomeric dielectric film for high sensitivity and durable sensing over wide pressure range. ACS Applied Materials & Interfaces. 2016;**8**(31):20364–20370. DOI: 10.1021/acsami.6b05177

[52] Mannsfeld SC, Tee BC, Stoltenberg RM, Chen CV, Barman S, Muir BV, Sokolov AN, Reese C, Bao Z. Highly sensitive flexible pressure sensors with microstructured rubber dielectric layers. Nature Materials. 2010;**9**(10):859–864. DOI: 10.1038/nmat2834

[53] Li B, Fontecchio AK, Visell Y. Mutual capacitance of liquid conductors in deformable tactile sensing arrays. Applied Physics Letters. 2016;**108**(1):013502. DOI: 10.1063/1.4939620

[54] Wong RD, Posner JD, Santos VJ. Flexible microfluidic normal force sensor skin for tactile feedback. Sensors and Actuators A: Physical. 2012;**179**:62–69. DOI: 10.1016 /j.sna. 2012.03.023

[55] Won DJ, Baek S, Kim H, Kim J. Arrayed-type touch sensor using micro liquid metal droplets with large dynamic range and high sensitivity. Sensors and Actuators A: Physical. 2015;**235**:151–157. DOI: 10.1016/j.sna.2015.09.044

[56] Keplinger C, Sun JY, Foo CC, Rothemund P, Whitesides GM, Suo Z. Stretchable, transparent, ionic conductors. Science. 2013;**341**(6149):984–987. DOI: 10.1126/science. 1240 228

[57] Yang CH, Wang MX, Haider H, Yang JH, Sun JY, Chen YM, Zhou J, Suo Z. Strengthening alginate/polyacrylamide hydrogels using various multivalent cations. ACS Applied Materials & Interfaces. 2013;**5**(21):10418–10422. DOI: 10.1021/am403966x

[58] Bai Y, Chen B, Xiang F, Zhou J, Wang H, Suo Z. Transparent hydrogel with enhanced water retention capacity by introducing highly hydratable salt. Applied Physics Letters. 2014;**105**(15):151903. DOI: 10.1063/1.4898189

[59] Chen B, Lu JJ, Yang CH, Yang JH, Zhou J, Chen YM, Suo Z. Highly stretchable and transparent ionogels as nonvolatile conductors for dielectric elastomer transducers. ACS Applied Materials & Interfaces. 2014;**6**(10):7840–7845. DOI: 10.1021/am501130t

[60] Chen B, Bai Y, Xiang F, Sun JY, Mei Chen Y, Wang H, Zhou J, Suo Z. Stretchable and transparent hydrogels as soft conductors for dielectric elastomer actuators. Journal of Polymer Science Part B: Polymer Physics. 2014;**52**(16):1055–1060. DOI: 10.1002/polb.23529

[61] Nie B, Li R, Cao J, Brandt JD, Pan T. Flexible transparent iontronic film for interfacial capacitive pressure sensing. Advanced Materials. 2015;**27**(39):6055–6662. DOI: 10.1002/adma.201502556

[62] Robinson SS, O'Brien KW, Zhao H, Peele BN, Larson CM, Mac Murray BC, Van Meerbeek IM, Dunham SN, Shepherd RF. Integrated soft sensors and elastomeric actuators for tactile machines with kinesthetic sense. Extreme Mechanics Letters. 2015;**5**:47–53. DOI: 10.1016/j.eml.2015.09.005

[63] Li R, Nie B, Zhai C, Cao J, Pan J, Chi YW, Pan T. Telemedical wearable sensing platform for management of chronic venous disorder. Annals of Biomedical Engineering. 2016;**44**(7):2282–2291. DOI: 10.1007/s10439-015-1498-x

[64] Yuk H, Zhang T, Lin S, Parada GA, Zhao X. Tough bonding of hydrogels to diverse non-porous surfaces. Nature Materials. 2016;**15**(2):190–196. DOI: 10.1038/nmat4463

[65] Yang CH, Chen B, Lu JJ, Yang JH, Zhou J, Chen YM, Suo Z. Ionic cable. Extreme Mechanics Letters. 2015;**3**:59–65. DOI: 10.1016/j.eml.2015.03.001

[66] Yang CH, Chen B, Zhou J, Chen YM, Suo Z. Electroluminescence of giant stretchability. Advanced Materials. 2015; DOI: 10.1002/adma.201504031

[67] Ni N, Zhang L, Wang Y, Shi L, Zhang C. Transparent capacitive sensor for structural health monitoring applications. International Journal of Applied Electromagnetics and Mechanics. 2016;**52**(3–4):1577–1584. DOI: 10.3233/JAE-162215

[68] Cohen DJ, Mitra D, Peterson K, Maharbiz MM. A highly elastic, capacitive strain gauge based on percolating nanotube networks. Nano Letters. 2012;**12**(4):1821–1825. DOI: 10.1021/ nl204052z

[69] Ho DH, Sun Q, Kim SY, Han JT, Kim DH, Cho JH. Stretchable and multimodal all graphene electronic skin. Advanced Materials. 2016;**28**:2601–2608. DOI: 10.1002/adma.201505739

[70] Shen Y, Lin YH, Nan CW. Interfacial effect on dielectric properties of polymer nanocomposites filled with core/shell-structured particles. Advanced Functional Materials. 2007;**17**(14):2405–2410. DOI: 10.1002/adfm.200700200

[71] Hongying Z, Wang MY, Jian Z. Soft compressive sensor design and analysis. In: 2015 IEEE 7th International Conference on Cybernetics and Intelligent Systems (CIS) and IEEE Conference on Robotics, Automation and Mechatronics (RAM); 15 July 2015; IEEE; 2015. pp. 255–260. DOI: 10.1109/ICCIS.2015.7274583

[72] Park J, Lee Y, Ha M, Cho S, Ko H. Micro/nanostructured surfaces for self-powered and multifunctional electronic skins. Journal of Materials Chemistry B. 2016;**4**(18):2999–3018. DOI: 10.1039/c5tb02483h

[73] Pang C, Koo JH, Nguyen A, Caves JM, Kim MG, Chortos A, Kim K, Wang PJ, Tok JB, Bao Z. Highly skin-conformal microhairy sensor for pulse signal amplification. Advanced Materials. 2015;**27**(4):634–640. DOI: 10.1002/adma.201403807

[74] Li T, Luo H, Qin L, Wang X, Xiong Z, Ding H, Gu Y, Liu Z, Zhang T. Flexible capacitive tactile sensor based on micropatterned dielectric layer. Small. 2016;**12**(36):5042–5048. DOI: 10.1002/smll.201600760

[75] Gerratt AP, Michaud HO, Lacour SP. Elastomeric electronic skin for prosthetic tactile sensation. Advanced Functional Materials. 2015;**25**(15):2287–2295. DOI: 10.1002 /adf m.201404365

[76] Kwon D, Lee TI, Shim J, Ryu S, Kim MS, Kim S, Kim TS, Park I. Highly sensitive, flexible, and wearable pressure sensor based on a giant piezocapacitive effect of three-dimensional microporous elastomeric dielectric layer. ACS Applied Materials & Interfaces. 2016;**8**(26):16922–16931. DOI: 10.1021/acsami.6b04225

[77] Wan S, Bi H, Zhou Y, Xie X, Su S, Yin K, Sun L. Graphene oxide as high-performance dielectric materials for capacitive pressure sensors. Carbon. 2017;**114**:209–216. DOI: 10.1016/j.carbon.2016.12.023

[78] Laflamme S, Kollosche M, Connor JJ, Kofod G. Soft capacitive sensor for structural health monitoring of large-scale systems. Structural Control and Health Monitoring. 2012;**19** (1):70–81. DOI: 10.1002/stc.426

[79] Laflamme S, Kollosche M, Kollipara VD, Saleem HS, Kofod G. Large-scale surface strain gauge for health monitoring of civil structures. In: Gyekenyesi AL, Yu T-Y, Shull PJ, Diaz AA, Wu HF, editors. SPIE Smart Structures and Materials+Nondestructive Evaluation and Health Monitoring; 26 April 2012; San Diego, California. International Society for Optics and Photonics; 2012. p. 83471P. DOI: 10.1117/12.913187

[80] Kharroub S, Laflamme S, Song C, Qiao D, Phares B, Li J. Smart sensing skin for detection and localization of fatigue cracks. Smart Materials and Structures. 2015;**24**(6):065004. DOI: 10.1088/0964-1726/24/6/065004

[81] Laflamme S, Ubertini F, Saleem H, D'Alessandro A, Downey A, Ceylan H, Materazzi AL. Dynamic characterization of a soft elastomeric capacitor for structural health monitoring. Journal of Structural Engineering. 2014;**141**(8):04014186. DOI: 10.1061 /(ASCE)ST.1943-541X.0001151

[82] Laflamme S, Cao L, Chatzi E, Ubertini F. Damage detection and localization from dense network of strain sensors. Shock and Vibration. 2015;**2016**:2562949. DOI: 10.1155/2016/2562949

[83] Downey A, Laflamme S, Ubertini F. Distributed thin film sensor array for damage detection and localization. In: Lynch JP, editor. SPIE Smart Structures and Materials+Nondestructive Evaluation and Health Monitoring; 20 April 2016; Las Vegas, Nevada. International Society for Optics and Photonics; 2016. p. 98030R. DOI: 10.1117/12.2219301

[84] Ni N, Zhang L, Zhou J, Wang Y, Liu F. Dielectric elastomer cantilever beam sensor. In: Bar-Cohen Y, editors. SPIE Smart Structures and Materials+ Nondestructive Evaluation and Health Monitoring; 8 Mar 2014; San Diego, California. International Society for Optics and Photonics; 2014. p. 90561M. DOI: 10.1117/12.2045054

[85] Yoon SG, Chang ST. Microfluidic capacitive sensors with ionic liquid electrodes and CNT/PDMS nanocomposites for simultaneous sensing of pressure and temperature. Journal of Materials Chemistry C. 2017;**5**:1910–1919. DOI: 10.1039/C6TC03994D

[86] Peng P, Rajamani R, Erdman AG. Flexible tactile sensor for tissue elasticity measurements. Journal of microelectromechanical systems. 2009;**18**(6):1226–1233. DOI: 10.1109/JMEMS.2009.2034391

[87] Gisby TA, O'Brien BM, Anderson IA. Self-sensing feedback for dielectric elastomer actuators. Applied Physics Letters. 2013;**102**(19):193703. DOI: 10.1063/1.4805352

[88] Kofod G, Stoyanov H, Gerhard R. Multilayer coaxial fiber dielectric elastomers for actuation and sensing. Applied Physics A: Materials Science & Processing. 2011;**102**(3):577–581. DOI: 10.1007/s00339-010-6066-5